The Evolution of Everything

Big History seeks to retell the human story in light of our position as a late-arriving species on a very ancient planet. This book provides a deep, causal view of the forces that have shaped the Universe, the Earth, and humanity. Starting with the Big Bang and the formation of the Earth, it traces the evolutionary history of the world, focusing on humanity's origins. It also explores the many natural forces shaping humanity, especially the evolution of the brain and behavior. Moving through time, the causes of such important transformations as agriculture, complex societies, the Industrial Revolution, the Enlightenment, and Modernity are placed in the context of underlying changes in demography, learning, and social organization. Humans are biological creatures, operating with instincts evolved millions of years ago, but in the context of a rapidly changing world, and as we try to adapt to new circumstances, we must regularly reckon with our deep past.

Brian Villmoare is Associate Professor in the Department of Anthropology at the University of Nevada, Las Vegas (UNLV). His research interests range from broad questions of evolutionary theory to high-resolution studies of the hominin face. His current research projects include studying the role of selection and genetics in evolutionary change and extinction, the specific evolutionary constraints and selection pressures responsible for hominin craniofacial form, and paleoanthropological fieldwork in Ethiopia. In 2013 he participated in the discovery of the oldest fossil specimen of our genus, *Homo*.

The Evolution of Everything
The Patterns and Causes of Big History

BRIAN VILLMOARE
University of Nevada, Las Vegas

CAMBRIDGE
UNIVERSITY PRESS

Shaftesbury Road, Cambridge CB2 8EA, United Kingdom

One Liberty Plaza, 20th Floor, New York, NY 10006, USA

477 Williamstown Road, Port Melbourne, VIC 3207, Australia

314–321, 3rd Floor, Plot 3, Splendor Forum, Jasola District Centre, New Delhi – 110025, India

103 Penang Road, #05–06/07, Visioncrest Commercial, Singapore 238467

Cambridge University Press is part of Cambridge University Press & Assessment, a department of the University of Cambridge.

We share the University's mission to contribute to society through the pursuit of education, learning and research at the highest international levels of excellence.

www.cambridge.org
Information on this title: www.cambridge.org/9781108495653

DOI: 10.1017/9781108862530

First published 2023

A catalogue record for this publication is available from the British Library.

Library of Congress Cataloging-in-Publication Data
Names: Villmoare, Brian, 1967– author.
Title: The evolution of everything : the patterns and causes of big history / Brian Villmoare, University of Nevada, Las Vegas.
Description: Cambridge, UK ; New York, NY : Cambridge University Press, 2023. | Includes bibliographical references and index.
Identifiers: LCCN 2021029788 (print) | LCCN 2021029789 (ebook) | ISBN 9781108862530 (epub) | ISBN 9781108495653 (hardback) | ISBN 9781108797320 (paperback)
Subjects: LCSH: Human beings–Origin. | Human evolution. | Genetics–Methodology. | Evolution (Biology)
Classification: LCC GN281 (ebook) | LCC GN281 .V53 2021 (print) | DDC 599.93/8–dc23
LC record available at https://lccn.loc.gov/2021029788

ISBN 978-1-108-49565-3 Hardback
ISBN 978-1-108-79732-0 Paperback

Contents

Figure Credits

The publishers, organisations and individuals listed below are gratefully acknowledged for giving their permission to reproduce figures in which they hold the copyright.

Copyright holder/credit	Figure number(s)
Artists Rights Society	
Otto Dix/Artists Rights Society (ARS), New York/VG Bild-Kunst, Bonn	23.2 (middle right)
BundesArchiv	
© BundesArchiv	12.4 (top)
Getty Images	
© Mark Garlick/Science Photo Library/Getty Images	2.1
© Handout/Getty Images News/Getty Images	2.3
© Auscape/Universal Images Group/Getty Images	3.1
© QAI Publishing/Universal Images Group/Getty Images	3.2
© Arctic-Images/Stone/Getty Images/CC-BY-SA-3.0	4.2
© Mint Images-Frans Lanting/Getty Images	4.3 (top)
© Mark Ireland/Moment/Getty Images	4.3 (bottom)
© Daniela Duncan/Moment/Getty Images	5.3 (top left)
© Hans Lang/imageBROKER/Getty Images	5.3 (bottom left)
© Peter David/The Image Bank/Getty Images Plus	5.3 (bottom right)
© BSIP/Universal Images Group/Getty Images	6.1, 6.3
© Stefanie D. Hueber, Georg F. Weiller, Michael A. Djordjevic, Tancred Frickey/Hox protein classification across model organisms by CLANS analysis, Hueber et al./Wikimedia Commons/CC-BY-CA-4.0	6.5
© Dinoguy2/Creative Commons (Corrected Miglewis)	8.1 (top)
© Corey Ford/Stocktrek Images/Getty Images	8.1 (bottom), 8.2 (left)
© Schafer & Hill/The Image Bank/Getty Images	8.3 (top left)

Copyright holder/credit	Figure number(s)
© DEA/G. Cigolini/De Agostini/Getty Images	8.3 (top right)
© Layne Kennedy/Corbis Documentary/Getty Images	8.3 (bottom left)
© Hoberman Collection/Universal Images Group/ Getty Images	9.5 (bottom left)
© Ronan Bourhis/Universal Images Group/Getty AFP	9.5 (top right)
© David Haring/DUPC/Oxford Scientific/Getty Images Plus	10.1 (left)
© Manoj Shah/Stone/Getty Images	10.1 (right), 10.2 (right)
© Anup Shah/Stone/Getty Images	10.2 (left)
© Tambako the Jaguar/Moment Open/Getty Images	10.2 (middle)
© Duncan1890/DigitalVision Vectors/Getty Images	11.1
© Auscape/Universal Images Group/Getty Images	11.3 (top)
© DEA/N. Cirani/De Agostini/Getty Images	11.3 (middle)
© Lionel Bonaventure/Staff/AFP/Getty Images	11.3 (bottom)
© Universal History Archive/Contributor/Universal Images Group/Getty Images	12.4 (bottom)
© guenterguni/E+/Getty Images	13.1 (top)
© Rüdiger Katterwe/EyeEm/Getty Images	13.1 (bottom)
© Takuya Kabe/EyeEm/Getty Images	14.1 (right)
© Hadi Zaher/Moment Open/Getty Images	15.3 (bottom left)
© Photostock Israel/Science Photo Library/Getty Images	15.3 (bottom right)
© AGF/Contributor/Universal Images Group/Getty Images	16.1 (bottom left)
© Universal History Archive/Universal Images Group/Getty Images	16.6 (bottom left), 18.2 (bottom right)
© Ed Freeman/Stone/Getty Images	19.4 (top left)
© Alexander Spatari/Moment/Getty Images	19.4 (top middle)
© Buyenlarge/Archive Photos/Getty Images	19.4 (top right)
© Print Collector/Hulton Archive/Getty Images	19.4 (bottom left)
© DEA/M. Ranzani/De Agostini/Getty Images	19.4 (bottom middle)
© Mondadori Portfolio/Hulton Fine Art Collection/ Getty Images	19.4 (bottom right)

Copyright holder/credit	Figure number(s)
© Science & Society Picture Library/SSPL/Getty Images	20.2 (top)
© Sovfoto/Universal Images Group/Getty Images	21.2 (top left)
© AFP/Stringer/AFP/Getty Images	21.3 (left), 23.8 (top left)
© Archive Photos/Archive Photos/Getty Images	21.3 (right)
© Bettmann/Bettmann/Getty Images	22.2 (middle right), 23.3 (bottom)
© New York Times Co./Archive Photos/Getty Images	23.1 (top middle)
© Apic/Hulton Archive/Getty Images	23.1 (top right)
© Felix Garcia Vila/iStock/Getty Images Plus/Getty Images	23.2 (top right)
© Fine Art/Corbis Historical/Getty Images	23.2 (left, second down)
© ullstein bild Dtl./ullstein bild/Getty Images	23.2 (bottom right), 23.3 (top left), 23.5 (bottom left), 23.6 (bottom left)
© Roger Viollet/Roger Viollet/Getty Images	23.5 (top middle)
© Photo 12/Universal Images Group/Getty Images	23.5 (top right)
© Universal History Archive/Universal Images Group/Getty Images	23.6 (top middle)
© DEA Picture Library/De Agostini/Getty Images	23.6 (bottom right)
© Zh. Angelov/Stringer/Hulton Archive/Getty Images	23.7 (left)
© PhotoQuest/Archive Images/Getty Images	23.8 (top middle)
© Paul J. Richards/Staff/AFP/Getty Images	23.8 (bottom left)
© Shah Marai/Stringer/AFP/Getty Images	23.8 (bottom right)

Journal of Heredity

Alfred Blakeslee, Journal of Heredity	5.1

Montana State University

© David Ward, Montana State University/ Handoucebmp yellowstone Algais/bmp yellowstone	7.2

NASA/JPL/CalTech

© NASA/JPL/CalTech	2.2, 15.2 (top)

Copyright holder/credit	Figure number(s)
Reporters without Borders (Reporters sans frontières)	
© Reporters without Borders	24.2
University of Wisconsin and Michigan State Comparative Mammalian Brain Collections, funded by NSF and NIH	
© University of Wisconsin and Michigan State Comparative Mammalian Brain Collections	14.2
Wikimedia Commons	
© Selbst erstellter Screenshot von XEphem 3.7.3	1.1 (top)
© Selbst erstellt	2.2
© Astroskiandhike	3.3
© Yassine Mrabet/Miller-Urey Experiment/Wikimedia Commons	4.1
© Frank Schulenburg/Black-tailed deer(*Odocoileus hemionus columbianus*) at Point Reyes National Seashore, California, United States/Wikimedia Commons/CC-BY-SA-3.0	5.3 (top right)
Mouagip/Codons sun/Wikimedia Commons	6.4
Woudloper/The geological clock: a projection of Earth's 4,5 Ga history on a clock/Wikimedia Commons	7.1
© Simon Villeneuve/test de traduction/Wikimedia Commons/CC-BY-SA-3.0	7.3
Charles D. Walcott/*Opabinia regalis* Walcott, 1912/Wikimedia Commons/{{PD-UK-unknown}}	8.2 (right)
© Whndout bmp jpg Figure 8 3 Trilns/bmp jpg	8.3 (bottom right)
© Dwergenpaartje/A Schizochroal eye of the trilobite *Phacops rana*/Wikimedia Commons/CC-BY-SA-3.0	8.4 (top left)
© USDAgov/Common house fly, *Musca domestica*/Wikimedia Commons/CC-BY-SA-3.0	8.4 (top right)
© wilkdxplorer/Squid eye/Wikimedia Commons/CC-BY-2.0	8.4 (bottom left)
© Kamil Saitov/A human eye with its unique iris pattern/Wikimedia/CC-BY-4.0	8.4 (bottom right)

Copyright holder/credit	Figure number(s)
Whndout bmp JPG Figure 9 4 Tetris/bmp JPG	9.3 (bottom left)
© Maija Karala/Whndout bmp jpg Figure 9 4 Tetris/bmp jpg	9.3 (top right)
© Conty/Comparison between the fins of A., Crossopterygian fishes and the legs of B., tetrapods/Wikimedia Commons	9.4
© NordNordWest/Spreading of Homo sapiens/ Wikimedia Commons	12.1
© Tuvalkin/World map in Mercator style projection showing human skin color according to Biasutti 1940/Wikimedia Commons/CC-BY-SA-3.0	12.2
Virginia Archives/Racial Integrity Act of 1924/ Wikimedia Commons	12.3
© Alexander Vasenin/Moon jellyfish (Aurelia aurita) at Gota Sagher (Red Sea, Egypt)/Wikimedia Commons/CC-BY-SA-3.0	14.1 (left)
© John A Beal, PhD/Lateral Portion of Frontal, Parietal, Occipital, and Superior Portion of Temporal Lobe Resected/Wikimedia Commons/ CC-BY	14.3 (top)
© Thomas Schultz	14.3 (bottom)
Icelandic Hurricane/Projected path of Tropical Storm Norman (2006)/Wikimedia Commons	15.2 (bottom)
© Igor Chuxlancev/Ants build their-body bridge to cross the big canyon/Wikimedia Commons/CC-BY-4.0	15.3 (top)
© George Dehio/A plan of Lincoln Cathedral in England by G Dehio/Wikimedia Commons/{{UK-PD}}	15.4 (left)
© Mopane Game Safaris/Day with bushman in Grassland Bushman Lodge, Botswana/Wikimedia Commons/CC-BY-SA-4.0	16.1 (top)
© Idobi/Hadzabe Men Practicing Bowing/ Wikimedia Commons/CC-BY-SA-3.0	16.1 (bottom right)
© Daniel Schwen/El Castillo (pyramid of Kukulcán) in Chichén Itzá/Wikimedia Commons/CC-BY-SA-4.0	17.3 (top left)
© Patrick J. Finn/ASI monument number N-MP-220/Wikimedia Commons/CC-BY-SA-4.0	17.3 (top right)

Copyright holder/credit	Figure number(s)
© John Trumbull/The Death of General Warren at the Battle of Bunker's Hill, June 17, 1775/ Wikimedia Commons/{{UK-PD}}	19.3 (bottom)
© Hélène Rival/Flintlock musket signed Pillon, fully carved and engraved. Weapons Department in the Museum of Art and Industry in Saint-Étienne, France/Wikimedia Commons/CC-BY-SA_3.0	20.1 (top)
© E.L. Hoskyn/Steam powered weaving shed/ Wikimedia/{{UK-PD}}	20.2 (right)
© KiloByte/"Cocaine toothache drops", 1885 advertisement of cocaine for dental pain in children. United States./Wikimedia {{UK-PD-unknown}}	20.3 (top left)
Hardie, D. W. F./Photograph of Widnes in the late 19th century showing the effects of industrial pollution/Wikimedia Commons/{{UK-PD}}	20.4 (top left)
© Lewis W. Hine for the National Child Labor Committee/Child laborer/Wikimedia Commons	20.4 (top right)
© Harper's Weekly/The Haymarket Riot/Wikimedia Commons/{{UK-PD}}	20.4 (bottom)
Hans Gude/Fra Hardanger/Wikmedia Commons/ {{UK-PD}}	20.5 (top left)
Caspar David Friedrich/Wanderer Above The Sea of Fog/Wikimedia Commons/{{UK-PD}}	20.5 (top right)
J. M. W. Turner/The Fighting Temeraire/Wikimedia Commons/{{UK_PD}}	20.5 (bottom)
© Rybinsk State Historical-architectural Art Museum and National Park Photofiles	21.2 (top right)
© Alexander Wienerberger/Diocesan Archive of Vienna (Diözesanarchiv Wien)/BA Innitzer	21.2 (bottom)
© James Gordon/Palmyra, Syria/Wikimedia Commons/CC-BY-2.0	22.1 (top)
Alice Harris, Daniel Danielson, others/Mutilated Congolese children and adults (c. 1900-1905)/ Wikimedia Commons/{{UK-PD}}	22.2 (bottom left)
© Unknown author/Scribner's Magazine 1916/ 1916 advertisement for the United Fruit Company Steamship Line/Wikimedia Commons {{UK_PD}}	22.3 (bottom left)

Copyright holder/credit	Figure number(s)
National Air and Space Museum/Bessie Coleman/ Wikimedia Commons/{{UK-PD}}	23.1 (top left)
© Cornellrockey/US Army WWII field artillery/ Wikimedia Commons/{{UK-PD}}	23.1 (bottom left)
© Toni Frissell/Members of the 332nd Fighter Group attending a briefing in Ramitelli, Italy, March, 1945/Wikimedia Commons	23.1 (bottom right)
Claude Monet/Le Bateau-atelier/Wikimedia Commons/{{UK-PD}}	23.2 (top right)
Edvard Munch/Madonna/Wikimedia/{{UK-PD}}	23.2 (bottom left)
© Palczewski, Catherine H. Postcard Archive. University of Northern Iowa/Anti suffragette propaganda from the early 20th century/ Wikimedia Commons/{{UK-PD}}	23.3 (top left)
Library of Congress	23.4 (left)
© Unknown author/A portrait of Atatürk from the 1930s/Wikimedia Commons {{UK_PD}}	23.4 (middle)
© Georg Pahl/Bundesarchiv, Bild 102-14597/ Wikimedia Commons/CC-BY-SA 3.0	23.5 (top left)
© Bundesarchiv, Bild 102-14468/Georg Pahl/ Wikimedia Commons/CC-BY-SA 3.0	23.5 (bottom right)
© Unknown author/Women at the Siege of the Alcázar in Toledo/Wikimedia Commons/ {{UK_PD}}	23.6 (top left)
© Unknown author/Guernica, Ruinen/Wikimedia Commons/CC-BY-SA 3.0	23.6 (top right)
© RAWA/Taliban beating woman in public/ Wikimedia Commons/CC-BY-2.0	23.7 (right)
© Denver News/Ku Klux Klan members and a burning cross, Denver, Colorado, 1921/Wikimedia Commons/{{UK-PD}}	23.8 (top right)

Individuals

Randall Bytwerk	16.6 (top right), 16.6 (middle left)
Ernst Haeckel	9.1
Ludwig Hohlwein	16.6 (bottom right)
William Kimbel/IHO	10.3, 10.4, 11.2

Copyright holder/credit	Figure number(s)
Marta Lahr	16.4
Vladimir Nikolayevich Ivanov	13.3 (right)
Jill Pruetz	13.2
Max Reimer	16.6 (top left)
Scott Turner	15.4
Brian Villmoare (author)	1.1 (bottom right), 1.2, 5.2, 6.2, 9.2, 10.5, 13.3 (left), 13.4, 15.1, 16.2, 16.3, 16.5, 17.1, 17.2, 17.5, 17.6, 18.1 (top left), 18.5, 18.6, 19.1, 21.1, 22.2 (top left), 24.1, 24.3, 24.4, 24.5. 24.6

Acknowledgments

Given the range the topics in this book, I was obliged to rely on the knowledge and advice of experts from across the disciplines. Any errors of commission, omission, or oversimplification, of course, are mine alone. For opinions, discussions, and critiques, I would like to thank Ramon Arrowsmith, Levent Attici, Alyssa Crittenden, Erin DiMaggio, Jennifer Fish, Paula Huntley, David Klein, Pierre Lienard, Bill Jankowiak, Rebecca Martin, Tim McHale, Hoski Schaafsma, Chet Sherwood, David Tanenhaus, Bowen P. Weisheit, Jr., Lars Werdelin, and Erin-Marie Williams. I would also like to acknowledge Dominic Lewis and the editors and staff at Cambridge University Press, my agent, Barbara Rosenberg, the invaluable assistance of Bri Heisler, and, especially, Amy Llanso, for her technical editorial assistance over many drafts.

Chapter 1: Introduction

This textbook is a bit different from most that you will be exposed to in your collegiate career. One major difference is the breadth – whereas most textbooks focus on a single subject (sociology, chemistry, history, biology, economics, etc.), this textbook encompasses almost all of the traditional fields of academic knowledge. One natural result of this breadth is that you will not learn as much about any one of those disciplines. However, one of the goals of this textbook is to show you how these fields are interrelated, and that by seeing the world through multiple disciplines, you can gain a much greater understanding of the various factors that are responsible for the world and Universe that you see today.

One of the growing fields of study in academics has been the "Big History" movement. The purpose of Big History is to give students a broader perspective of the past. Big History is typically taught in history departments, but this textbook will show you how the events of the past are the result of so many factors that the past must be viewed through multiple disciplines.

The main underlying theme of this book is the need to use science, in various forms, to understand the *causes* of past events, even in the recent past. We all understand that we use scientific disciplines like physics and astronomy to understand the origins of the Universe, geology to understand the formation of the Earth and movement of the continents, and biology to understand the evolution of life. But we can also use science to understand much more recent phenomena.

When we examine a historical event such as World War II, we frequently look at it from the perspective of the individuals involved in the conflict. For example, we may read about D-Day from the viewpoint of a soldier on a beach in Normandy, facing a barrage of incoming machine gun fire. Or we may read about the women building tanks and airplanes on the assembly lines, and their material contribution to the war effort. Or we may read an account by African American veterans from the 761st Tank Battalion, who had to fight racism in segregated combat units. This form of history is called "narrative," in which we try to understand a broader event by seeing it from various perspectives, with the hope that, by reading enough narratives and seeing as many perspectives as possible, we will acquire a thorough knowledge of the events. Today this is probably the most common form of history, and one reason is that it is so personally compelling. Narrative allows us to "feel" what it might have been like to be involved in those historic events, and it is natural that we would enjoy this form of history.

But narrative, for all its emotional reward, will not necessarily help us understand the broader forces that drove the events. To understand the causes of past events, we need to step back and examine the events from a distance. In the case of World War II, personal narratives may not help us understand why the Allies won the war. To do that, we must take a more scientific perspective on past events. At the outset of World War II, Japan and Germany had some of the strongest industrial economies in the world. This explains why they were able to build enormous armies, air forces, and navies, and why they were able to be so successful in the early years of the war. But it is important to remember that the United States was not in the war for the first three years, when the Axis powers had so much success. Even before the war, US industrial output was greater than that of all the Axis countries combined. And the US population alone was almost as great as that of Germany and Japan (and the population of the Soviet Union, by 1942 a member of the Allies, was greater still). Once the United States was in the war, it was largely just a matter of time before the economic force of the United States, along with the enormous populations of the USA and the Soviet Union, wore down the Axis. The United States was able to manufacture more tanks, airplanes, and trucks than all the Axis countries combined, whereas the Axis countries ran out of resources and so were literally unable to continue the fight. In fact, a significant aspect of Allied strategy centered on reducing Axis industrial output (bombing factories, mining harbors) while maximizing their own. This is in no way to trivialize the contributions and sacrifices of the individuals engaged in the conflict but just to point out that understanding the bigger picture of a large-scale event like World War II means stepping back and trying to understand the larger forces that determined the outcomes. Scientific historians do exactly this, by examining relationships among, say, resources and power. Do countries with a lot of natural resources and large populations tend to be more powerful than technologically similar countries with fewer resources and smaller populations? Even though it is a historical question, it can be examined in a scientific framework.

This highlights a difference in types of causes. Broadly speaking, scientists talk about two types of causes: proximate and ultimate. In general, a *proximate cause* is the event that led immediately to the consequence, whereas the *ultimate cause* is the underlying driver of the circumstances. In World War II, we might say that the bombing of Pearl Harbor on December 7, 1941 by the Japanese Imperial Navy was the cause of the USA entering the war. But that was just the precipitating event. The underlying cause was the expansion of world power by Japan, which had conquered several Asian regions and regarded the United States as a rival in the Pacific Ocean. Under those circumstances, a conflict was more or less inevitable; the question was largely what form it would take. So the attack was the proximate cause, but the ultimate cause was Japanese expansionism.

This applies to biology as well – two male elk may fight for the right to mate with the herd of females (proximate), but the underlying reason is that natural selection has acted to make male elk aggressive, as more aggressive males are more likely to reproduce (ultimate).

What Is Science?

This textbook employs an explicitly scientific perspective on events in the past. So it is probably worthwhile to explore what science is, and why its perspective is so unique. Most of us (even scientists ourselves) have an image in our heads of what science is and what scientists look like. The stereotypical image is of someone in a white lab coat in a laboratory, and the fact is a great deal of science is done that way. Lab coats are useful for keeping your street clothes clean while working with chemicals or dissected animals, and laboratories are useful spaces for keeping materials uncontaminated and out of the elements while working with them, and for keeping unpleasant smells out of the rest of the building. Sometimes we have a positive view of science, particularly when it produces medical breakthroughs, but sometimes we have a negative view of science, for example, when discussing the potential of cloning humans.

But that is not really what science is. Science is simply a way of understanding. In its most basic form, science simply tries to describe the material world in as objective a way as possible. A scientific statement about the world has several important characteristics:

1. Any scientific statement must invoke only forces and causes that can be observed by anyone using empirical methods. Science cannot attribute

actions, for example, to "spirits" that can only be seen by one privileged spiritualist.

2.   Any statement that one scientist makes about the world must be testable by other scientists. For example, if a scientist attributes a flood to a meteor, based on the presence of a crater, other scientists must be able to examine the crater themselves; they cannot simply take the word of the first scientist.

3.   Parsimony: if there are competing explanations, the one with the fewest assumptions will be accepted. For example, if there are two hypotheses on the origin of the Egyptian pyramids, and one requires the presence of alien spacecraft to move the stones, it will be regarded as less likely than an explanation that simply requires the presence of large numbers of workers pushing stones up ramps. The first hypothesis requires assuming something we have no concrete evidence for (aliens) and is therefore less parsimonious.

There are several important implications of this viewpoint. The first is that scientists do not speak from a position of authority. Unlike a spirit medium or palm reader, who claims to have access to privileged knowledge or powers, a scientist does not claim to have unique abilities. Within science, while there are certainly scientists who are widely respected, no true scientist would ever claim to have a perspective that is not subject to challenge. In fact, this is largely how science progresses: one scientist makes a claim that is tested by other scientists, and if the claim is found to be incorrect another may be proposed. If it is not rejected ("falsified," as scientists often say), then the hypothesis is provisionally accepted. But scientists know that at any time, any scientific hypothesis is subject to refutation. Even if a young scientist is challenging an older, more famous scientist, the claims are examined on their own merits rather than the reputation or position of the older scientist. In this way, science is one of the most democratic forms of knowledge and study – anyone (with sufficient study) can do it, and no one is above challenge, no matter their reputation. In fact, many young scientists have made their names by successfully challenging the long-held theories of senior scientists.

The history of science is replete with hypotheses that have been accepted for a time and then modified or rejected altogether. But with every rejected hypothesis we are refining our knowledge of the Universe – we know what is *not* the answer, so we know to look elsewhere.

Objections to Science

Although today we often regard science as presenting a relatively authoritative perspective on the Universe, historically there have been many objections to science. One of the most common challenges to science is when a scientific discovery or model contradicts a religious text. The most well known such conflict is the debate between the scientific perspective on the evolution of species (including humans) and Creationism. Some interpretations of the Book of Genesis in the Old Testament have humans as the product of divine creation. Under this model, humans were created as a fully developed species and have no particular biological relationship to other living things on Earth (which were also created). This creationist model (which is not universally accepted by all religious authorities) is in complete contradiction with virtually any modern biological model, under which all species, including humans, evolved from ancestral forms through the forces of evolution.

There have been various responses by society to this conflict. In some of the more religious regions in the USA, one response has been to outlaw the teaching of biological evolution or force the teaching of creationist-based models, although these laws have largely been overturned in the courts during the last half of the twentieth century. Other responses have been to challenge the position of scientists in society, so that they are not seen as reliable and objective.

Another objection to science is that sometimes the implications of applying mechanistic biological models derived from animals to humans challenges our position in the natural world. Many people prefer to see humans as the pinnacle of the natural world, and if we are the product of the same forces that have produced poison ivy, shrimp, algae, and cockroaches, then we may not have any way to distinguish our species. In some ways this critique is accurate, because as far as science is concerned humans are just another species. Science does not "order" nature in any way that treats humans as, in any sense, better. Rather, we are simply different. When we study human evolution in later chapters, we will detail some of the ways in which we are different, and for some of those ways it would be hard to argue that we are superior.

Most of us will have heard the story of how our parents met: how, if one or another parent hadn't gotten off the bus one stop early, or hadn't decided at the last minute to come to a party, or any variety of circumstances had been slightly altered, our parents would not have met, and we, as individuals, would literally not exist. The same applies in the natural world – species are the result of random events. One implication of this is that humans, as a species, might well not have existed if any number of circumstances over the last several hundred million years had changed. Although most of us can readily understand how we, as individuals, might not have ever existed, many of us have trouble with the idea that humans, as a species, might not have existed. But this is probably the most important implication of natural selection.

The idea that humans, and the Earth itself, is the product of the random forces of geology, astronomy, biology, and physics, makes some people uncomfortable. If we are, in effect, lonely flecks of space dust in an indifferent Universe, where do we derive purpose and morality? Unfortunately, science offers no solutions here. Science simply attempts to explain the material facts of the Universe, present and past. As a consequence, science offers no statements of morality or purpose – that is the realm of philosophers, theologians, and even artists. But if philosophers and theologians are to offer us useful morals and purposes, they must have a firm understanding of the facts of the Universe, and here science plays its role – as the underlying foundation of all spheres of thought.

The Scope of This Book

One of the premises of this book is that the scientific perspective is appropriate for investigating almost any aspect of the past and present, as long as it is a question of fact. This book starts with the Big Bang, some 14.5 billion years ago, and, covering major events and epochs in the past, passes right up to today. The first part of the book covers topics that traditionally fall under hard science – physics, astronomy, geology, and biology – whereas the second part, after the arrival of humans, encompasses disciplines that are considered "softer" sciences, such as archaeology, psychology, demography, history, political science, and economics. Each of these disciplines has their own topics, methods, and research traditions, and in this book you will learn a great deal about what these many thousands of scientists and researchers have discovered over the last several hundred years. This book is a synthesis, so I am necessarily brief when delving into any one topic, but I hope that you will find enough of interest in these many topics that you will pursue some of them further.

Many of the ideas presented here are still subject to vigorous debate among scientists, and since science is a field that thrives on disagreement, sometimes there will not be a definitive answer. For many working scientists (such as myself), the debates are one of the main things that makes the field interesting. Science is not simply learning a set of facts but debating and arguing about how to interpret them. You should feel free to challenge any ideas you see in this book – look for objective evidence and see what your own conclusions are. But remember the key to scientific claims: any evidence must be available for all to see, and the explanation with the fewest assumptions is the most scientific.

The Question of Scale

One issue we have when examining questions is picking an appropriate *scale*. For example, when we want to know about the power of a star, we need to look at the atomic or subatomic scale where nuclear fusion occurs. Very specific forces apply at those tiny scales. But when we want to predict how stars move through space, we pull back and look at how large bodies are affected by gravity. At this level, many subatomic forces no longer apply.

In biology, scale is also important. When we want to understand how a single species of dinosaurs behaved, we might focus on a few well-preserved specimens. But when we ask about the cause of the extinction of all dinosaurs, we don't necessarily study a single individual fossil or even a single species. In that case, we look at the appearance and disappearance of hundreds or thousands of species.

This same principle applies when examining other types of questions. Although we often read histories about individuals, science tends to be interested in broader "why?" questions. For example, the US Civil War is often presented in terms of the actions of individuals, or armies, and the ways in which one individual may have changed the course of a battle, or, potentially, the outcome of the war. One historical question is, for example: could the Confederacy ever have won? A scientific way of examining this question is to ask if a country with an agriculture-based economy has ever conquered a country with a large industrial base. How often has that happened; and if it has never happened, would we say that the Confederacy is unlikely to have won, no matter the actions of a few individuals? That question would be addressed by studying broad historical patterns rather than a single conflict. So, although it is, in a sense, a historical question, it might be answered in a scientific way by picking an appropriate scale of analysis.

Scale again is a factor when talking about events in time. Some events take place over enormous timescales. From the Big Bang to the origin of the Earth was roughly 10 billion years. Time scales of that magnitude are essentially incomprehensible for a human. We are born, live, and die within 1/10,000,000,000 of that timeframe, and so can have no way to truly understand how time passes over such a long period. On Earth, we often talk about species existing for millions or tens of millions of years, and even here we have no ability to understand this on a direct level. We apply numbers to these time intervals, but we have evolved to exist minute by minute or day to day, because we need to avoid threats and find food and water in those timeframes. But for much of world's history and prehistory, very little large-scale change occurred day to day or even year to year. In fact, large-scale changes occurred that would have been imperceptible in the lifetimes of any animals. Today, that process continues: even as historical events occur daily, the Earth is changing in large and imperceptible (or nearly imperceptible) ways that may have dramatic effects on the future of humanity.

Box 1.1 The Bubble

This book may significantly challenge the way you view the world. This book presents explanations that may surprise you, or contradict what you have learned previously, whatever your political, philosophical, or religious perspective. In fact, that is one of the purposes of this book. An academic education *should* challenge you. Hopefully, as you read this book you will reexamine the way you view the world. Sometimes you will find your perspective justified, but sometimes you may find yourself comparing two seemingly compelling ideas that are incompatible with each other. That is another purpose of this book – to force you to think rigorously, not just about the Universe but about how *you view* the Universe. This may occasionally make you uncomfortable, but that is OK – that is what education is about – getting you out of "the Bubble." The hope is that, by the end of the book, you will have adopted an intellectually rigorous, and sometimes skeptical, perspective that you can employ for the rest of your academic career.

Box 1.2 Progress in science

Science progresses by proposing hypotheses, which are subsequently rejected or refined, or accepted. One of the most famous examples is our understanding of the solar system. The earliest recorded solar systems placed the Earth at the center of the solar system (the "geocentric" model). This model was first formally proposed (as far as we know, based on historical documents) by the ancient Greeks: Animaxander, then Aristotle, and later Ptolemy (although Pythagoras was an early dissenter). This model was broadly consistent with the traditions of several religions, and was therefore not reexamined until almost 2,000 years later.

The track of the Sun and Moon followed predicted paths, but the orbits of the planets proved hard to explain – if they were simply orbiting around the Earth they should traverse the sky much as the Sun and the Moon. But they don't – they make "loops" as they pass through the night sky. There were some ad hoc explanations (see the image of planetary epicycles in Figure 1.1), but Copernicus observed that it was much simpler to explain the celestial patterns if the Sun was at the center of the Solar System. The "loops" were made by planets orbiting the Sun, as the inner planets, orbiting in faster, smaller orbits,

Box 1.2 (cont)

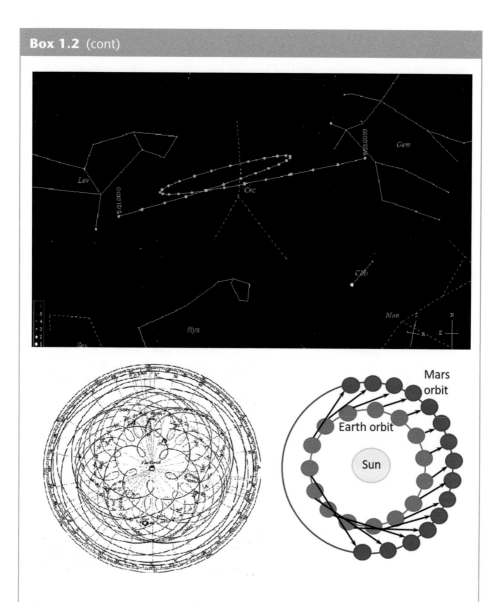

Figure 1.1 Two contrasting models for the apparent movement of Mars. As Mars passes through the sky (top of figure), it loops back over its earlier path. There are two ways that this can be explained. One is that, if Earth is the center of the solar system, Mars must be making loops (epicycles) as it orbits around Earth (lower left). This was the preferred model of the sixteenth-century Catholic Church. The second possible explanation is that Earth and Mars are both orbiting the Sun, and that our perspective on Mars changes since we are in a faster-moving, inner orbit (lower right). As we pass by Mars, it appears to reverse course. Which of these two models do you think are more parsimonious (fewer ad hoc assumptions), and therefore preferred, under the scientific model? Since there seemed to be no way to explain epicycles for orbiting planets, Galileo preferred the second model.

Box 1.2 (cont)

pass by the outer planets. By the turn of the seventeenth century, telescope technology had improved enough that Galileo could observe that the planets were composed of matter much like the Earth, suggesting that the "uniqueness" of the Earth was less than had originally been thought. His persecution by the Roman Inquisition for promoting the "heliocentric" model, with the Sun at the center of the solar system, is well known.

Copernicus and Galileo both argued for a model in which the planets (including Earth) orbited around the Sun in circles. Even seventeenth-century astronomers could see that a circular orbit could not explain the patterns of planetary movement, and it took two mathematicians, Johannes Kepler and Isaac Newton, to determine that the planets, under the attractive forces of gravity, moved in ellipses around the Sun. Subsequent discoveries (particularly general relativity) have refined the explanations, and discoveries in astronomy have moved apace.

The history of planetary astronomy is illustrative of how science proceeds. In the case of the geocentric model, subsequent discoveries and observations made it necessary to throw out the whole model (it was completely "falsified"). But in the case of the identification of elliptical orbits, science refined rather than rejected the heliocentric model. And this is how science progresses today. Sometimes models are completely discarded, but more often existing ideas are refined with the discovery of new observations, or new mathematical models to better explain the observations.

Box 1.3 What does science do and not do?

Science does a lot of things very well but there are some things it does not do at all. It is, at its most basic, simply a conceptual way of examining the Universe. But it is only interested in material facts: How old is the Earth? Where did humans come from? Why did the dinosaurs go extinct? Why is there no gorilla as large as King Kong? What causes malaria? How does rattlesnake venom kill? Are there other inhabitable planets? These are all questions that science can address. Some are very hard to answer, even with modern technology (are there other inhabitable planets?) and some are simple (what causes malaria?). These are questions of fact.

Box 1.3 (cont)

But science cannot tell you whether Vincent van Gogh was a better painter than Edouard Manet (although, in my opinion, he was). We can systematically question people about their preferences, or examine how much the paintings sell for, but this only tells us about the popularity of the painters not which was a "better" painter. Judgments on artistic merit fall under the philosophical field of aesthetics, and there people can debate the merits of either painter, and perhaps (or not) come to a consensus on judgment. But whether or not one painter is "better" is simply not a question that science can address since it does not deal with a question of objective fact. There is nothing to measure or assess objectively.

Similarly, science does not deal with questions of morality. The decision on whether an act is moral or immoral is based on a set of rules established by other people. Morality deals in "oughts" – you ought *not do* this act (murder), but you ought *do* another (throw a life preserver to a drowning swimmer). Most societies probably agree with those rules but not all, and so morality is a product of society. It is not something that is inherent to nature. In fact, animals perform acts all the time that we might well regard as immoral if done by a living human.

Sometimes difficulties arise when we see animals, particularly animals we are closely related to or are closely involved with, perform acts we see as immoral. In the early 1970s Jane Goodall, observing chimpanzees in the Gombe Preserve in Tanzania, watched as they behaved in ways that we would consider immoral in humans. They engaged in acts that would be labeled as murder in humans, and even warfare. Now, most of us would agree that it makes no sense to judge their acts in a moral framework, since they are not humans. But even when those acts are performed by humans, science makes no statements about the morality of those actions. Science has no "oughts."

Science can, however, help provide clear facts that help us to make ethical decisions. It can illuminate the kinds of social circumstances that might lead to violence (do societies with fewer economic opportunities tend to be more violent?), just as it might help understand the social implications of various remedies to violence – for example, does capital punishment deter murder in a society? Although the essential issue of capital punishment is clearly moral, since it deals with killing, the simple objective question of a specific social response (say, a reduced murder rate) to capital punishment *is* a potential scientific question.

Box 1.4 Pesudoscience

Medical Pseudoscience

Sometimes ideas are presented as science yet are without any of the characteristics of true science. Typically, although not always, it is because somebody wants to sell you something. Having the aura of science can lend a product credibility. This is known as *pseudoscience*, and it is commonly seen in advertisements for products that claim to have medicinal powers, such as copper bracelets, therapeutic magnets, crystal therapy, "essential oils," or water with a changed molecular structure (e.g., "memory water").

Typically, the advertisements are full of scientific jargon that is designed to fool the consumer into thinking the product has been scientifically tested. Often the advertisements will show people with white lab coats holding clipboards or test tubes, along with anecdotal claims by "customers" who have had their pains and aches relieved by the product. But if you keep in mind the actual scientific method, you can determine which products have no basis in science.

The test:

1.   Have the products been examined by a respected independent laboratory for effectiveness?
2.   Did the laboratory follow normal scientific protocols for medical tests – in other words, did the laboratory test for the "placebo effect"?
3.   Are the results of the test open for examination by anybody? Could we interview the test subjects themselves, or do we have to take the word of the company?

It is not always easy to determine if a laboratory is legitimate, and in the United States there are few regulations on products that do not make specific medicinal claims. It is important to keep in mind that just because a few people claim to have been "cured" by the product does not mean it works. If 12 out of 50,000 people claim relief from arthritis pain from copper bracelets, in all likelihood the relief came from another source (the placebo effect perhaps, or another medicine, or just natural healing). But the advertisements may only show the claims of the twelve who found relief.

In the United States there is long tradition of pseudoscientific fraud, going back more than a hundred years, to the salesmen of various tonics that claimed to cure any number of maladies (the original "snake oil" salesmen). The tonics in the nineteenth century were typically just a mixture of opium, cocaine, or alcohol that would temporarily relieve pain, but sometimes they were genuinely dangerous with arsenic and mercury commonly used. Today, with the internet, it is easier than ever for fraudulent hucksters to find people who are genuinely in pain and prey on their hope for relief or cure.

Box 1.4 (cont)

Nonmedical Pseudoscience

But the methods of pseudoscience are also used to convince people of a variety of other types of nonscientific beliefs. For example, claims about the powers of pyramids are sometimes framed using scientific language. Many of the claims about UFOs, especially for evidence of previous visits found in ancient rock art, are pseudoscience, since there is no way for any legitimate scientist to test the interpretations that the figures in the rock art represent ancient aliens or astronauts. This criticism also applies to claims of scientific legitimacy by advocates of extrasensory perception, crop circles, palmistry (palm reading), cryptozoology, and dowsing, among many, many others.

Sometimes you may hear of engines "powered by water" that were suppressed by oil companies or the government. Since the idea of an engine powered by water violates known physics and engineering properties, and there are no examples for scientists to examine, this falls squarely under the category of pseudoscience. Conspiracy theories, in general, fall under the category of pseudoscience, since, by their very nature, the evidence cannot be examined (and there are many – the "Moon-landing conspiracy," the "suppressed cancer cure" conspiracy, and "cold fusion" are just a few). In the case of the water-powered engine, promotion of the idea was typically investor fraud, and several "designers" have been found guilty of fraudulently taking money from investors knowing full well the engine did not work.

Most of the broadly followed religions today existed prior to the advent of science, but some more recent religions and cults have characteristics of pseudoscience. Scientology has many of the characteristics of pseudoscience since it makes specific claims (through its "Dianetics" program) that cannot be independently examined by scientists. Similarly, "creation science" presents nonscientific ideas in a scientific framework, with the hope that the reader will be fooled into thinking that the ideas are scientific (see the section on 'Creationism' in Chapter 5).

In general it is a good idea to be skeptical of any scientific claim if the presenter is selling you something or wants you to join an ideology. If a claim is genuinely scientific, it will have evidence that is open for all to see, and hopefully they will have no interest in separating you from your money. Of course, one of the best ways to know if someone is presenting you with pseudoscience is to learn the actual science, so hopefully, after reading this book, you will be better prepared to defend yourself!

Box 1.5 Math is the language of nature

Most students know that math is necessary for science. This fact actually serves as an obstacle for many students who might be interested in science but don't feel very strong in their math skills. Long ago, I was one of those kids – I loved reading about scientific discoveries like fossils and distant galaxies and lost cities in the jungle. But I was a poor math student, starting in elementary and middle school, so I accepted that I would have to be a casual consumer of science, reading about the latest finds in *National Geographic* rather than focusing on science and math coursework. This was my pattern, even through college, where I took only one 100-level science course in four years (I got a B−).

It wasn't until I was thirty years old, going back to school, that I decided to give it another try. This time, something happened, and suddenly my brain was picking it all up, making connections, seeing the patterns. This is when I realized the importance of math for thinking about all aspects of life and it is part of the inspiration for this textbook.

There is no math in this book, so don't worry about that. But it is worth exploring, just in a superficial way, why math is critical for a scientific understanding of the Universe and humanity.

The most common use of math in most sciences is the application of statistics. You may have heard that statistics is about probability and that is true, but mostly it is about whether or not we are fooling ourselves. As we will discuss later, human brains are very good at seeing patterns. In fact, they are so good that they often see patterns that aren't real.

Let's take an example from epidemiology. Say you hear about an old friend living near a nuclear power plant who suddenly gets cancer (a benign tumor). And you also hear about two other people who live on the same block as your friend who also have a form of cancer. Let's say that no one on your block has cancer at all, but you are aware of someone who lives between you two who has cancer, the only one on their block. This kind of "coincidence" would make some people think that the proximity to the power plant is causing the cancer (after all, we know that radiation can cause cancer).

Let's plot that out, since we are scientists. In Figure 1.2 the horizontal axis (X-axis) is the distance of a block to the power plant. The vertical axis (Y-axis) is how many people on that block have cancer.

When the initial data are plotted out, it certainly looks like there is a pattern (Figure 1.2a). From this, you might be justified in demanding the closure of the plant, or having the power company pay for new homes further away from the plant.

But what happens when we sample every block in the area? Let's say we go door to door for blocks around and find out every incidence of cancer in the surrounding blocks. If we do the pattern looks something like Figure 1.2b.

Box 1.5 (cont)

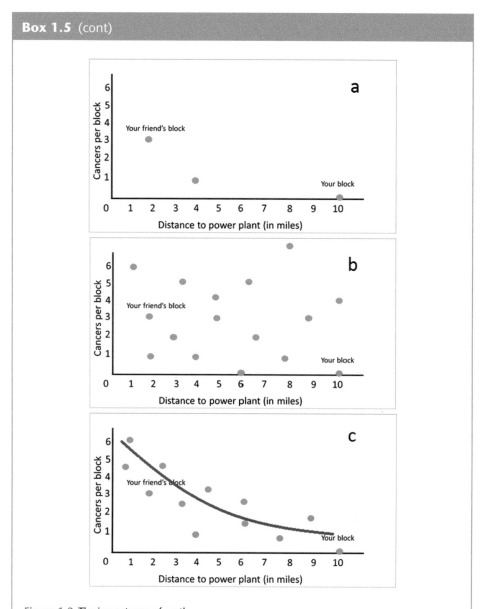

Figure 1.2 The importance of math.

So, it turns out it *was* a coincidence that your friend and their neighbors had cancer – by chance, there was a high rate of cancer on your neighbor's block, and a low rate (none) on your block. The cancer rates are randomly scattered on this graph. There is *no* relationship between proximity to the power plant and the appearance of cancer. This is exactly why statistics was invented – to check whether our guesses are correct.

Box 1.5 (cont)

But if we had seen a pattern like in Figures 1.2c, with a lot more incidents of cancer close to the nuclear power plant, and fewer the further you went out, we would definitely have reason to suspect a radiation leak. The important thing was that we went and examined the real data, and didn't just trust our initial instinct. We *tested our hypothesis*.

This is a pretty straightforward example, and you wouldn't even have to use any math to get to the answer, just plot the data. But sometimes it is a tougher thing to estimate by just looking and we have to use math.

What may surprise you is that most statistics are pretty simple. In this case, the statistic is called a correlation coefficient, and is roughly on the level of a ninth-grade algebra problem (all you need are some averages and square roots). So it is not particularly complex, and there is no reason for students to be intimidated (most of us use statistical software anyway).

Fairly simple math also helps explain why Godzilla and King Kong cannot exist. If you look at an animal like a deer, which is a herding grazer, and compare it with an elk, which is also a herding grazer, you will notice that although the elk is less than twice as tall, it is more than four times as heavy. Why is this? The answer is that animals are three-dimensional things, so increasing size increases size in multiple dimensions. Look at a cube of three feet along each side. It is $3 \times 3 \times 3$ feet = 27. If you double that, it is not $27 \times 2 = 54$, it is $6 \times 6 \times 6 = 216$. This simple law means that getting larger makes an animal very heavy, and being massive is hard work. A gorilla is only about 30 percent taller than a chimpanzee, but it is three times as heavy (maybe 400 pounds for an adult male). This is why you could never have a gorilla as large as King Kong – a 25-foot gorilla (about five times as large as a normal gorilla) might weigh 25 tons (400 lbs $\times$ 5 $\times$ 5 $\times$ 5). A gorilla skeleton would have trouble supporting that! If you look around at nature and find really big animals, much of their bodies are legs. This is why *Tyrannosaurus rex* (at a mere 15 tons) had huge legs and tiny arms. And to be *really* big we have to look for aquatic animals, where water supports the weight.

The world provides us with information like this all the time and it is up to us to make sure we can interpret it correctly. I once met Robert Alvarez, one of the most famous geologists of the last forty years (and discoverer, with his father, Walter, of the crater formed by the huge meteor that struck the Earth 65 million years ago, wiping out the dinosaurs). We were talking about math and he said: "math is the language of nature." What he meant is that nature is always telling us its secrets, but often we can only really understand them if we bother to understand the math.

Part I: Introduction to the Scientific Perspective on the Past

Chapter 2: The Origins of the Universe

Most of us take for granted the presence of our planet, our solar system, and even distant stars. That is because we are, by the standards of the Universe, pretty small creatures who don't live very long. So, in our lifetimes, and even the lifetimes of our parents, grandparents, great-grandparents, and back thousands of years, not much has changed in the night sky. But the Universe did not always look like this.

The Big Bang

About 13.8 billion years ago, everything you see, on Earth and in the night sky, even extending beyond what our most powerful telescopes can see, was compressed into a tiny subatomic space. Currently, estimates are that the Universe was 1×10^{-33} centimeters, which is much, much smaller even than the components of atoms: electrons, neutrons, and protons. That tiny point was, at that moment, essentially pure energy – there were no "objects." Even subatomic particles did not yet exist. This tiny ball of near-incomprehensible energy expanded out over tiny fractions of seconds, and currently scientists can model the state of the Universe as far back as 1×10^{-11} seconds, using specialized, expensive, and very large lab equipment (particle accelerators). In this time, the Universe spread out in size, rapidly doubling, then doubling again and again and again, within tiny fractions of seconds, so that by 0.00000001 seconds after the Big Bang the Universe was already the size of our solar system, and the energy was generating subatomic particles. The temperatures over this time period were similarly enormous but dropping rapidly as the Universe expanded, starting at perhaps 1×10^{35} degrees centigrade right after the Big Bang, and dropping to 1×10^{10} degrees by 1/10 of a second (by comparison, our Sun is only 5,700 degrees). By the end of that 1/10 of a second, the Universe was already a light year across, and here, the energy was no longer as dense, and the generation of subatomic particles ceased (see Box 2.7).

This 1/10 of a second threshold is critical, because, from this point on, the number of subatomic particles in the Universe is fixed – there will be no new generation of these particles, and they are simply shuffled among different

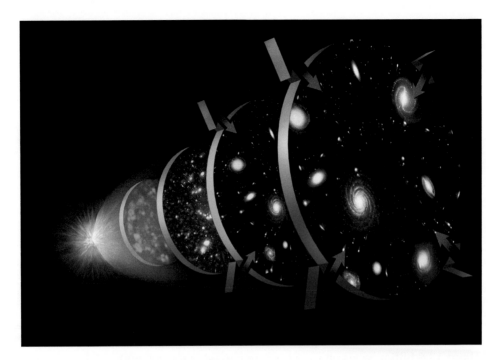

Figure 2.1 In the initial expansion after the Big Bang, within tiny fractions of a second, subatomic particles had formed out of the raw energy of the Big Bang, but the subsequent formation of atoms, clusters of atoms, stars, and galaxies took more than a billion years. The Universe continues to expand and today is more than 93 billion light years across.

atoms in the Universe for the next 13.8 billion years. From here on, physics follows the relatively familiar patterns that we still see across the Universe – subatomic particles, such as neutrons, protons, and electrons, forming atoms, and those atoms undergoing changes to become other atoms, to produce the Universe we see today.

The Universe has continued to expand over the last 13.8 billion years (see Figure 2.1), and today it is so large that even our largest telescopes cannot see to its edge. The observable Universe is 93 billion light years across, but that is simply how far we can see, not how large it is. In essence, we can see stars 46.5 billion light years away in any direction, even as we know it is larger – potentially much larger.

Expansion of the Universe

You may have noticed something a little funny about the math in that last paragraph. If the Universe is at least 93 billion light years across, and we can see stars 46.5 billion light years away, that would seem to mean that the light waves we see left their stars 46.5 billion years ago. Yet the Universe is only 13.8 billion years old. How could this be? The answer is that the Universe is still expanding. The most distant light waves we see were emitted billions of years ago, yet we are continuing to drift away from the source, very rapidly. For example, the very distant galaxy GN-z11 is roughly 32 billion light years away, but the light we see from it is only 13.4 billion years old. Roughly 400 million years after the Big Bang, that galaxy emitted light that we are only picking up today. And over those 13.4 billion years since it emitted its light, we have drifted an additional 18.6 billion light years apart. It is a little like tossing a baseball between two cars that are moving away from each other – when the baseball was tossed, the cars might have been 20 feet apart, but as they pulled away the distance increased so that, when caught, the cars might be 50 yards apart.

This expansion is an especially interesting phenomenon. One fact that still confounds astronomers is that the expansion is *accelerating*. Normally, when we think of the laws of motion, objects, once in motion, tend to stay at their velocity unless something applies force to it. For example, comets keep a roughly constant pace, as do planets, satellites, and other similar objects. But stars and galaxies are actually increasing their speed away from each other, dispersing at an ever greater rate. In fact, the expansion of the Universe is itself faster than the speed of light. To date, astronomers still argue about the source of energy causing this acceleration, because they cannot see any specific source that could cause such acceleration. (If you want to become a famous astronomer, solve this problem!)

Another interesting implication of this acceleration is that events that occur too far away will never send light that reaches us. Since the expansion is accelerating, events beyond a certain distance (14.5 billion light years) will emit light that can never keep up with the expansion. Imagine those two cars again. If you are in one car and throw an 80 mph fastball at the other car, but it accelerates up to 90 mph away from you, the baseball will never reach it. So we will never know about events that are too far away.

One thing that is difficult for some students to understand is the idea that it is not the objects in space spreading out into space, but it is space itself expanding. The Universe is not simply the objects in the space, it *is* the space. There is no "place" outside the Universe – there is nothing beyond the edge of the Universe, because there is no "beyond." This also means that there is no "center" to the Universe – no place where the energy or matter is denser from the center point of the Big Bang. Every place is expanding away from every other place, not just the objects or energy but the space in which the objects are situated.

The Formation of Stars and Galaxies

After the Big Bang, the Universe was full of very simple atoms – mostly hydrogen (which has only one proton and one electron, and no neutrons). Astronomers estimate that 75 percent of the mass of the Universe today is composed of hydrogen atoms, but initially hydrogen comprised almost all of the mass. Much of the story of the Universe over the next 13.8 billion years is the story of what happens to these hydrogen atoms.

Over time, these hydrogen atoms tended to clump together. Even though these atoms are very tiny, they do have mass, and anything with mass will pull on other objects, simply through gravity. As these clumps got larger and larger, they started to have stronger gravitational forces and pull in more and more hydrogen. Once these clumps of hydrogen became large enough, they started to burn, and these burning hydrogen balls are what we today call stars.

So stars are great balls of burning gas, but they do not burn in the conventional sense. When we see wood or hydrogen gas burning, we are witnessing a specific type of chemical reaction called oxidization, in which a fuel chemically combines with oxygen, releasing heat. When oxygen combines slowly with a fuel, say iron, we get rust, and a very slow form of oxidization. And when we have a fast oxidization, we get what we call fire.

However, in stars we have another process entirely. Here, two different atoms are compressed together enough that they form another element. This process is called *fusion*, and it is a type of nuclear process. In fact, it is the same type of process that is used in most modern nuclear weapons (the so-called hydrogen bombs). The process starts with atoms of the simplest element, hydrogen (which were created in the Big Bang), being forced together to form the next simplest element, helium. The force that pushes these hydrogen atoms together is their own gravity, compressing the atoms.

One way to think about this force is to imagine yourself diving to the bottom of the ocean. The deeper you go, the greater the pressure of the water, and because humans have spaces of air in their bodies (especially in their lungs, and in the sinuses of their heads), they cannot dive beyond a certain depth before the pressure crushes them. The force of the water under the effect of gravity is, in effect, greater at greater depth. It is much this way in a ball of gas, in which the center is under greater pressure. Normally, the nuclei of atoms repel each other, but if the ball of gas is large enough, with enough gravitational force, hydrogen atoms will be forced together, and the nuclei undergo nuclear fusion. Fusion releases a tremendous amount of energy, and when you look up at the Sun you are witnessing exactly this fusion reaction.

But once the atoms are fused into helium the process does not stop (Figure 2.2). In fact, the fusion process continues, forcing together simpler

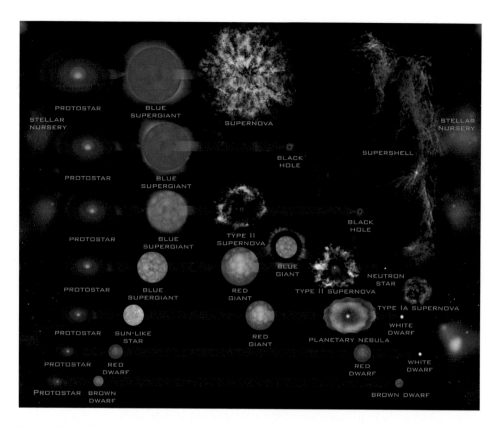

Figure 2.2 Stellar evolution. Notice how much the size of the star determines its fate. All stars of the same size will go through the same stages of stellar evolution. Note that our Sun will end up as a white dwarf (and subsequently a black dwarf, undergoing heat death when all the energy is exhausted), whereas more massive stars will become supernovas. Very small stars (i.e., brown dwarves) never undergo nucleosynthesis, and so do not change their states over time because there is insufficient gravity to generate fusion. A brown dwarf is larger than the planet Jupiter but smaller than the Sun.

atoms, and creating more complex, heavier atoms. In most stars, the process generates the elements you are familiar with – lithium, beryllium, boron, carbon, and nitrogen, right on down the periodic table until they get to iron. Iron is so dense that, in most stars, the gravity is not strong enough to squeeze it. So the end point of most stars is a big ball of iron floating in space. However, if the star is large enough, the gravity can be so strong that even a ball of iron can be compressed. When this occurs, the star is said to become a supernova, which is the release of energy from the fusion of iron. This explosion, through fusion, generates force that can fuse iron into the heavier elements, and there are many (zinc, lead, gold, tungsten, and uranium among them). The supernova is an explosion that spreads these elements across the Universe, where they later become other stars, planets, asteroids, or other celestial bodies. The fact that these heavier elements are only produced by supernovas is why they are less common on Earth, whereas iron is very common as it is the end product of most stars. The Universe produces a great deal of iron but not as much gold and silver.

Stars live and die – they have a finite lifespan. That lifespan is determined by their mass (see Figure 2.2). Large stars, under much greater gravitational forces, live much shorter lives, as their atoms are forced to fuse at a great pace. Smaller stars burn their fuel at a more stately pace, not as hurried by the powerful gravitational forces of larger stars.

Once a star has enough gravitational mass it will find itself attracting, and being attracted to, other stars. As these stars are tugged together, they will combine their forces of attraction and their inertia, to form an orbit. Over time, as more and more stars are pulled together, they will make an enormous rotating mass called a galaxy. Many of the "stars" you see in the night sky are actually these groups of stars, burning brightly but distantly.

Our Solar System and Earth

The Universe has been seeded with elements from these stars living, then dying, in massive explosions, over billions of years. These particles, adrift in space, ultimately clump under gravitational force, much as the hydrogen does, but if they do not form large bodies with large gravitational masses they will never undergo fusion. This is how planets are formed (even at 318 times as massive as the Earth, Jupiter is not large enough to undergo fusion). Using the exact forces that bring stars and galaxies together, these clumps of matter start rotating in orbits. This is precisely what happened to our solar system about 5 billion years ago. Clumps of matter near our star, the Sun, were far enough away not to be pulled into the star but not far enough away to resist its gravitational pull. This matter orbited the Sun, clumping under its own gravitational force into medium-sized bodies that we know today as the planets. One of those clumps, forming roughly 4.5 billion years ago, is our planet Earth.

Box 2.1 Matter and energy

You have probably seen the famous equation: $e = mc^2$. The variables of that equation are pretty simple: energy = mass × speed of light squared. This means, of course, that there is a certain amount of energy stored in mass, and the greater the mass (of an object) the more energy it has in it. This is the reason why nuclear reactions, even of relatively small amounts of mass, release a tremendous amount of energy. Mass is lost, and converted into energy.

Box 2.1 (cont)

But you can, algebraically, rework this equation in a way that is also important: $m = e/c^2$. This says that mass = energy divided by the speed of light squared. The implication of this overall equation is that, much as mass can be converted into energy, energy can be converted into mass.

In an important sense, matter and energy are interchangeable. So when we talk about the Big Bang, we argue that, at the origin, there was nothing but raw energy. Part of the process of the Big Bang was the conversion of this energy into matter. Much of the work of physicists over the last fifty years has been to learn just how energy is converted into matter. This work is expensive and time-consuming, typically taking place in high-energy particle accelerators, where the conditions at times just after the Big Bang are simulated.

One important implication of this equivalency is that you can think of matter, in the form of subatomic particles, as lumps of "condensed" energy. We perceive of it as something different than energy, since it takes a quite distinct form from other forms of energy (light, heat, etc.), but the underlying physics tells us that everything in the Universe composed of matter is ultimately composed of energy. Ultimately, there are only different forms of one "stuff."

Box 2.2 How do we detect elements in space?

You will sometimes read about the chemical composition of various celestial bodies, or how the Universe is mostly hydrogen. Normally, when we evaluate chemical composition here on Earth, we have to physically or chemically inspect the element to know. So how do astronomers on Earth, millions or billions of light years from what they are observing, know what it is made of?

The answer is something you have probably already been exposed to – a neon sign. In a neon sign, the tube of neon gas has electricity running through it. This electrical energy has the effect of making the gas vibrate at a very high speed. When the gas vibrates, it releases the energy as light. For sign makers, and astronomers, the important thing is that different chemical elements will vibrate at different frequencies, making different colors. For example, in neon signs, the element neon glows orange, hydrogen glows red, helium glows yellow, carbon dioxide is white, and mercury is blue. When you look at a sign with different colors, you are looking at several different tubes with different gases.

Astronomers will analyze the light from a star or other celestial body (see Figure 2.3) and look at all the frequencies of light being emitted, so they

Box 2.2 (cont)

can determine which elements are present. It is fairly simple in theory, but obviously it requires some pretty sensitive instruments to be able to determine what light is being emitted by extremely distant stars.

Figure 2.3 The Crab Nebula. This is the remnant of a supernova that became visible to Earth in 1054. The supernova was visible to the naked eye on Earth for roughly two years before receding to its present size. Several supernovas have been highly visible to humans over the last few centuries, and are documented in writings from China, Europe, and the Middle East. They may also be recorded in rock art from North America and Australia. The supernova that resulted in the Crab Nebula (SN1054) is estimated at roughly 6,500 light years away, so it would have actually occurred at around 5,500 BCE, with the light not arriving on Earth till 1054.

Box 2.3 Size matters in space

Stars don't start out as stars. They start out as random atoms of hydrogen floating through space as the result of the Big Bang. Each atom, tiny as it is, has a gravitational force, so, over millions of years, these atoms will drift toward

Box 2.3 (cont)

each other, ultimately forming great big balls of hydrogen atoms. But what makes these balls of gas become stars – nuclear reactors generating millions of degrees of heat?

It is their own gravitational force. When a ball of gas gets massive enough, the force of gravity at the center compresses the hydrogen, and the nuclear fusion reaction starts. The pressure is similar to what you feel when you dive to the bottom of a deep swimming pool – the deeper you go, the more pressure you feel on your ears and sinuses. In fact, you can only go so deep before it gets dangerous, not because of a lack of air, but because the pressure is too much for the human body. Inside a star, the same principles are operating.

The idea that pressure alone can generate combustion is used by humans on Earth. When you see diesel tractor-trailers driving down the highway, you are seeing the result of this process. Diesel fuel in those engines is not ignited by a spark plug as it is in gasoline engines. Rather, the diesel (and air) is injected into the cylinder, and the piston compresses it, producing so much pressure that the fuel ignites on its own. If you remember your high school science class, your teacher probably told you that increasing pressure on a gas generates heat. In a diesel engine, the increase in pressure on the air and fuel generates so much heat that the fuel ignites, driving the engine and truck forward.

But back to our big balls of gas floating in space – sometimes a ball of gas does not get big enough to generate those enormous forces. In our own solar system, Jupiter is a giant ball of gas, and even though it is equivalent in mass to 318 Earths, it is still not enough. So, instead of combusting, it is just a giant gas planet. Sometimes these balls get large enough to start a small amount of fusion, and generate just a bit of energy – these are called "brown dwarves." If in millions or billions of years they attract more hydrogen atoms, they may become massive enough to ignite in an overall fusion reaction. But until then, they just sit, on the cusp.

The lifetime of a star is also largely determined by size (see Figure 2.2). A truly big star has so much gravitational force that the fusion reactions will happen very rapidly, and the star may only last several hundred years. A smaller star, such as our Sun, has a slower burn rate, and the steady process of fusion in such a smaller star's furnace can last billions of years. This is lucky for us, because it took almost 4 billion years of life on Earth before humans arrived.

Box 2.4 Fission and fusion

When we talk about stars, we are talking about nuclear reactions. When we talk about nuclear power plants, we are talking about nuclear reactions. And when we talk about nuclear weapons, we are talking about nuclear reactions.

But all nuclear reactions are not the same. In fact, there are two basic types – fission and fusion – and they are essentially the opposites of each other.

The earliest nuclear weapons, like those used in World War II, as well as all nuclear reactors used to generate power, are fission devices. They generate energy by splitting one atom into two atoms (many times over and over again). This division of the atom releases a great deal of energy. Splitting atoms is not that easy, so we tend to use uranium, which is among the easiest elements to split. The reaction can be slow and controlled, as in a nuclear reactor, or fast, as in a nuclear explosion, depending on how fast the atoms are split. In a nuclear fission explosion there is a very rapid chain reaction, as one atom being split causes enough energy for others to split, and so on, in a quick chain reaction that causes a rapid release of energy. In a nuclear reactor, atoms are split just a few at a time, preventing a rapid chain reaction, so that it can be controlled.

But stars use fusion, in which two smaller atoms are fused into a large one, essentially the reverse of fission. This process actually releases much more energy than fission and, for us on Earth, it is much less controllable. When you hear about the "hydrogen bomb," that means a fusion bomb in which hydrogen atoms are fused to make helium. This is precisely the process that stars use to generate their energy. But it requires an enormous amount of pressure to get those atoms to fuse (see Box 2.3), however, so it is very difficult to do. In fact, in a fusion bomb, a fission explosion is used to generate the force necessary to fuse the hydrogen atoms into helium. So a hydrogen bomb is essentially two bombs – a fission bomb and a fusion bomb.

But we have not figured out, yet, how to use fusion to generate power. And there are good reasons why we would want to. As you probably know, the byproduct of fission from a nuclear reactor is very poisonous to humans and essentially all other life-forms. And it is poisonous for thousands of years. The byproduct of hydrogen fusion is helium, which is harmless (and very useful). But, despite some fraudulent claims, a way to generate enough force for a fusion reaction without a fission explosion has never been found, so currently this method is obviously too risky. If we can ever figure out how to generate energy from fusion, we will have a fairly clean source of energy, and considering how common hydrogen is, we would have fuel forever.

Box 2.5 Why spheres?

Have you ever wondered why stars, planets, and many galaxies are in the shape of a sphere? Why are they not saucer shaped, or donut shaped, or cubes for that matter?

The answer has to do with something unique about a sphere. In a sphere, all points in the surface are equidistant from the center. For a large object in space, this means that gravity effects the surface equally. Imagine an irregularly shaped object – as gravity pulls on the surface of the object, the surface material might be drawn toward the center. Spots further from the center (high spots) will be pulled closer. But as the object becomes a spherical shape, the gravitational forces start to equal out, so that every part of the surface is being pulled equally. This is a stable state that many celestial bodies ultimately become.

Something similar happens with soap bubbles. In a soap bubble, it is not gravity but surface tension that is pulling on the surface. When you make an irregularly shaped soap bubble with your bubble wand, surface tension pulls more on some parts than on others, until the bubble becomes a sphere. Once it is a sphere, surface tension is the same across the entire bubble, and it wants to stay in that shape. Until you pop it, of course.

This makes the sphere a very natural shape, and you will see it across nature, in space, in geological formations, and in plants and animals.

Box 2.6 Before the Big Bang

Astronomers seem to have a pretty good understanding of the Big Bang, and what happened right afterwards, even up to 10^{-43} seconds after the Big Bang (a length of time so small that astronomers had to make up a new word for it – a *Planck*, named after the famous physicist).

But what about *before* the Big Bang? Knowing what happened before the Big Bang is a big problem. At the instance of the Big Bang, everything, including space and time, was compressed into a tiny subatomic space. All this information was so compressed that it seems like there is no way to know anything about what there was before the Big Bang. No information seems to have been able to escape that compression of energy, matter, time, and space.

Of course, that hasn't stopped people from trying to figure it out. But the problem is that, without any real data, they are all guesses. Some scientists have argued that the Universe is in a never-ending cycle: it expands after a Big Bang, then compresses billions of years later via gravity into another subatomic space, which produces another Big Bang, and the cycle starts again. The problem with this idea is that the Universe appears to be accelerating as it

Box 2.6 (cont)

expands, not slowing. If we were in a cycle of Big Bangs, we would expect an explosion, then a slowing as gravity took effect, then a compression. The acceleration implies that the Universe will keep getting larger and never compress back into a smaller space.

Other scientists argue that there was nothing "before" the Big Bang. They argue that the Big Bang created the various forces of the Universe, and much as the Big Bang created gravity, or magnetism, it created time. Therefore, it doesn't make sense to talk about "before the Big Bang," because there could be no "before" the creation of time. We perceive of time as something that has always existed, but this may simply be a limitation of the biology of our brains. Perhaps we perceive of time in a way that is somehow inaccurate.

Many of these ideas get philosophical pretty quickly because of the lack of data, so many scientists choose not to even try to guess. It is, for now at least, a unique and unanswerable question.

Box 2.7 A light year

Astronomers use a measure of distance called a light year. As you might expect, it is the amount of distance that a light beam travels through empty space in a year (light gets slowed down passing through water, glass, or other materials).

You have probably heard that nothing can travel faster than light, so when a spaceship in a science fiction movie goes into "hyperspace," just keep in mind that you are watching *fiction*. That limitation is one of the predictions of Einstein's theory of special relativity, and it still holds for objects in space.

The confusing thing is that the Universe is itself expanding faster than the speed of light. That means that some stars are moving away from each other faster than the speed of light. So, for example, as the Universe expands, and all objects move away from each other, the *relative* speed of our Sun moving away from the very distant galaxy GN-z11 is actually greater than the speed of light. Today we are seeing the light it emitted about 13 billion years ago in our direction. But at some point, long ago, its *relative* motion away from us exceeded the speed of light, which means that we will never see any light emitted in our direction after that moment. The light will be racing toward us at the speed of light, but since we are moving away faster than that we will never see it – it will just be racing through empty space forever. And if we ever invented a spaceship that could somehow travel just at the speed of light, it would never get to GN-z11, because the galaxy would keep receding away from our spaceship faster than we could go.

Chapter 3: The Structure and History of the Earth

Formation of the Earth

Around 4.6 billion years ago, the space dust and rock orbiting around our Sun had consolidated into a fairly sizable ball. Gravity was doing most of the work here – pulling the space material into a ball (a process called *accretion*) but also keeping this growing ball orbiting the local star. This ball was heated by the newly formed local star, the Sun, and from the forces of its own gravitational pressure. In fact, it was so hot that it was still entirely molten. During the first few hundred million years of the formation of the Earth, the rest of our solar system was also still young, and space debris floated in and out of our orbit, not yet forming into the rest of the planets of our solar system. This was a time when the newly forming planets were repeatedly struck by this debris, which was often moving at extremely high speeds.

One product of these collisions is our Moon. Roughly 4.5 billion years ago, a meteor struck the new planet and blasted a chunk of it off into space. This chunk started orbiting its parent, and we now call it the Moon. Over time, the forces of gravity inside the Moon pulled it into a sphere (see Box 2.5). Likewise, the Earth's gravity forced it to return to a sphere. This means that the Moon is composed of the same material as is the Earth, a fact that we discovered when astronauts landed on the Moon almost fifty years ago.

The Earth has several layers – the outer *crust*, which is about 25 miles thick in the continents, is all we ever see, but it actually represents less than 1 percent of the Earth's mass. Far below that is the inner *core*. The core is made up of the heavier elements on Earth – iron and nickel, which have settled through the other molten materials to the center of the Earth under the forces of gravity. The inner core is actually thought to be in solid form, even though it is still hot enough to be liquid. The pressure of the outer layers are so great that the molecules in the core are forced together, and once packed that tightly they become a solid. The outer core, under slightly less compressive pressure, is a metallic liquid, that plays a critical role for life on Earth (see Box 3.2).

But between those two layers, the largest portion of the Earth, above the core, and below the crust, is a thick, slow-moving fluid called the *mantle*.

Originally, the entire Earth was a molten lump in space, but over time the outer surface of the Earth cooled (space is, after all, pretty cold) and formed the rocky outer crust. One way to imagine this is to think of a volcano – it erupts, and then the lava cools into rock. The Earth was essentially all lava – it was completely molten, and when it cooled on the outer surface it left a hard, rocky outer section. This means that the crust and the mantle are largely composed of the same materials, but the outer crust is in solid form whereas the mantle is still hot enough to be fluid.

The Earth has undergone many large-scale changes since this period, but there are still remnants of the rock formation from this time period. Geologists have discovered ancient rocks that date to the end of this cooling period, some more than 4 billion years old (see Box 3.3). The period of the formation of the surface lasted for about 700 million years, and is known as the *Hadean Eon*. The Earth during most of this time was uninhabitable, as it was far too hot for life to survive.

The Appearance of Life

Toward the very end of the Hadean, the Earth's surface had cooled enough for water to form. I discuss the origins of life on Earth in Chapter 4, but one thing to keep in mind is that water is probably absolutely necessary for life to form. And it is during this time, after about 3.8 billion years ago, that we see the first evidence of life. This period of the Earth is called the *Archaean Eon*, and it lasted more than a billion years, until 2.1 billion years ago. This is when we see the first continents form, and geologists have found evidence for the movement of continents as far back as 3.8 billion years ago.

The life on Earth at this time was very simple – probably just bacteria – and it survived through photosynthesis, using the energy from the Sun to live and reproduce. There was almost no oxygen in the atmosphere for much of this time period, so it would have been extremely inhospitable for humans. The surface would have been recognizable to us because oceans had formed, and land masses as well, so the overall form of the surface would have been somewhat familiar to the modern eye.

In many ways, the Archaean is the start of modern geological processes. During this time period continents formed and drifted, much as they do today. Also, with the formation of large bodies of water, modern erosional processes and rock formation started. Geologists have identified *chert*, which is formed when silicon-based molecules precipitate in water, that dates from this era. The formation of chert today is identical. This is when we first see sedimentary rocks (such as sandstone), which form when sediments settle in the bottom of large bodies of water. So during the Archaean, the geology of Earth as we think of it had established its patterns: the movement of water was driving much of

Figure 3.1 The bands of rusted iron in this rock formation document the appearance of large amounts of free oxygen on Earth between 3.1 and 2.1 billion years ago. The source of this oxygen was simple single-celled photosynthetic bacteria called cyanobacteria. Without the large amounts of free atmospheric oxygen generated by the cyanobacteria, no later complex life could have evolved. It took more than 2 billion years of photosynthesis to build up enough atmospheric oxygen for this later complex life to evolve.

the geological processes on the surface, even as the powerful forces under the crust were driving the movement of the continents (see Box 3.6).

During the Archaean primitive single-celled life expanded, and photosynthetic bacteria spread around the world. Photosynthesis produces oxygen, and for over a billion years oxygen was being generated, but it could not readily build up in the atmosphere because of the presence of elements on the surface of the Earth that readily bind with oxygen, notably iron. In fact, this is the period of time in which we see large bands of oxidized iron (see Figure 3.1), evidence that oxygen was present, but also evidence that it was not available for the atmosphere.

Eventually, however, there was no more iron on the surface to soak up the oxygen, and the transition to the next major period in the evolution of the Earth, the *Proterozoic Eon*, was characterized by the build-up of atmospheric oxygen. Oxygen is necessary for more complex life, so this build-up was a critical step in the evolution of multicellular life. This accumulation of atmospheric oxygen is known as the *Great Oxygenation Event*, and it was an essential precursor for all life that followed. This oxygen in the atmosphere created environmental patterns and temperatures that would be recognizable today, and there are researchers who argue that, during this time period, the surface temperatures might have dropped low enough that the environment could have created glaciers across the landmasses.

The Final 500 Million Years

For us, the last 10 percent of the Earth's history is the most interesting, as it is when complex life comes to dominate the biosphere, but it is worth remembering that it is a relatively brief period compared to the overall history of our planet. The Proterozoic ends with the appearance of complex, multicellular life at the *Cambrian Explosion*. We discuss the Cambrian and later eras over Chapters 3–9.

Box 3.1 Why does geology matter to everyday life?

It might seem that geology is only important for a few practical reasons, such as finding and mining minerals found beneath the outer surface of the planet, or making sure the ground is strong enough to support a large building before construction. But the Earth, and how we are populated across it, is a major driving factor in historical and political trends, and has been since before humans first walked the Earth.

One major factor is the distribution of resources. During World War II, the countries that controlled iron and oil had major advantages over those that did not, and the control of these resources drove many of the events of the war. Prior to the outbreak of the war against the USA, Japan had, through imperial expansion, invaded Malaysia, Indonesia, Korea, and parts of China. A primary goal of this expansion was to acquire access to resources critical for building up a large military. Germany, similarly, lacked primary access to iron ore and oil, and the Nazi regime invested considerable effort into capturing the oil fields of the Caucasus region and transporting iron ore from conquered Norway. The United States possessed deposits of both iron and oil, as well as other critical materials such as bauxite (for aluminum), copper, and manganese, and was able to produce the mechanized weapons that won the war, relatively unimpeded. Today, Saudi Arabia owes its political power to the fact that it has enormous oil deposits, and the adjacent country Yemen, which has little international power, has scant oil resources. But before the twentieth century, when oil was unimportant and relatively unexplored, Saudi Arabia was a comparatively undeveloped, premodern country.

Geology has driven some major biological events in prehistory. Prior to 2 million years ago, South America and North America were separated by several hundred miles of ocean. South America was populated by marsupials (pouched

---

Box 3.1 (cont)

mammals like kangaroos), as it once was joined to Australia (and Antarctica), whereas North America was populated by placental mammals, due to its connection with Asia at the Bering Strait. Between 1 and 2 million years ago, continental drift pushed South America against North America. North American mammals migrated south and South American mammals migrated north. Over the next million years or so, the placental mammals slowly outcompeted the marsupials, and today only the opossum survives as a marsupial in the Americas. The drift of the continents, and the level of the ocean, were important determinants for the biological populations of the Americas.

But, even long before the evolution of mammals, the Earth was uniquely suited to life, and this is because of its geological structure. The Earth is large enough to retain a fluid inner core, which generates the magnetosphere that helps protect the Earth from the Sun's radiation. This protective barrier helps retain liquid water on the Earth's surface, which is essential for life (see Box 3.2).

As we explore historical events in later chapters, it is important to understand that there are often deeper causal factors that have driven many of these occurrences, and geology is one of the most underestimated factors that has driven world history and prehistory.

Box 3.2 Why does the Earth have an atmosphere but the Moon and Mars do not?

As we look up into the night sky it is natural to wonder about life on other worlds. Our nearest neighbor is Mars, and fiction writers have offered a wide array of ideas about Martian life, sometimes hostile, sometimes benign. Mars is a good candidate for life – it has been around roughly the same amount of time as the Earth, and it inhabits the "Goldilocks Zone" for planets, in which they are not too far from a star (too cold) nor too close (too hot). But the surface of Mars looks desolate and forbidding – a waterless, airless desert on which life could never start. Despite this appearance, the main reason why Mars has no life actually has nothing to do with its surface – it is what is under the ground that really matters.

Most of the interior of the Earth is a viscous fluid that moves very slowly, and due to differences in heat between the layers the liquid moves in large flows, up toward the surface, then back down. The cyclical motion of iron in the fluid metal outer core has the effect of generating a powerful magnetic field that extends

Box 3.2 (cont)

from the core of the Earth out tens of thousands of miles into space (Figure 3.2). For most people, the visible effect of this powerful magnetic field is apparent when you pick up a magnetic compass. The needle points toward the north because that is the orientation of the magnetic field generated by the motion of liquid metal inside the Earth. The polarity of this magnetic field is effected by the Earth's rotation, so the poles of the magnetic field are at the poles of the Earth.

As useful as compasses are, the most important job of this magnetic field, extending tens of thousands of miles out into space, is to act as a shield against the powerful solar radiation the Sun generates. This shield is known as the *magnetosphere*. Without the magnetosphere, the Sun's powerful radiation would strip away our atmosphere, including most of our liquid water. Since the presence of water is a critical precondition for the evolution of life, the magnetosphere is one of the most important factors that a planet must have if life is to evolve.

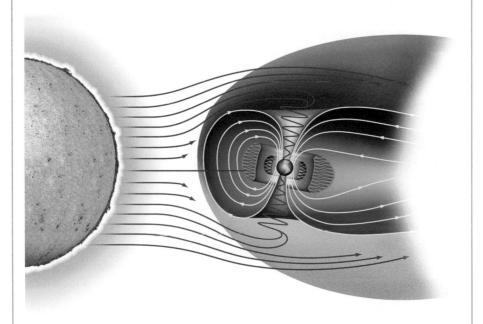

Figure 3.2 The magnetosphere. Probably the most important result of the Earth having a liquid outer core is that movement in this layer, which has a large amount of iron in liquid form, generates a powerful magnetic field. The magnetic field is convenient for us, because we can use compasses to identify the north and south poles of the Earth, but more importantly, it shields the Earth from the Sun's powerful radiation. Without this protection, the Sun would have stripped away Earth's atmosphere and liquid water. Mars has no fluid inner core and mantle, and generates no magnetosphere, which is why it has hardly any atmosphere and no liquid water.

Box 3.2 (cont)

But Mars has no liquid center, and no magnetosphere. This is due to its small size. Mars is about the same age as Earth, but it is roughly the half the mass of the Earth. Since smaller objects cool faster, its liquid core cooled enough to solidify more than 3 billion years ago. This means the planet can generate no magnetosphere, and without a magnetosphere Mars has no atmosphere and very little water.

But Mars was not always like this. Our planetary neighbor has been the subject of direct exploration since the 1970s, when the *Viking* lander and *Viking* orbiter both started sending high-resolution images from the surface of the planet. Over the last forty years, orbiters photographing the surface of Mars have identified ancient river beds, in which liquid water was once flowing, and other geological features that are best explained by the widespread presence of liquid water across the planet. Today, any water on Mars is in a solid form, mostly in frozen icecaps at the north and south poles. But billions of years ago Mars had a full atmosphere similar to ours and large bodies of liquid water.

The discovery of water on Mars has led astronomers and biologists to consider whether life could have existed on that planet. So far, there has been much speculation but very little actual evidence that life has ever existed. And if life did exist, it would have been similar to the early single-celled primitive life that existed on Earth 3 billion years prior to the evolution of multicellular organisms. But, if there were life, it would have died out billions of years ago, when the core of Mars cooled and solidified, and the magnetosphere disappeared, allowing the harsh radiation of the Sun to strip away the life-giving atmosphere. As we continue to explore Mars, we may be able to ultimately determine whether the small red planet next door ever did support life.

Box 3.3 How do we date old rocks?

Geologists, paleontologists, and archaeologists all study events that happened thousands, millions, or billions of years in the past. Knowing the dates of the events is often critical, but how do these scientists date these ancient occurrences?

There are two basic principles for dating rocks. The first is called *relative dating*, which uses the principle of *superposition*. Under superposition, older rock layers are always below younger layers, because younger layers are laid on top of older layers. There is not always a direct relationship between the thickness of layers and their age, because sometimes upper layers get stripped away by erosion, but younger layers have no way of getting underneath the

Box 3.3 (cont)

older layers. So if you have an absolute date for one layer, you know that everything below it is older and everything above it is younger

The other category of dating rocks is known as *absolute dating*. In absolute dating, the actual ages of the rocks themselves can be determined. Only certain types of rocks can be dated directly, and most methods for dating rocks and bones use the principle of *radioactive decay*. Some atoms are inherently radioactive, which means that they are always ejecting subatomic particles. When one of these elements kick off a particle, it becomes a different element. If there is a rock composed of one of these unstable elements, over time more and more of the rock will change from the original element to the new one.

Scientists know how fast this radioactive decay occurs (the *half life* of the unstable element), so when they want to date a rock, they look at how much of the original (parent) atoms are present relative to the number of new (daughter) atoms. The age of the rock can be determined by measuring the ratio of parent atoms to daughter atoms.

Of course, only a few rocks contain sufficient amounts of radioactively unstable elements, so scientists often spend an enormous amount of time looking for such rocks. When dating the oldest rocks on Earth, geologists look for zircons, which are small crystals that have tiny amounts of uranium when they originally form. Over time, this uranium decays into lead. So, to date the crystal they look at the ratio of uranium to lead (*uranium-lead dating*). Currently, the oldest date from zircons is 4.4 billion years ago.

Other atomic elements are also used for dating fossils and bones. In order to date the earliest human bones in East Africa, rocks in the associated layers of sediment are dated with *argon-argon dating*, in which an unstable isotope of argon decays into another, more stable isotope of argon. The principles are identical to the uranium-lead dating but with different atomic elements.

A similar principle applies to dating organic material such as bones with *carbon dating*. Some of the carbon in your bones is in a particular form, carbon-14 (^{14}C), in which there are two more neutrons than in most radioactively stable and common form of carbon carbon-12 (^{12}C). Every time you eat, you replenish the ^{14}C, which comes into our bodies through plants, or animals that have eaten plants. But when an animal dies, it stops absorbing any ^{14}C, and over time that ^{14}C begins to decay into the more stable ^{12}C. Scientists wanting to know how long ago the animal died can look at the ratio of ^{14}C to ^{12}C and know how old the bones are.

The problem with carbon dating is that ^{14}C has a very short half life (about 5,700 years), and after about 60,000 years there is so little ^{14}C left that it

Box 3.3 (cont)

cannot be measured. So if scientists find a bone or plant fragment with no measureable ^{14}C, all they can say is that it is older than 60,000 years. This means that it cannot be used on such things as dinosaurs or early humans, which are millions of years or hundreds of millions of years old. Fortunately, the rock-dating methods, such as uranium-lead or potassium-argon, have half lives in the billions of years; so as long as there are radioactively unstable rocks in the sediment layers with the fossils, they can be dated.

Sometimes scientists combine relative and absolute methods. For example, if paleontologists discover a fossil, and it was found in an undatable layer of rock but it falls between two layers that can be dated, they know that the fossil could be no older than the deeper layer and no younger than the upper layer. This bracketing is common when paleontologists discover early humans in East Africa, where there are many datable layers that are often only a few tens of thousands of years apart.

Box 3.4 The "Goldilocks Zone"

Earth is a nice place to be. It is not raging hot like Mercury (which gets to 400 degrees centigrade at noon), nor is it a frozen rock in space like Pluto (which never gets warmer than −220 degrees centigrade). For living things, Earth is just the right temperature – warm enough for water to be liquid, yet not so warm that water simply vaporizes. And since life cannot exist, as far as we know, without water, Earth is the perfect place for life to evolve. It is just the right distance from the Sun. This distance, which creates the not-too-hot and not-too-cold conditions is called the *Goldilocks Zone* (it is "just right").

But what about other planets? In our solar system, Mars and Earth fall in this perfect distance from the Sun (this region is technically known as the *Circumstellar Habitable Zone*). For decades, astronomers have been looking at other stars with orbiting planets, hoping to find where else in the Universe life could exist. Quite a few candidates have been proposed. Several stars discovered by the *Kepler* space telescope have been seen as good candidates: the medium-sized star Kepler 62 has planets orbiting that appear to fall within the habitable zone, as well as Kepler 186 and Kepler 442 (the stars were named after the telescope that discovered them). But these stars and planets are still extremely distant – 490 light years is the closest, whereas 1,200 light years is the most distant observed so far by *Kepler* and related missions. So even if there is life on planets, we might never know, especially if the life never acquired radio wave transmission technology.

Box 3.4 (cont)

It would take far too long for space probes with our current level of technology to get to these planets to search for life. Currently our fastest probes travel at roughly 36,000 mph (58,000 kph). A light year is roughly 5.8 trillion miles (9.5 trillion kilometers), so with our fastest probes it would take 16,000 years to travel a single light year. Obviously, traveling to a star 490 or 1,200 light years away is not practical, because it would take more than seven million years for the probe to arrive. But even if it did, and started sending radio signals back to us, it would take another 490 years for the news to reach us on radio waves. That is a long time to wait!

Box 3.5 Our solar system: rocky planets and gas planets

You are probably familiar with the planets in our solar system: Mercury, Venus, Earth, Mars, Jupiter, Saturn, Uranus, and Neptune. One thing about the planets that you might have noticed is that the inner planets (Mercury, Venus, Earth, and Mars) are all rocky planets, whereas the outer planets (Jupiter, Saturn, Uranus, Neptune) are all gas planets. That is not an accident. The same powerful solar winds that stripped away the atmosphere from Mars (see Box 4.2) pushed the gas molecules in the inner solar system out past all the rocky planets. Once out at that greater distance from the Sun, the solar winds are not as strong, so the gas could accumulate into planets. In general, there is so much hydrogen in the solar system that the mass of the gas planets far exceeds the mass of the rocky planets; Jupiter alone is many times more massive than all the rocky planets combined. And when astronomers, using powerful telescopes, look out at more distant solar systems, that pattern appears to hold, with rocky planets close to stars and gas planets further out, suggesting that there is nothing particularly unique about our solar system.

Box 3.6 Continental drift explains everything

Once Earth's crust formed, and water began to accumulate, the geological patterns and forces took hold that govern the planet to this day.

The surface of the Earth is formed by this crust, but the inner 99+ percent of the Earth is fluid. The crust is thickest on land and thinnest in the deep ocean floor, so cracks or weaknesses in the crust tend to occur on the ocean floor. In fact, that is where new crust is formed, as the roiling molten mantle forces its way through the crust at the weakest points. This process, whereby molten mantle forces itself through cracks in the crust, creating movement of the crust,

Box 3.6 (cont)

is known as *plate tectonics*, and is one of the most important geological processes to understand because it explains so much of the broad-scale geological patterns we see on Earth.

The mantle is a thick fluid with strong currents generated by the heat of the core. As this thick fluid cycles, it generates powerful forces on the crust. These forces push on the crust, and create cracks as the large portions of the crust are pulled apart or pushed together. In places where these cracks open up, the mantle rises to fill the gap. Generally, this happens on the seafloor, where the crust is thinnest and weakest (although there are some areas of continental crust that are pulled apart – for example in East Africa). This mantle filling into the crack cools, forming new crust surface. This is why the sea floor is the youngest part of the Earth's surface. But the process is continuous, so as the new ocean floor grows along these cracks, it pushes the existing chunks of the Earth's crust outwards. This process is called *seafloor* spreading. The chunks of the existing crust that are forced apart by this process are called *tectonic plates* and they have been moving since the formation of the Earth's surface. Of course, if there is new crust being generated, somewhere the crust must be disappearing, and this is exactly what happens – as the edge of the seafloor plate is pushed into an existing continental plate it is forced underneath, because the continental plates sit higher on the surface of the Earth. As the oceanic crust is forced under the continental crust it is, in essence, recycled to become mantle again. This *subduction* generates enormous force, and it creates uplift on the continental crust. The Andes mountain range, along the west coast of South America, is an example of this process, with the Nazca Plate being forced into and beneath the South American Plate.

There is considerable evidence for the movement of the plates. For example, the complementary shape of Africa and South America is not an accident – they were both part of an ancient supercontinent called Pangea more than 200 million years ago. Another piece of evidence comes from biology. The ancestral mammalian populations of South America and Australia are marsupials (kangaroos, koalas, Tasmanian devils) because more than 150 million years ago, Australia and South America were joined together (with Antarctica). Roughly 85 million years ago they started to separate, with their marsupial populations, moving only a few inches a year. North America and South America were finally joined 2 million years ago, very recently by geological measures, but the movement of the plates is fast enough that it can be measured using modern technologies like satellite-based global positioning systems (GPS).

These plates continue to move, and some move quite quickly. The Australian plate moves fast enough that maps have to be corrected often – a recent

Box 3.6 (cont)

correction had to adjust GPS coordinates for the continent about 5 feet relative to the 1994 position. Since all the other continents are also moving, the maps of Australia have had to be changed several times since the mid-twentieth century.

In addition to explaining the shape and position of the continents, continental drift explains the where and why of major geological features. The boundaries of plates are under tremendous force, as the edges of these massive sections of crust collide or move apart. These lines of differential relative motion, known as *fault lines*, are where earthquakes tend to occur, as the plates slide against each other, releasing the pressure in tremors. This is also where mountains ranges appear: as two plates push into each other the material has nowhere to go but up. Plate tectonics also explains the locations of volcanoes: when plates pull apart, molten material can rise up; however, when plates collide, it also creates instability in the surface that can allow molten material to push up. In fact, the pressure of the colliding continental plates may even generate some of the heat responsible for volcanoes. Although volcanologists still debate some of the specifics of the causes of volcanoes, it is clear that they tend to appear at the edges of the plates (see Figure 3.3).

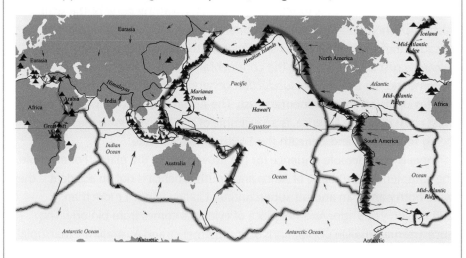

Figure 3.3 The tectonic plates and the "Ring of Fire." This image shows the boundaries of the plates (blue lines). The relative movement of these plates (blue arrows) accounts for most of the major geological changes we have seen over the last several billion years. For example, the appearance of strings of volcanoes in the Pacific (shown as red triangles), along with the rising islands, such as Hawaii, are explained by this process (the long, curved fault line bounding the Pacific is known as the "Ring of Fire"). Also, large mountain ranges such as the Rockies and the Alps are the result of continental plates being forced together. Earthquakes are generated by the slippage of two plates against each other. Plate tectonics is the most important major explanatory model for understanding broad patterns in geology.

Box 3.6 (cont)

The production of new mantle deep under the sea floor also generates local geological features. Sometimes the liquid mantle is a bit hotter in some spots, and this heat melts the crust, creating bulges that push upwards, creating strings of volcanoes rising from the ocean floor. The Hawaiian Islands are now dead volcanoes that are the result of the mantle pushing through the ocean floor.

So, although geology is complex, and can require years of effort to interpret even small areas of the Earth surface, in reality a few simple underlying processes are responsible for the patterns we see.

Box 3.7 Three kinds of rocks

If you go for a walk out in the forest or desert you will see a lot of rock – on the ground at your feet, in far-off cliffs, everywhere. The entire Earth is made out of rocks of one type or another. It may seem like a random jumble, but you can actually tell a great deal about the past just by looking at the rocks around you. Geologists are able to read the rocks effortlessly, but even without learning formal geology you can actually read the past from the ground.

Broadly speaking, there are only three kinds of rocks. The simplest kind of rock to understand is *igneous rock*. Igneous rocks are the result of volcanic activity, and are best understood simply as lava that has cooled and hardened. Basalt is an example of an igneous rock, as are pumice and granite – they can range from quite soft to very hard. One thing to keep in mind is that the Earth's crust was formed of igneous rocks, so any rock you find was, at one point, igneous rock, even if it has later changed to one of the other two forms.

If rocks are subjected to weathering and become sand or small pebbles or dissolve and then reprecipitate (e.g., sandstones) they may become *sedimentary rocks*. Sedimentary rocks form from small particles of rock that have eroded, typically due to weathering or wave action, which then settle at the bottom of bodies of water. Sedimentary formations form as layers and are often very visible when exposed (as in the Grand Canyon). These layers of sand or small rocks are compressed by the weight of the upper layers, plus the water, and over millions of years harden into sedimentary rocks. They are bound together by the chemicals in the water, but sedimentary formations can be soft relative to igneous rocks and can erode quite quickly (although some, such as chert, are very hard). Sedimentary rocks are the rocks in which fossils are frequently found, so paleontologists will often be found searching through

Box 3.7 (cont)

the layers of these rocks. In fact, some sedimentary rock layers can be composed entirely of the bodies of animals, as in limestone, which is made up of the shells of small marine organisms that accumulated at the bottom of the ocean over millions of years. Because sedimentary rocks form in large bodies of water, there are no such rock formations older than the start of the Archaean Eon, at 3.8 billion years old, when water started to accumulate on Earth; however, they are typically much younger.

Perhaps the most interesting and complex rock type is *metamorphic rock*. Metamorphic rock starts as either igneous or sedimentary rock (or even another metamorphic rock), but under enormous pressure (from the weight of upper rock layers) and heat (from nearby hot liquid magma deep in the Earth) the rock transforms (but does not melt). One example of this transformation is when limestone, which is a sedimentary rock, over millions of years, under the forces of heat and pressure, transforms into marble. Another example is the formation of schist from a mudstone. Geologists can look at metamorphic formations and interpret the forces that have created it, for example through the warping the original sedimentary formations. When metamorphic formations from deep in the Earth are exposed, through uplift and erosion, it allows geologists to interpret some of the forces deep below the present surface. Because the rocks were originally in another form, they often leave evidence of the various stages in the transformations. For example, a metamorphic rock may leave evidence of its original igneous form in a chemical signature, then the subsequent transformation to a sedimentary layer in the form of the layered bands, and finally the forces that transformed it into a metamorphic rock through the warping of the banding patterns.

So, when you are out for a stroll, and happen to glance over at a nearby exposed rock, keep in mind the history present in that one single stone!

Box 3.8 Where did water come from?

Life on Earth is completely dependent on water, so from the perspective of any living thing, the appearance of large quantities of water on the surface of the planet is probably the most important geological change in the planet's history. But where did water come from? The planet originated as a molten rock floating through space, so the change to a water-covered surface capable of supporting life is not necessarily intuitive.

Box 3.8 (cont)

Currently there are three hypotheses for the appearance of water on the planet. The first is that the chemical elements that form water were present in the rocks below the crust; they formed water deep underground and percolated up over millions of years. Water is simply hydrogen and oxygen, so the elements are simple, and more importantly, common throughout the Universe. They could have been present deep under the crust as separate elements that combined in the mantle. Evidence for this idea is water-worked diamonds found near the surface of the Earth. Diamonds only form under extreme heat, and so are billions of years old. Some diamonds, rounded and eroded, show evidence of having traveled with water as water rose to the surface.

Another hypothesis is that water is the result of *outgassing* from volcanic activity. Volcanoes produce many chemicals when they erupt – methane, for example, contains hydrogen, and carbon dioxide contains oxygen. The hydrogen and oxygen in those chemicals, in the presence of the heat from the volcano, could combine to form water.

The last hypothesis is that water came to Earth on a comet that hit the Earth. The Earth has been hit many times by meteorites and comets, and some comets are known to be large, icy blocks. For example, the comet Hartley 2 is known to be covered with ice. A large enough comet, or several, could have deposited ice that melted to water after impact.

It is worth keeping in mind that these theories are not necessarily mutually exclusive – there could be multiple sources of water on Earth. And without this water, life on Earth would certainly never have evolved.

Chapter 4: Life

All life on Earth descends from a common ancestor. We know this because all life is related – we all use the same genetic code (DNA), and the other building blocks of life (the amino acids) are identical. In fact, it is possible to determine relatedness between any two living things by just looking at how different their DNA is. So, even though you look nothing alike, you are related to a housefly, a dandelion, a mushroom, and even the bacteria that causes ear infections. All of this life comes from a single, ancient ancestor. Sometime, between 3.5 and 4 billion years ago, life originated, and as far as we know it only happened once. All life after that point is a result of slow changes over time through evolution. Because this was such a profound moment, it has long fascinated researchers.

We don't know precisely where or when life originated, but we do know something about early life from fossils found in very old layers of rock. The earliest life that became preserved in fossils was a very simple single-celled organism that converted the Sun's energy into growth using photosynthesis, much as plants and algae do today. This life was quite widespread around the world by more than 3 billion years ago, and paleontologists find it in places as far apart as Greenland and Australia.

But how did this life originate? To understand how life originated we have to think about what the world was like during the transition from the Hadean to the Archaean. At that time, long before the appearance of any plants, the world was hot and rocky, with large bodies of water. The Earth did not have a large envelope of gaseous oxygen (in the form of the O_2 molecule), which is today a product of billions of years of photosynthesis. Rather, there was scant O2 but large amounts of methane, ammonia and water. Methane and ammonia are organic molecules, which means that they have carbon, but in these molecules is also nitrogen and hydrogen (these same chemicals are found in all living organisms today). The conditions around 3.6 billion years ago – a mixture of relatively simple molecules sitting atop a relatively hot planet Earth – are known as the *primordial soup*. At some point, these chemicals started forming self-replicating structures. But these earliest structures left no trace, at least as far as we know.

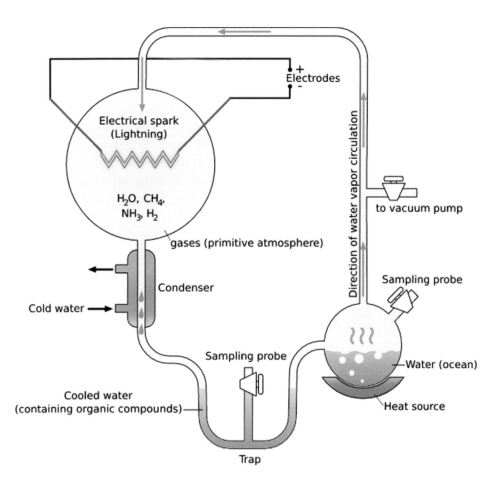

Figure 4.1 The Miller–Urey experiment (1952). In this experiment, the scientists attempted to replicate the conditions of early Earth in an attempt to determine how life could have originated. They generated water vapor (by heating water, on the right side), then added the vapor to ammonia, methane, and hydrogen. Using a spark generator, they passed electricity through the gasses. The condensed gas was then collected. They discovered that many of the amino acids used to generate all life were found in the resulting condensate, suggesting that relatively straightforward chemical and physical processes were responsible for the generation of life on Earth.

Since there are no fossils of the very first organisms, scientists have tried to reconstruct how life must have originated using their understanding of the chemistry on early Earth. By the middle of the twentieth century our understanding of early Earth was sufficient to allow us to try to figure out how life could originate from something nonliving, and scientists attempting to understand how life originated have conducted a series of experiments attempting to replicate the earliest appearance of life. The most famous of these experiments is the *Miller–Urey* experiments of 1952 (Figure 4.1). In this series of experiments, the researchers put ammonia, methane, water, and hydrogen in a flask, heated the flask, and then ran an electrical current through the gas. This roughly replicated the conditions on early Earth, in which a hot surface temperature was likely often subject to electrical storms, injecting electrical energy into the primordial soup. They checked the condensed gas the next day and found amino acids. Amino acids are the ultimate product of DNA (see Chapter 6), and the bodies of all living things are composed of amino acids. Of the twenty amino acids that make up all life on Earth, Miller and Urey found six. Some fifty years later, researchers decided to reexamine the samples to see if the amino acids were correctly identified and, with more modern and sensitive tests, discovered not only the original six amino acids but an additional three in smaller quantities.

But it is important to emphasize that even though Miller and Urey found a way to generate some of the components of life in their experiments, they did not actually generate life. Subsequent research by Miller and Urey, as well as quite a few other scientists, has not yet been able to make an actual life-form from simple chemical precursors. However, researchers have made significant progress toward understanding how early life formed. Recently, using similar methods to Miller and Urey, NASA researchers have been able to generate some of the actual components of genetic code from simple chemicals (see Box 4.5).

Box 4.1 What is life?

Consider a living animal and a dead one. What is the difference between the two? Well, it can't be the components, because they are made of exactly the same materials, organized in exactly the same way. And it is not necessarily the presence of energy, because a body can be heated, or exposed to radiation or electricity, yet it will remain resolutely dead.

This is because life is a process. When something is alive it is *doing* something. There have been a great many attempts by scientists over the years to define life, and a few of the definitions include the presence of:

- Homeostasis – regulation of the internal elements of the living thing.
- Organization – being composed of one or more cells.
- Metabolism – converting chemicals and raw energy into forms that can be used by the organisms (and the generation of waste).
- Growth – increase in size or cell number.
- Adaptation – response to the external environment over time.
- Response – rapid response to stimulation from the environment.
- Reproduction – the ability to produce more cells or duplicate itself.

Box 4.1 (cont)

The problem with this list is that it provides a number of criteria by which we might identify something that is living, but it isn't really a broad, general definition of what life itself is. But let's think about what happens when an organism stops living.

When something dies it not only stops its growth and movement, but, over time, it tends to return to the simple elements of which it is composed. It you have ever seen a dead animal, you know that after a fairly short time it will decompose and no longer be organized into the structures of which it was composed when it was alive. So, in effect, it is going from a highly organized state (with cells, organs, etc.) to a less organized state. This process, by which objects go from a highly organized state to a less organized state, is known as *entropy*.

In physics, the *Second Law of Thermodynamics* states that, over time, any closed system will change from a highly organized state to a less organized state (structural equilibrium), and the energy in the objects in that system will spread out, resulting in thermal equilibrium. This is occurring in our Universe, as the highly organized and energy-dense (hot) celestial bodies spread out, distributing themselves evenly throughout the Universe and cooling slowly. In your own house you can observe the thermal element of this phenomenon: if you fill up your bathtub with hot water and leave it for a few hours, the heat will leave the water and spread around your house, raising the temperature of the air a few fractions of degrees. At the end of the day, the water will be the same temperature as the room, having reached thermal equilibrium. If you leave a glass of water out long enough, it will evaporate, and the water molecules will be distributed evenly in the form of humidity.

Another form of this process is at play all over Earth in the form of erosion – the physically organized layers of rock are being eroded so that, some day, all the current mountains will be eroded flat (reducing the potential energy), and the separate layers of stratigraphy will be combined in an undifferentiated mass (reducing the physical organization). This force is inexorable – any amount of organization or energy will be reduced by the forces of the Universe. But things that are alive go against this process, taking simple chemicals and organizing them into physical structures. And they consolidate energy, in the form of their organized materials (think of the energy-dense fats or sugars in all living things).

In 1875 Ludwig Boltzmann offered what I regard as the best overall definition of life:

> *The general struggle for existence of animate beings is not a struggle for raw materials – these, for organisms, are air, water and soil, all*

Box 4.1 (cont)

> *abundantly available – nor for energy which exists in plenty in any body in the form of heat, but a struggle for [negative] entropy, which becomes available through the transition of energy from the hot sun to the cold earth.*

For him, life was defined as the use of energy to defeat the universal forces of entropy. The reason this is such a good definition is that it truly captures the essence of life – it is a struggle against the Universe, which seeks to reduce energy and order. Life seeks to capture energy and use it to create order. And this may be what is so unique and special about life – it resists the raw destructive forces of entropy.

Box 4.2 Why does life need water?

When astronomers search for planets that might have life, they are often searching for evidence of water. When water was discovered on Mars, it dramatically increased the probability that life could have once existed there. Water is known to exist on several planets in our solar system, as well as on some of the moons of Jupiter and Saturn. Because water appears to be so common in our solar system, scientists expect that it is relatively common on other planets in other solar systems throughout the Universe.

One of the first questions that might come to mind is: Why is water necessary for life? The first part of the answer is that life needs fluid. Chemicals do not combine very well, nor nearly as completely, in their solid form. Think for a minute about two chemical elements that need to combine to make life. How do they find each other? If they are just solid pieces of rock, they probably won't. They may just sit on the surface of the planet for billions of years. But if they are mixed into a fluid, they will inevitably find each other.

Imagine a big glass of water. If you put a drop of blue food coloring in on one side, and a drop of red on the other, those two colors do not stay separate for long. After a while, the colors will combine. This is because of the way molecules in a fluid state move – they have a great deal of energy, and just keep bouncing around off each other. Given enough time, every molecule will bounce off every other molecule.

But why water and not some other fluid? What is so special about H_2O? Well, some scientists suspect that it is possible for life to form with other fluids, but water does have special advantages. The first advantage is that it appears to be

Box 4.2 (cont)

relatively common throughout the Universe. Hydrogen and oxygen are very common, being relatively simple atoms, and we would expect to find it pretty much anywhere we look in the Universe. It readily exists as a gas, liquid, or solid, which means that it is physically stable once formed. It is also pretty stable chemically – water can dissolve many solids without reacting to their constituent chemicals. And that ability to dissolve many substances is also important – the water molecule has polarity: the hydrogen atoms, which are positively charged, are on one side of the molecule, while the negatively charged oxygen atom is on the other. When something is placed in water, the two different electrical charges tug differentially on the different components of whatever substance is in the water, and can literally pull it apart. This ability to dissolve substances means water is excellent for nature to experiment with mixing various chemicals.

So water is probably the best candidate for a fluid that is a necessary precondition of life, as it allows for the continuous mixing of various chemicals, and this mixing is probably what led to life itself. But once life has formed, especially complex life, having water in the bodies of living things provides a way for the stable transport of other chemicals to get around the body. All of these factors are why astronomers invest so much time, energy, and money in the quest for water. Without water, there may be no life itself!

Box 4.3 Where did life come from?

Although scientists have demonstrated that natural processes can produce many of the precursors for life (see main text and Box 4.4), life has never been generated experimentally. This is probably because scientists have not been able to precisely reproduce the circumstances under which life originated – some combination of simple chemicals plus energy (heat or electricity). Earth also had an enormous amount of time and space to experiment – most of the Earth was probably under water for hundreds of millions of years, subject to energy provided by lightning, volcanoes, and underground heat from the mantle radiating up, in which there were just lots of opportunities for chemicals to mix under various circumstances.

However, some scientists argue that life probably came from somewhere else. Under this model (known as *Panspermia*), life may have originated in distant galaxies billions of years ago. Over time, meteors crashing into life-

Box 4.3 (cont)

bearing planets would knock off chunks of the planets. These chunks, now meteors in their own right, could carry simple life (such as bacteria) wrapped in ice across the Universe. These meteors could travel into new solar systems, randomly seeding the planets in the "habitable zone" with life.

Although the research is still in early stages, there are some interesting results and discoveries. Experiments by NASA and other groups have demonstrated that bacteria can survive extended periods of time in space. In the 1980s a meteorite was discovered in Antarctica that appears to be from Mars. This meteorite has very unusual microscopic formations that have been interpreted as being the result of microbial life. To date, there is no scientific consensus on these formations. As mentioned previously (Box 3.2), Mars currently lacks a magnetosphere and an atmosphere, and life has not been found by various exploration modules. However, at one time Mars did have a magnetosphere, so life could well have existed. This debate is ongoing – so stay tuned!

In 1969 the *Murchison Meteorite* fell on southern Australia. This meteorite has been of particular interest to researchers studying the origins of life. Careful study of the components of the interior of this meteorite has found a variety (well into the thousands) of organic compounds normally associated with life, as well as amino acids, which are part of the structure of proteins that make up all living things (actually, some of the amino acids are identical to those produced by the Miller–Urey experiments). But perhaps the most fascinating discovery was uracil. Uracil is part of the genetic code known as RNA, which is closely related to DNA. So, whether or not the compounds found on the Murchison Meteorite are the product of extraterrestrial life, it certainly appears that the chemical processes associated with the appearance of life are not uncommon throughout the Universe. Interestingly, that meteorite appears to have been eroded by liquid water at some time in its past.

To date, there is no irrefutable evidence that life on Earth had an extraterrestrial origin. One reasonable conclusion, however, based on evidence like the Murchison Meteorite, is that the chemistry associated with the origins of life, such as complex organic compounds and amino acids, was not unique to early Earth. Most scientists interpret this to mean that life has probably evolved *independently* on many, many planets throughout the Universe. However, considering how long Earth was populated by single-celled cyanobacteria (roughly 3 billion years), before even simple multicelled organisms appeared, we are likely to miss most of the life in the Universe. Until a distant life-form learns how to transmit radio signals, we will probably never see them.

Box 4.4 Can we make life?

All life appears to be the result of a single event, some 3.8 billion years ago. Everything that lives today, or has ever lived, traces its origin back to that single moment when life started. That is why all living things – bacteria, plants, animals, fungi – share the same genetic code; our DNA is made up of the same components, and they produce the same amino acids and proteins using the same processes.

Naturally, understanding that first moment of life has fascinated scientists since they first started examining life systematically. Even Aristotle, some 2,500 years ago, had ideas on the origins of life (although precious little data to work with). This natural curiosity led a graduate student in chemistry, Stanley Miller, and his professor, Harold Urey, to conduct a famous experiment on the origins of life in 1952. They took the chemicals that they thought would have been common on early Earth: water, ammonia, methane, and hydrogen, and placed them in a sealed, half-liter, round-bottom flask. The flask was heated so that the water would evaporate and a spark was run through the vapor, to simulate lightning. Finally, the vapor was allowed to condense into a tube. When they examined the condensed water, it was found to contain more than half of the amino acids that are normally the result of genetic transcription.

Other experiments have identified other parts of our basic genetic code that are easily produced using chemical precursors. In the early 1960s Spanish biochemist Joan Oro showed that hydrogen cyanide and ammonia could, when combined in water, produce adenine, which is one of the four chemical bases of DNA, and a building block of the genetic code in all organisms. In succeeding decades there has been considerable research into the chemical origins of life and later experiments have produced other results that support the idea that the chemistry of early Earth was sufficient for life to evolve.

But researchers have not yet been able to "produce life in a test tube." Why not? One factor is likely to be that we do not yet understand the precise geological and atmospheric conditions when life appeared. Also, life has only appeared once on Earth, and whatever the chemical conditions generated early life, it was clearly an unusual event. If life were easy to produce, we would see multiple strains of life, unrelated to each other. That is not the case – there is only one. But one advantage that nature has is time and space. Miller and Urey conducted their experiment in a half-liter of water over a few weeks. Life on Earth had a much bigger laboratory - it evolved somewhere over the entire expanse of the planet, over hundreds of millions of years. Some relatively unlikely chemical event occurred, but much like going fishing every day for

---

Box 4.4 (cont)

years, rather than just one morning, increases your chance of catching a fish, having trillions of gallons of water over hundreds of millions of years, subject to thousands of volcanoes and millions of lightning storms, increases the chance of a rare event occurring.

Box 4.5 RNA World

The discovery of the Murchison Meteorite, and NASA experiments in which part of the universal genetic code was synthesized from relatively simple precursor molecules, has led researchers to hypothesize about the genetics of the earliest life-forms. Today, all life-forms have DNA, a complex strand of genetic code that forms in the familiar double helix, and these long molecules contain the entire genetic code used to generate the anatomy of all living things. The chemical organization of the bits of data in the coils of DNA is such that each of the two strands of the double helix is a mirror of the other, so the data of one strand are duplicated in the other. This would seem to be inefficient, but the "doubleness" of the DNA actually protects it. By having a mirror, the bits of DNA are bonded to other DNA, with no molecules available for bonding with random chemicals floating by. It makes the DNA strand, and the data stored in it, stable. Perhaps even more importantly, the mirrored side provides the template for repair. This stability, and the fidelity of the genetic code, is important, since our DNA is used millions or billions of times over our lifetimes. Any error would be replicated many, many times over.

However, there is a simpler genetic code that all living things have, called RNA. It is used by our cells to make short-term copies of genetic material when a particular gene is activated. Because it is only used temporarily, it is a transitory molecule, and the long-term stability is not as important, since the molecule is repeatedly remade by our cells each time a gene is activated. The sugar backbone of RNA is slightly different than that of DNA, which makes it less chemically stable than DNA. But, importantly, the same information stored in DNA can be stored in RNA, even if it lacks the advantages of the double helix.

The fact that uracil, one of the main chemical components that store data in RNA, can be synthesized from precursor chemicals found in meteorites, has led scientists to hypothesize that the earliest life-forms used RNA, rather than DNA, to store genetic code. This hypothesis is called the "RNA World" hypothesis.

Box 4.5 (cont)

Under this hypothesis, RNA served as DNA does today, as the main storage for the genetic code of early, simple organisms. At some point, DNA, with its protective self-bond, developed from RNA. DNA is extremely similar to RNA – the main difference in the chemical composition being the substitution of thymine for uracil.

Today, some viruses use RNA to store their genetic code, rather than DNA (see discussion in Box 4.6 on whether viruses are "alive"). Viruses, as genetic parasites, use the genetic replication mechanism of their host cells (who have DNA in their cells) to duplicate their genetic code, and they appear to be able to convert their RNA to DNA to co-opt this replication mechanism. So the RNA to DNA conversion may not be difficult, from an evolutionary perspective. Any organism that converted from RNA to DNA would have had a substantial advantage, since the genetic code in DNA is so much more stable, so all RNA-based organisms would have been outcompeted. Unfortunately, neither RNA nor DNA preserve in the fossil record, so we have no way of knowing when the conversion of RNA to DNA might have taken place, but right now it makes sense as an intermediate step on the way to the evolution of DNA as the universal genetic code.

Box 4.6 What does it mean to be alive?

We all think we know what we mean when we say something is alive. It grows, it reproduces, it responds to the environment, etc. But sometimes there are things that appear on the edges of the definitions, things that may not fit neatly into the boxes. The most well-known "things" that sit at the edge of life are viruses. Viruses possess some of the traits we associate with life, namely that they reproduce and that they contain a genetic code. But they don't have the ability to do anything on their own – for example, they don't eat or respire, nor do they engage in photosynthesis. In fact, they don't process energy or matter in any way. Nor do they regulate their internal chemistry as prokaryotes do. Finally, they can't reproduce on their own. They use the DNA replication mechanism of the host cell (bacterial, plant, animal, etc.) to duplicate their own genetic code (which can be DNA or RNA).

Because they lack many of the characteristics of other types of organisms, some scientists do not consider them as "living." But even though they do not have these characteristics, they obviously do reproduce with a genetic code that evolution has the opportunity to shape. In fact, virologists use the genetic

---

Box 4.6 (cont)

code in viruses to determine the evolutionary relationships among the various strains.

Another "thing" that sits even further outside the definition of "living" are prions. All organisms are composed of proteins – proteins are the ultimate product of DNA activation, so all of our tissues are composed of proteins (and fats). Prions are proteins that are similar to proteins in tissue, but they are misfolded and will replicate under specific chemical circumstances. The most famous prion is probably the one responsible for bovine spongiform encephalopathy, also known as mad cow disease, but other neurological diseases have been attributed to prions (the human variant of this disease is known as Creutzfeldt–Jakob disease). This protein, once in the brain, duplicates, absorbing the healthy brain proteins and converting them to prions. Because the prion is not living, even as much as a virus, it cannot be "killed." So, for example, cooking meat with prions will not eliminate the threat of the prion in the same way that cooking will kill viruses and bacteria. This makes prions very dangerous.

But prions possess no genetic code; they are simply complex proteins that happen to generate more of themselves in certain chemical circumstances. They are in no sense of the word alive, even if they seem to behave in a way that exhibits growth in the way we associate with viruses and bacteria.

Early Life

The earliest life known from the fossil record is in the form of single-celled organisms that we would today call bacteria and archaea. They were *prokaryotes*, which means that they were among the simplest organisms, without many of the cellular structures seen in the cells of plants and animals (known as *eukaryotes*). Notably, the cells lacked a nucleus. In eukaryotes, the nucleus surrounds the cell's genetic code (the DNA), and protects it from exposure to radiation (which can cause mutation) and from parasites (such as viruses).

These simple organisms still exist today and are known as *cyanobacteria* (Figure 4.2). They engage in photosynthesis, as do plants, and convert sunlight into energy for reproduction. The earliest fossil evidence of these bacterial organisms is in the form of *stromatolites* (Figure 4.3), which are large mound-like accumulations of these bacteria that form after thousands of years of bacteria living in the same location. These structures are as old as 3.6 billion

Figure 4.2 Green algae. The earliest life on Earth that was widely successful was cyanobacteria, also known as green algae. This is a very simple organism – a bacterium – that also engages in photosynthesis. For almost 2 billion years, this was the dominant life-form on Earth, and the long-term success of cyanobacteria changed the geology of the Earth itself, generating oxygen that is reflected in ancient layers of oxidized iron. This oxygen is also what made more complex life possible later and is the oxygen in our atmosphere today.

years old and are found in the fossil record in Australia and South Africa. For decades scientists assumed that stromatolites were an extinct form of photosynthetic bacteria, until in the 1970s a large number were discovered in a shallow bay on the western coast of Australia, still forming identical mound-like structures. Additional stromatolites have been subsequently identified in warm oceans around the globe.

For much of the history of Earth, the only form of life was cyanobacteria. From roughly 3.6 billion years ago until fewer than a billion years ago, this multicellular life had no competition. From its origin, it spread around the globe, reproducing and engaging in photosynthesis. For our purposes, this generation of oxygen was a critical element in the later evolution of multicellular life. For almost three billion years, cyanobacteria were generating oxygen around the world. Today we see evidence of this build-up in the iron bands in the Earth that have become rust (oxidized), and, up until 1.8 billion years ago, the oxygen formed bonds with elements and compounds in the earth and ocean and did not accumulate in the atmosphere. Oxygen readily forms molecular bonds, and this is one of the most important characteristics of

Figure 4.3 Stromatolites. The earliest fossil evidence of cyanobacteria are stromatolites, which are rocky mounds built of layer and layers of cyanobacteria (top image). Although these were identified as the earliest life on Earth by the early twentieth century, they were thought to be extinct until populations were discovered on the coast of Western Australia (bottom image). Later, other populations of stromatolites were discovered in different shallow, temperate ocean waters. As far as we know, stromatolites have continuously existed on Earth for more than 3.5 billion years.

oxygen. For example, it is one of the reasons we can absorb it so easily by breathing. We have iron in our blood (in the hemoglobin molecules) that bonds with oxygen coming in through our lungs, and can transport it to our cells very rapidly (iron is red in the presence of oxygen, which is why rust is reddish and our blood is red). Some animals use copper to bond with atmospheric oxygen to transport it to their cells, which is why some arthropods have greenish blood (copper is green when oxidized – probably most noticeable on the Statue of Liberty, whose outer skin is completely composed of copper sheets).

Over time, the chemicals in the Earth became saturated with oxygen, and at that point the oxygen became free (that is, not having bonded with any other compound or element), existing in large quantities on the surface of the planet. This started with the oceans at around 2.4 billion years ago, and the atmosphere by roughly 0.85 billion years ago, and it is after this point that we have a concentrated build-up of atmospheric oxygen on Earth. The additional free oxygen in the oceans enabled the evolution of eukaryotes (organisms with nuclei) roughly 2.4 million years ago, but it was not until after 850 million years ago, with a build-up of atmospheric free oxygen, that multicellular organisms (metazoans), with their complex and oxygen-dependent structures, could evolve.

Chapter 5: Evolution

In Chapter 4 we looked at the origins of life itself. All subsequent forms of life on Earth are descendants of this ancient, simple life-form. But how did the appearance of a simple, one-celled organism living in the "primordial soup" lead to all the amazing forms that evolved over the next 3.5 billion years – from sunflowers to jellyfish to grasshoppers to slime mold to humans? The answer is *evolution*. The very term "evolution" is, in some quarters, regarded as a controversial word, and the study and teaching of this subject has been outlawed many times in many places over the last 150 years. But, among biologists, evolution is universally accepted as the "grand unifying theory" of biology. One of the preeminent biologists of the twentieth century, Theodosius Dobzhansky, once observed that "nothing makes sense in biology except in light of evolution." In spite of this fact, some groups, both in the USA and abroad, continue to attempt to prevent the topic of evolution from being taught in schools.

So what is evolution? The definition of the word is simply this: a change in genetic variation in a population over time. This is obviously a pretty broad definition, and in fact evolution comes from four main sources:

1.   The simplest source of genetic change in a population comes from *gene flow* – interbreeding between two groups in a species that previously had relatively isolated gene pools. This happens all the time in the natural world. If the ice in the north pole starts to melt and polar bears no longer have large ice floes on which to live, they will migrate to dry land and interbreed with the local brown ("grizzly") bear population. The resulting offspring are a new genetic combination, and the interbreeding will have changed the gene frequencies of the bear population. This also happens with humans, who have frequently migrated throughout the last 100,000 years (and especially the last 500 years), regularly interbreeding with local populations wherever they have traveled.

2.   Another easy-to-understand source of change in genetic variation is *mutation*. If one individual undergoes a genetic mutation, and then reproduces over succeeding generations, the novel genetic mutation is passed to offspring

and a change in gene frequencies has occurred. One example might be red hair in humans. That was probably the result of a random genetic mutation more than 20,000 years ago. It didn't disadvantage the original person who acquired the mutation, and so that person reproduced and passed their gene down through the generations. This passing of the mutation changed the genetic frequencies for hair color in that human population.

3. A more difficult concept is *genetic drift*. Sometimes genetic frequencies in populations change simply at random. This is because not every individual from a population will always reproduce at exactly the same rate as every other individual. For example, consider a population of eighteen rats. "Fancy" rats come in three basic colors: brown, black, and white, and assume for the moment that they are equally represented (a third of each color: six brown, six white, six black). If, in one generation, several of one color don't happen to reproduce (just a chance event), the next generation will reflect a lower percentage of that color. This is a change in gene frequencies from one generation to another, but it is not driven by anything other than chance.

4. The final source of change in genetic frequencies in populations is *natural selection*. Evolution by natural selection is a change in genetic frequencies in a population that is not simply random – it is driven by something. Some individuals reproduce, and some don't, because some are prevented from reproducing. Often, this takes the form of some animals not surviving long enough to reproduce, and their gene combinations are therefore eliminated from the gene pool, which changes the overall population's genetic frequencies.

So, how does this occur? Natural selection is reliant on one basic fact: not every individual is the same in a population. This is known as population *variation* (Figure 5.1). If you look at a population of organisms, from fungus to elm trees to chimpanzees, you will notice that all of these living things have elements that are variable *within* the species. For example, some elm trees grow more leaves than others, or start their lowest limbs higher, or have deeper roots. Some chimpanzees are bigger than others, or have larger teeth. All of these things are differences within populations. Sometimes, under some conditions, various elements of organisms give them an advantage in survival. A classic example is the thickness of fur coats in coyotes (see Figure 5.2). If a coyote lives in the Arctic, a thick coat is an obvious advantage during a particularly cold winter. Coyotes with thinner coats may simply die (and therefore be prevented from passing their genes on to later generations). Alternately, if the coyote lives in the American Southwest, a thinner coat may provide an advantage

| 1 | 0 | 0 | 1 | 5 | 7 | 7 | 22 | 25 | 26 | 27 | 17 | 11 | 17 | 4 | 4 | 1 |
| 4'10" | 4'11" | 5'0" | 5'1" | 5'2" | 5'3" | 5'4" | 5'5" | 5'6" | 5'7" | 5'8" | 5'9" | 5'10" | 5'11" | 6'0" | 6'1" | 6'2" |

Figure 5.1 Bell curve. Although we sometimes look at our eye color or dimples and try to find similarities with our relatives, most features of our anatomy are not so discretely patterned. The vast majority of anatomical features on living things fall along a bell curve. Such features as height, weight, strength, skin color, shoe size, resistance to disease, and body odor fall along a continuum. This continuum is a distribution called the "normal distribution," although it is informally known as the "bell curve" for its overall shape. By looking at this historical photograph, you can see that the height of the draftees fell along a bell curve, with the average height somewhere around 5"7' and 5"8', and most of the other draftees a few inches taller or shorter. At the ends (what is known as the "tails" of the distribution) are a few who are very short or very tall. By looking at a distribution like this you can get a sense of the variation in the group, and what is common and uncommon in the population.

during particularly blistering summers. This variation is what nature has to work with, and closely related animals can acquire significant differences if they live in dramatically different environments. Another straightforward example is the difference in color between polar bears and brown bears. They share a common ancestor from some half-million years ago, and the ancestor was almost certainly brown. But once bears moved out onto the polar ice, lighter-colored bears were better able to ambush prey, and over time lighter and lighter-colored bears were better able to pass on their genes. In other words, the light-colored fur was *selected* for.

The same process is responsible for the enormous variety of dog breeds. People selected for the traits they wanted (size, aggression, loyalty, fur color, ear shape, snout length, etc.) by only allowing some dogs (the ones with the desired trait) to breed. Repeated over hundreds of generations, humans have been able to produce dogs a varied as Great Danes and schnauzers, all working from the ancestral wolf population. When humans perform the selection it is known as *artificial selection,* and humans have repeatedly done this to animal and plants (such as the grain crops harvested today) throughout the last 10,000 years.

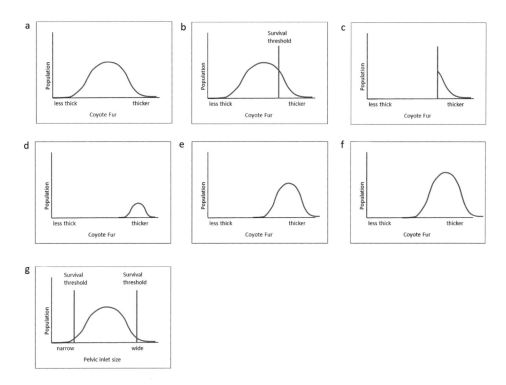

Figure 5.2 Directional selection. In this example, there is a population of coyotes whose thickness of fur falls along a normal distribution. We see the cold here as the force of selection (5.2b), so that less furry animals are less likely to survive and reproduce, whereas furrier ones are more likely. During the winter, the less furry coyotes die off (5.2c), and the subsequent generations are furrier than the older generations (5.2d–f).

In Figure 5.2g we see how stabilizing selection works. Instead of selection pushing the population in one direction, as in the example above, selection trims both sides of the distribution. So, in the example above, the human pelvic inlet can be neither too wide (which would affect locomotor efficiency) nor too narrow (which would impede childbirth). This is how natural selection finds "optimal" shapes for anatomy – by trimming both sides of the distribution.

The Evolution of New Species

When two populations of a species live in different environments, they will tend to acquire adaptations specific to their conditions. There is only one living species of wolf, *Canis lupus*. In North America it exists in very divergent areas, so that the Mexican wolf, which lives in the US Southwest and Mexico, and the northern gray wolf, which lives in snowy northern regions, have different coat colors that provide them with camouflage in their specific environments, as well as differences in coat thickness specific to their exposure to temperature. But wolf populations have not been separate for very long (only a few tens of thousands of years), so these two wolf types readily interbreed when in contact.

However, sometimes two populations are separated for so long that they will or can no longer interbreed. The red fox (*Vulpes vulpes*) and the gray fox (*Urocyon cinereoargenteus*) are both North American foxes, but they diverged several million years ago, and during that separation they acquired enough differences in anatomy and behavior that they will no longer interbreed. In fact, during this time there have been significant genetic changes, and these two species no longer have the same number of chromosomes, so even if they did attempt to mate there would be no fertile offspring.

When there is genetic change between two populations within a species it is known as *microevolution*, and when two populations diverge enough so that they become distinct species it is called *macroevolution*. But the actual processes of evolution are identical – gene flow, mutation, genetic drift, and selection. It is just a question of the degree of evolution. When enough evolutionary differences have accumulated, and the two populations no longer interbreed, evolution has generated a new species. It is often a question of time – the longer two populations of a species are isolated from each other, the more differences will accumulate. Sometimes we will capture a set of populations in the middle of this divergence. In North America, we usually consider wolves (*Canis lupus*) and coyotes (*Canis latrans*) to be different species, since they are fairly different in anatomy and behavior, and will not typically interbreed. But under some circumstances they do interbreed (under conditions of reduced population, or in zoos) and the offspring are perfectly viable. Wolves and coyotes diverged between 100,000 and 800,000 years ago, so we are probably observing the speciation of these two groups, even though they retain the ability to interbreed.

Box 5.1 Forms of selection

Natural selection is dependent on population variation. Think of pretty much anything you can measure on a tree, or mushroom, or grasshopper, or human. If you plot these measurements they will fall along a bell curve, so that most individuals fall close to the mean (average) but some individuals fall far on either side. In directional selection, one end (or tail) of the bell curve is selected against. For instance, fur thickness in coyotes can be measured, and when it is it falls along a bell curve. Most individuals fall toward the middle, but some are much furrier and some much less furry. If a harsh winter arrives, the less furry coyotes are more likely to die. This pushes the mean value of the population toward the furrier side, and the next generation will, on average, be furrier. This type of selection, from one side of the bell curve, is called *directional selection*, and over time it will tend to drive the population average of a given

Box 5.1 (cont)

trait (as well as the underlying genetics) in a particular direction, so that the mean value changes over time (see Figure 5.2 a - f).

However, some selection forces do not change the mean value but maintain it. In *stabilizing selection* (also known as *balancing selection*), selection acts on both tails of the distribution, keeping the bell curve narrow (Figure 5.2 g). Under stabilizing selection there is an optimum, and any organism that strays too far from the mean is selected against.

One famous example of this is the human pelvis. The human pelvis does two very important jobs. The first is probably obvious – it is where the legs attach, so it must have a shape that allows for efficient locomotion. The ability to walk and run is critical in nature, and when food is sparse an animal that can forage more efficiently (burning fewer calories per mile) will be more likely to survive and reproduce. Generally, narrower hips tend to be more efficient – Olympic marathon runners, both male and female, have relatively narrow hips because humans with narrow hips tend to keep the body mass more easily centered, with less side-to-side movement. Over a marathon, the seemingly slight advantage afforded by narrow hips adds up, so you will rarely see competitive marathon runners with wide pelvises. Something similar was at work in our deep evolutionary past, except that, instead of running for fun, we were covering large distances looking for food.

However, the human pelvis has another, equally (or more) important job – it provides the birth canal. Humans are born with extremely large heads, and, unlike in other animals, birth is dangerous because the human head must fit through a relatively small opening in the pelvis. Before modern medicine, death in childbirth was one of the leading causes of death in women. But one way natural selection made birth easier was to give humans wider pelvises (and in humans, females do have slightly wider pelvises than males, on average).

So there are two forces that natural selection is imposing on humans. One force is making human pelvises narrow, so they can cover large distances while looking for food. This helped humans (in our deep past) avoid starving to death. The second force is trying to make human pelvises wider, so that birth is easier. So, essentially, these two forces of natural selection are working against each other. This means that there is a relatively narrow optimum shape for the human pelvis – anything wider or narrower would have been selected against in our evolutionary past.

Stabilizing selection is at work across all of nature – bird wings cannot be too small (insufficient lift) or too large (makes the bird slower, and costs many calories of food to support), squirrel fur cannot be too dark (too easy to see

---

Box 5.1 (cont)

against snow) nor too light (too easy to spot against the tree trunk), and so forth. Stabilizing selection explains why nature is able to find so many optimum shapes, sizes, colors, etc. across all features of organismal anatomy.

Box 5.2 The evolution of antibiotic-resistant bacteria

Humans have caused evolutionary change in many organisms, and one important example is in bacteria. Bacteria can be killed with antibiotics, and the development of antibiotics has saved tens of millions of people from death or disfigurement from serious diseases, including some of the most historically destructive diseases in history (e.g., bubonic plague, leprosy, syphilis, tuberculosis). But today a new problem has emerged – bacteria that were once easily treatable but now are resistant to antibiotics, notably MRSA (methicillin-resistant *Staphylococcus aureus*), which has proven to be a particularly virulent and deadly infection.

So how does evolution lead to these resistant strains of formerly easily treatable bacteria? The answer, once again, lies in population variation. Within a population of bacteria, there is variation in the resistance to antibiotics (and this resistance, once again, falls on a bell curve). For most strains, almost all of the bacteria are killed by the correct antibiotic, but there are some that will not be so easily killed. Normally, for someone with a healthy immune system, these few remaining bacteria are easily handled (their resistance to antibiotics does not make them any more resistant to the immune system). But for someone with a compromised immune system, these bacteria will reproduce and be the foundation for a new population of bacteria. And since these bacteria were highly resistant to the antibiotic, their offspring will also be highly resistant. The reapplication of the same, formerly effective antibiotic will no longer kill the bulk of the bacterial population. We now have a resistant strain of bacteria.

These resistant populations of bacteria are frequently found in hospitals, where patients often have compromised immune systems, and are able to reproduce in these patients. However, the widespread use of antibiotics in agricultural livestock has also provided opportunities for resistant strains of bacteria to evolve. Many diseases are able to leap across species to humans (diseases such as tuberculosis, influenza, smallpox, typhus, measles, bubonic plague, malaria, and Covid-19, among many others, originated in animal

Box 5.2 (cont)

populations). If a resistant strain evolves in, for example, pigs, humans may be exposed to a new, antibiotic-resistant strain.

Livestock are often kept in conditions where cleanliness is hard to maintain, and infections can easily spread among crowded animals. These conditions create circumstances in which antibiotics are necessary for the health of the animals. But given the enormous scale of industrial livestock agriculture, this also creates ample opportunity for resistant strains to be generated in cattle, pigs, and chickens. Because of this, disease scientists have recently called for the reduction or elimination of antibiotics in animal husbandry. However, it is much cheaper to use antibiotics than create new, cleaner, less crowded stockyards, so the economics of eliminating antibiotics for industrial agriculture has made this goal politically difficult to attain.

Box 5.3 Sexual selection

We often think of natural selection as being something external to the species. The cold selects for furry rabbits, or the snow selects for white fur in Arctic animals, or the speed of antelope selects for fast cheetahs, or dry ground selects for deep roots in desert trees like acacias. In these cases, the selection comes from something external to the species, and the species changes in response. But, actually, many of the strongest forces come from within species, and the strongest is probably *sexual selection*. This is because, although the cheetah is, in a sense, in competition with the antelope for speed, for many contests the cheetahs are in competition with each other.

This is especially the case for competition for reproduction, and in mammals it tends to be stronger in males than females. You may notice that, in some species, the male is larger, or more colorful, or more ornamented, than the females. Males are under strong selective pressure that gets expressed in this difference between males and females (see Figure 5.3).

Sexual Competition

The driver of all of this is the difference in reproductive potential between males and females. Because we, as humans, tend to form long-term pair bonds to raise offspring, we think of males and females as producing the same number of offspring, but this is not the case in most animals. In elk (*Cervus elaphus*), females form groups (often twenty or more) for protection against predators, and this group will stay together and reproduce with one or two related males. Assuming that both males reproduce equally, they will sire ten

Box 5.3 (cont)

Figure 5.3 Sexual selection. This type of display can be costly for peacocks, who have to grow and maintain metabolically expensive anatomies that can slow them down when attempting to elude predators.

The black deer male is much larger than the female. In this form of sexual selection, the selection does not take the form of female choice, as the females will mate with whichever male is available. Rather, when one alpha male drives away the other males, he is the only male left for mating.

In 5.3c the male crab spider is mating with the female. In many species of spider, the female eats the male after mating (or tries to), as he no longer serves any purpose to her or the young spiders.

In anglerfish the small males attach themselves to the female and become parasites. They live off of the blood of the female, and sometime multiple males will attach to one large female.

offspring. The female elk, however, can only produce one each year. But there are equal numbers of male and female elk (as in most animals) – what happens to the males who are not producing offspring? They do not reproduce at all.

So we have, for elk, instances in which some male elk produce many offspring, and some produce none. This is very unlike the situation for females; all females will be able to reproduce each season. What decides which males get to reproduce with the group of females? In elk, it is physical combat, in which males compete for the opportunity to reproduce. The winner drives away other males, essentially isolating the group of females so that he can maximize his reproductive opportunities. For males, this means that only the strongest males, who can defeat the other males, will ever get a chance to reproduce, and this serves as a selective force against smaller male elk, driving

Box 5.3 (cont)

males to be much larger, with larger racks of horns as weapons. But there is no incentive for females to get larger – they never have to engage in combat for the opportunity to reproduce with a large, healthy male, and a single, or few, mating opportunities is enough to assure maximizing the female's reproductive potential for that generation.

This *male–male competition* is one of the two most common forms of sexual selection, and is very common in mammals.

Mate Choice

The male–male competition model of sexual selection applies to animals in which females cluster together and one strong male can drive off competing, weaker males. But there are other forms of sexual selection. Sometimes, female choice drives the selective patterns. In solitary birds, the male is almost always more highly decorated than the female. Probably the most spectacular example is the difference between the peacock and the peahen (collectively known as peafowl). Males of the genus *Pavo* have a spectacular display of tail feathers whereas females do not. Intuitively it would seem to be a disadvantage to have such a display – predators are more likely to see a male, and the large feathers are heavier and metabolically expensive to grow. So there must be an offsetting advantage. The advantage is that females will *choose* to mate with the male that has the more spectacular display. If a male has a big tail-feather display, *and* is able to survive, this appears to indicate that the male is genetically strong and healthy. In other words, a good genetic investment for the female.

It is worth thinking about the concept of genetic investment in this context. The cost of reproduction, at least in terms of the actual mating effort (coitus), is minor – a few calories at most. If the mating effort produces no offspring, or offspring with genetic flaws, that is OK for the male because the investment was so small, and he can pursue other opportunities with other females. But for the female, the investment is potentially huge. A female only has so many reproductive opportunities in a lifetime, because reproduction represents, at least in vertebrates, a significant cost in calories to gestate the young, or the egg, and then the effort to raise the offspring to independence. In some animals, like elephants, it can only happen once every five years at best. If a female reproduces with a male that has a genetic flaw, that large investment may not pay off, and the female will have wasted a significant proportion of her lifetime reproductive capacity. So, for females, there is a significant incentive to make sure that the male with whom she is mating has the best

Box 5.3 (cont)

genes. For males, there is an incentive to display in a way that matches the expectations of the female, and males with better displays will get more mating opportunities. Because the stakes are so much higher for females, females tend to be far more choosy than males, whereas males may pursue more frequent mating opportunities, even if each of those opportunities is not with a genetically ideal female.

This difference in coloration or feather pattern is common in birds, but in some mammals, for example gibbons, the animals are so sparse across the environment that it is extremely difficult for one male to have many mating opportunities (unlike birds, gibbons cannot just fly across the jungle in search of mates). In these cases, there is very little competition for reproductive access. When competition is low, males and females tend to be very similar, and it can be difficult to tell male from female in gibbons without a very close look. Once gibbons acquire a mate, they tend to stay with that mate and form strong pair bonds, probably because of the difficulty of acquiring mates in the first place.

Animals in which the female can produce extremely high numbers of offspring show entirely different patterns. Female insects and spiders (arachnids) are often many times larger than the male because of the need for the female to consume and store enough food to grow large numbers of eggs. In wolf spiders, for example, females are several times larger than the males. Males will preferentially mate with the largest females because large females are more fertile, but also because the largest females can store more food and are less likely to engage in cannibalism (against the male) when mating. In some extreme species, like the triplewart seadevil (a type of anglerfish), the male is less than 1 percent the mass of the female, and essentially functions exclusively as a sperm delivery mechanism, as the male is parasitic (attached) to the female and cannot survive without being connected to her.

Human Sexual Selection

So what about humans? Well, in humans males are well known to engage in display behavior. This can take the form of physical display (body-building or clothing) but also resource display, in which males display their ability to acquire resources (expensive watches, cars, etc.). Males are larger than females, so some form of combat-type competition was probably present in our evolutionary past, and males still occasionally engage in display combat in front of females. Currently, humans live in mixed male–female groups in which both sexes are able to engage in mate choice. However, early in our evolutionary history, before 3 million years ago, human males were much

Box 5.3 (cont)

larger than females, so the reproductive pattern was likely to be similar to gorillas, elk, and lions, in which the social group was probably one large male and multiple females. We don't know why we changed from a harem to our current socially mixed pattern, but it probably happened sometime around 2 million years ago, and scientists (including myself) are trying to figure out how, when, and where the change occurred.

Box 5.4 Adaptive vs. nonadaptive variation

The importance of natural selection in tailoring organisms to their environments is a relatively straightforward concept. There are clear correlations between environmental variables such as temperature and body shape (rounder animals preserve body heat while linear animals shed it), or the relative presence of water and the ability to survive without it (e.g., some plants require a great deal of water, and they tend to appear in humid environments, whereas plants that require very little tend to appear in deserts).

This relationship between anatomy and environment, however, does not mean that *all* patterns of variation reflect adaptation. Much of the variation in living things is random, and is *not* the result of natural selection. One example comes from the lions of Africa. Most lions have large manes, which are the result of natural selection: manes help the males look larger, which can make them appear more threatening to other males when competing against each other to take over a pride of females. In Tsavo, a region in western Kenya, the male lions have small manes, or can even lack a mane. There appears to be no adaptive reason why the males there lack manes – there is nothing special in the environment that would make a mane disadvantageous. In all likelihood, it is a product of genetic drift: the original founding lion population of Tsavo had small manes, and that genetic make-up has just been retained over time. Similarly, the Cross River gorillas of Nigeria have reddish hair atop their heads. We can envision no adaptive advantage, and a similar explanation for the founding gorillas seems the likeliest explanation.

There are many similar instances in humans. Hair color, for example, is often nonadaptive. There appears to be no advantage for any particular hair color, but there also seems to be no special disadvantage. In native Australians very dark hair is the norm, but in the central desert some groups often have blonde

Box 5.4 (cont)

hair. The genetics of blonde hair are well understood, and it appears to have been a mutation thousands of years ago that has been carried forward. Again, no advantage or disadvantage, and therefore not an adaptation.

However, much human variation is adaptive. For example, skin color maps more or less directly onto latitude (prior to recent global migrations). This means that populations closer to the equator, which receive more sun, have darker skin. Darker skin is caused by the presence of the pigment melanin. Melanin in the skin blocks solar radiation from penetrating through the outer layers of the skin. And when less solar radiation penetrates the skin, there is a reduced chance of skin cancer. So skin color is a direct adaptation to solar radiation.

Similarly, body build in humans is an adaptation to temperature. Populations closer to the poles (Inuit, for example) have heavier body builds and shorter limb segments in order to retain body heat. Populations in hot desert areas tend to be slender with long limb segments, to help shed body heat (the Nilotic peoples of Sudan, for example).

Interestingly, when peoples migrate from their ancestral regions, they can have maladaptive anatomies. Australia, which has a majority population that migrated from an extremely northern region (Great Britain), is close to the equator (the northern third is above the Tropic of Capricorn). Today that formerly northern population has the highest rate of skin cancer in the world.

Box 5.5 Social resistance to evolution by natural selection

Up until 1859, there was no true scientific model for the origin of living things. Most religions offer explanations, their "origin story," but there was no satisfactory *scientific* explanation of biological origin before Darwin's *Origin of Species*. Darwin's explanation was famously thorough – he spent most of his 502 pages documenting the evidence in support of this theory while only mentioning humans in a single line (he later wrote *Descent of Man* to address human evolution). He examined evidence from artificial selection, natural variation, heredity, and fossils, and today, more than 150 years later, the book is still held up as a model for the presentation of a complex idea to a nonscientific audience.

The book sold out in a single day and electrified the literate world. On reading the book, most impartial observers simply acknowledged that the idea

Box 5.5 (cont)

was obviously true, and many scientists immediately took up the idea in their research (there were some notable exceptions). But this was not the case for the religious establishment across much of Europe.

Many religious figures pushed back strongly against the idea of evolution by natural selection, and there were multiple reasons for this. One major issue was that Darwin directly contradicted well-established religious dogma. In all of the religions that descend from ancient Judaism (modern Judaism, Christianity, and Islam), the Old Testament is held to be the central document of the religion. The Old Testament presents an origin story (Genesis) that is very clearly contradicted by Darwin's explanation. For some people who hew closely to their religion, any contradiction of the Old Testament is a threat to the entire religion. If the Bible is literally the "word of God," then it must, by definition, be true. But if some part of the Bible is incorrect, can it really be the work of an infallible creator?

In many ways, the response to Darwin is similar to the reaction of the Catholic Church to the publication of Galileo's *Sidereus Nuncius* (*Starry Messenger*) some 250 years earlier. Using his astronomical observations of the planets, Galileo famously promoted the idea that the Earth revolved around the Sun (*heliocentrism*), rather than the other way around (*geocentrism*). The position of the Catholic Church at that time was that the Sun and all planets revolved around the Earth, since the Earth was the center of Godly creation. The Catholic Church used several verses from the Bible to support this position. For example, 1 Chronicles 16:30 and Psalm 96:10 both say: "The world is firmly established; it shall never be moved." Similarly, see Psalm 104:5: "He set the earth on its foundations; it can never be moved."

The Catholic Church saw Galileo's heliocentric model in 1610 as a threat precisely for the same reasons as those faced by Darwin in 1859. If the religion is premised on the infallibility of its creator, but the primary document passed to humanity by this creator is in error, then either the creator is not infallible or the document itself is not literally true. Many modern religions accept this last interpretation, instead accepting the Old and New Testaments as a collection of stories that, while partially metaphorical, still communicate the essential ideas of the religion. But some literalist sects (in Judaism, Christianity, and Islam) instead reject scientific ideas, insisting that the Bible must be *literally* true in every specific detail. In the United States, so-called Fundamentalist Protestant sects adopt this position, and have attempted (sometimes successfully) to prevent the instruction of evolution by natural selection in public schools.

Box 5.5 (cont)

But there are other objections that are sometimes overshadowed by the issue of literalism. One major implication of Darwin's theory is that humans have evolved via the exact same processes responsible for the evolution of every other living thing. In our time we are becoming more accustomed to viewing humans as another species of animal, especially since we now know that we share so much of our genetic code with other living things. But Genesis specifically sets humans apart – in Genesis, humans are created separately from animals and in God's image. Darwin's model presents humans as just another animal. But even more, one implication of evolutionary theory is that no animal "has" to occur. Speciation is partially the product of random events in genetics, population parameters, and the environment, and if any of these conditions had been slightly different humans might not have evolved at all.

Most of us have heard a story from our parents or grandparents about how they met through some chance encounter, and had it not happened they would not have produced us. Some circumstance of events led to us, but if your grandmother hadn't gone to the grocery store that day, or if your father hadn't stopped to change that flat tire, you wouldn't be here. It is the same with species. If events had been slightly different, humans might not have evolved at all. This idea presents humans as products of chance, rather than the product of divine creation, and significantly challenges the idea of humans as in some way created in the image of God.

In later chapters we discuss the social roles of religion, but by 1859 Christianity had been a predominant force in European thought (in many areas – art, literature, philosophy, politics, and law to name but a few) for more than a thousand years. In some ways, it had long possessed a monopoly on thought. So, for Galileo and Darwin to directly challenge the Bible represented real threats to the religion. Viewed from this perspective, we should not be surprised that religious figures pushed back vigorously. In Galileo's case, the Roman Inquisition was powerful enough to represent a literal physical threat, but even in Darwin's time the pushback was strong. Famously Darwin, who was no fan of confrontation, refused to publish his idea for more than twenty-five years, because he was afraid of the social reaction. And although we rarely find anyone who supports the pre-Galileo geocentric model, Darwin's idea, striking so close to our own self-made identity as humans, continues to face resistance.

Box 5.6 Creationism

Modern arguments against Darwin's idea of evolution by natural selection take the form of creationism. Creationism attempts to reconcile our modern scientific knowledge of biology with biblical texts. These arguments are extremely varied for much the same reason that religions are so varied: different religions have different interpretations of religious texts. Some creationist arguments are extreme: so-called Young Earth creationists reject all scientific ages for the Earth, instead interpreting some genealogical accounts in the Old Testament to propose that the Earth is only 6,000 or so years old. Other creationists accept the scientific age of the Earth and the general model for evolution, instead arguing that humans are a special case and not subject to the general laws of nature responsible for other life.

Because there are so many, largely inconsistent, creationist theories, it is not really possible to keep track of and refute them all, and most scientists don't bother. Being a scientist is difficult enough without spending time refuting theories that are not based on empirical evidence. However, it is important to be aware of the general ideas proposed by creationists, and I outline some of the more well-known or interesting theories here.

Creationist geology – the Flood. In order to link events in the Bible to scientific evidence, some creationists have searched for traces of specific biblical events in the geological record. The flood of Genesis (of Noah and his family) is a prominent example, and for several hundred years writers have sought confirmation of the biblical flood in the Earth's record. The discovery of animals in the fossil record was taken as evidence for the flood, as was the movement of the continents. Under most of these models, events that would have taken millions or hundreds of millions of years happened in a short time. In fact, the age of the Earth is frequently questioned, and the use of radiometric dating has been challenged (one creationist has argued that radioactive decay was faster in earlier times, and that the ancient ages inferred from argon or uranium dating reflect this). Needless to say, there is no evidence for a worldwide flood some 6,000 years ago. In fact, as we discuss later, there are civilizations that long predate this hypothetical event, and evidence from archaeology and genetics demonstrate that humans have been living in some locations for tens of thousands of years without interruption.

Creationist biology – intelligent design and irreducible complexity. Some creationists take the position that natural selection could not have produced particular features of the anatomy. These anatomical features are so complex that a random process such as natural selection could not have produced them; rather they appear to have been designed. The eyeball is an oft-cited

Box 5.6 (cont)

example. The eye is an extremely complex organ, and if any one part of the eye is absent, or does not function perfectly, the entire organ is useless. The basic idea is that the eye works as a complete organ or not at all, so there could never have been an intermediate stage in evolution *toward* the eye. For example, there is no selective advantage to having a focusing lens if there is not already present a retina exactly the right distance away. However, the evolution of the eye is now well known. There are species of primitive animals (flatworms, for example) who have only a light-sensitive patch of skin on the head. In some animals, the importance of light perception is sufficient to put those receptors in a concavity in the skull for protection. Once in a depression, an animal that lives in the water could take advantage of the "pinhole camera" effect to focus this light by restricting the size of the opening. From here it would be a small step to enclose the water-filled depression and cover it with clear tissue. So, although in a modern eye, if any one part fails to function, the organ is useless, in evolutionary terms the intermediate steps did provide an advantage.

The failings of creationism all stem from one unavoidable problem: they start with the answer and work backward to try to find evidence that proves the answer. Science does the exact opposite: it collects data and then finds an answer that best explains the pattern. This is the true value of science: all ideas are open and subject to test. No ideas are sacred and unchallengeable.

Chapter 6: Genetics

One of the strongest pieces of evidence that all life on Earth originated from a single event is that every living thing shares the same underlying biological blueprint. The most important element of this blueprint is the set of instructions built into every organism. These instructions are known as the *genetic code*. Although each organism has a unique set of these instructions, so that every organism is unique, the actual elements of the code are shared among all organisms.

This code is a series of instructions for assembling each organism. This code is in the form of a particular type of chemical combination known as *deoxyribonucleic acid*, or *DNA* (see Figure 6.1). Information for the growth and development of every organism is stored in the DNA of that organism, and each cell of an organism has the DNA of that organism in it. Each cell knows what to do during development by having different parts of the genetic code read, and then following those instructions. For example, a skin cell knows to read the part of the DNA related to skin growth, so that when you cut your finger, the skin cells know to manufacture more skin cells during the healing process.

The most elemental unit of genetic code is the *nucleotide*, also known as a *base*. This is a single unit of information that is in some ways analogous to the "bit" that stores computer information. And the genetic information stored in bases is read by the cell in a way that is analogous to the way a computer reads the instructions of a program: a string of these bases makes up a gene, and when the gene is read, the body responds in some way – by growing skin cells, or releasing a hormone into the bloodstream, or stopping the growth of bone cells. These "bits" are always chemically paired with a genetic mirror of itself, so we call these *base pairs*.

The genetic code (*genome*) for organisms like plants and animals, with many different parts, can be very complex. Simple organisms, such as bacteria, tend to have very simple genetic codes, since they only need the code for a single type of cell. So, a bacterium might have a few hundred thousand base pairs (the smallest bacterial genome is 112,000 base pairs – *Nasuia deltocephalinacola*) to a few million, with a few thousand genes. But some plants and animals have enormously complex genomes with hundreds of millions or billions of

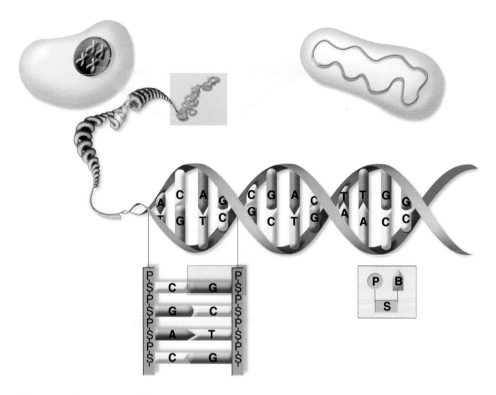

Figure 6.1 The structure of DNA. The individual units of DNA, the nucleotides, are in a long chain, with a sugar and phosphate backbone supporting them. The DNA is organized in a paired chain known as a double helix. Note that *adenine* is always opposite *thymine* and *guanine* is always opposite *cytosine*. So one side of the helix is a mirror of the other side. The bonds between these mirrored versions help keep the genetic code chemically stable, ensuring fidelity when the cells reproduce over the lifetime of the organism.

This twisted chain is further compacted by being coiled and organized into chromosomes. Each chromosome has different genes on it, and these genes tend to be inherited together. Humans have twenty-three chromosomes, but other animals have different numbers. Even our closest relatives, chimpanzees, have a different number (twenty-four) and this might well have prevented interbreeding between the two lineages millions of years ago when they diverged.

bases, and tens of thousands of genes. But there is not necessarily a direct relationship between the number of bases and the number of genes. Some genes are very long, whereas others are short, and, surprisingly, some bases do not appear to do anything; in other words, they are never read by any cell. In all likelihood, many of these dormant bases are part of the deep ancestral past of the animals or plant. So, in the case of humans, we lack tails (as do all apes), but somewhere in our genetic code is the series of nucleotide bases (the gene) that used to code for a tail. At some point (and we do not know why), the body decided to stop reading the gene for tails. Roughly 99 percent of the human genome does not code for any genes. This is the so-called junk DNA, but that is an inaccurate term, since it largely reflects the genetics of our deep evolutionary history that has been carried along. Further, the junk DNA may serve as a "spacer" between genes, and that may help regulate their expression.

We tend to think of humans as very complex animals, since our brains are so much bigger than those of other animals. This large brain enables a complex series of behaviors beyond that of any other animal on Earth. But, as far as nature is concerned, there is nothing particularly complex about our anatomy, nor about our genomes. We have a genome of 3 billion bases and some 20,000 genes. But this is roughly the same size of the genome in our closest cousins, the chimpanzees. Other animals have much larger genomes. The record (as of publication) is the marbled lungfish, which has 130 billion base pairs (and we don't yet know how many genes).

Why is it that the human genome is not especially complex, despite our having such a large and complex brain? Shouldn't the complexity of the anatomy be reflected in the genes? The answer is that nature likes to make copies of things, and we see this throughout the natural world. In the case of the human brain, there are probably a few genes that code for the different cells of the brain (such as neurons, glia, and axons, all of which I will discuss in Chapter 14), and a gene that instructs these genes to repeatedly make copies of those brain cells. So the instructions for a large brain might not be any more complex than the instructions for a small brain. The hard part is the initial step of making the instructions to make those brain cells.

Imagine a factory that makes cellphones. The hard part is putting the factory together, with the smartphone assembly line putting together the many different parts of a phone. But once the factory is built, making a lot of smartphones is not much more complex than just making a few – you just keep the assembly line running. The body does something like this with much of its anatomy. The brain is one example, but the spine is another. We have twenty-four vertebrae, and they are very similar to one another, with a few changes for the upper vertebrae to articulate with the ribs and the lower vertebrae larger in size to bear the weight of the upper torso. Nature can very easily add to or subtract from this number, depending on the demands of the animal's adaptation, by turning genes off or on. For example, a cat has seven lower vertebrae (ass opposed to primates' five), because it uses this part of the body to help it accelerate after prey. This evolutionary repetition is known as *serial homology*, and it enables the evolution of many complex life-forms.

Perhaps the most important example of serial homology is seen in the forelimbs and hind limbs of vertebrates. Early vertebrates only had forelimbs, which constrained their ability to adapt to land. But roughly 350 million years ago, a second set of limbs appears in the fossil record. These new limbs have the same elements as the forelimbs – one bone in the upper limb element (the humerus in the arm, and the femur in the leg), two bones in the lower element (the radius and ulna in the arm, and the tibia and fibula in the leg), along with the same numbers of wrist and finger/toe bones. This is because the genetic control of the hind limb is a copy of the genetic controls for the forelimb. Interestingly, this has only happened once in evolution (so far as we know), which is why nature has never made a six-limbed organism such as a Pegasus (the mythological flying horse).

Box 6.1 How is DNA organized in the human body?

DNA sits in every cell of every organism. In simple prokaryotic organisms, such as bacteria, the DNA floats freely in the cell. But in eukaryotic organisms, the DNA is in the cell nucleus. The nucleus is a valuable addition, because it separates the DNA from the process of protein production and thereby serves a regulatory role.

The actual nucleotides (bases) are units of information in four forms – adenine, cytosine, thymine, and guanine. Each nucleotide is held in place in the chain by a backbone of sugar and phosphorus (Figure 6.1). However, the genetic chain doesn't sit floating in the cell nucleus. Each nucleotide base actually bonds with another base according to very specific chemical rules: adenine to thymine and cytosine to guanine. And the mirrored bases have their own sugar/phosphorus backbone. This combination of bases and the sugar/phosphorus backbone in an extended chain is what forms the familiar double helix of DNA. One advantage of the double helix (as opposed to a single helix) is that the nucleotide bases are chemically protected from chemical elements floating throughout the cell. Chemicals will naturally bond with other chemicals under certain circumstances. If a nucleotide made a bond with a random chemical element floating in the nucleus, the genetic code would be compromised. With a double helix, the base is effectively already bound to another base, preventing additional, potentially compromising, bonds.

The strands of DNA are very long, especially if the organism's genome is complex. If the DNA of one human cell were stretched out, it would be longer than you are tall. This means that the DNA must be wrapped and coiled into a much smaller form that can fit into the cell nucleus. Once the DNA is twisted into a much smaller form, it also is broken into chunks. Those chunks of coiled and compressed DNA are called chromosomes. You are probably familiar with the human karyotype, in which human DNA is broken up into twenty-three lengths called chromosomes. You inherit one copy of DNA (twenty-three chromosomes) from your mother, and another

Box 6.1 (cont)

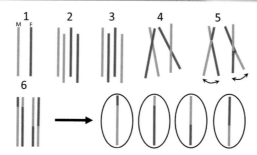

Figure 6.2 Meiosis. This is a simplified visualization of how DNA is copied for a sperm or egg cell, for a single chromosome.

Think of your own DNA – you receive a copy of each chromosome from your mother and father. In the image we are reproducing only a single chromosome, Normally we think of chromosomes as being in the form of an X, but for simplicity's sake, I present them here as a line (step 1).

The first step is the exact reproduction of each chromosome copy (step 2), which is followed by a moving of the copies so that you have a copy of your mother's chromosome next to your father's (step 3). Next, each copy of your mother's DNA overlays your father's (or vice versa) – this is called crossing over (step 4). Crossing over is completed when the overlain segments of DNA are exchanged (step 5), so that when they separate again (step 6) you now have totally unique combinations of your mother's and father's chromosomes. Each of those unique mixtures is now its own chromosome, and those are what are passed into the sperm or egg cell for reproduction. This means that, if you have a child, the child will receive a unique mixture of its grandmother's and grandfather's DNA via you.

Meiosis is one of the most important generators of variation for a species.

twenty-three from your father. So, in effect, you have two copies of each gene (see Figure 6.2 for the way the DNA is shuffled at each generation to create unique combinations of alleles).

Having two copies of each gene can be very useful, because, for example, if you inherited a defective copy of a gene for the growth of the eye's retina from your father, you would have a second, fully functional copy from your mother (each version of a gene is called an *allele*). Having two sets of genetic code shows up in other important ways as well. You are probably familiar with the idea of dominance and recessiveness. If you have inherited blue eyes from your father and brown eyes from your mother, the brown eye alleles are *dominant* and the blue eyes are *recessive*, meaning you will have brown eyes, even though you have the allele for blue eyes as well. And you can pass either to your offspring, so you, with brown eyes, could pass blue eyes to your child.

Box 6.2 How do the instructions in DNA generate anatomical features?

In a normal eukaryotic organism, like a plant or animal, the DNA sits in the cell nucleus. So how does that little string of instructions actually become our bodies? The first step that happens is that the DNA unwinds slightly so that it can be read. Normally, it is packed so tightly that there is no room for any parts of the cell to read it. But when a particular gene is going to be read, a chemical called an enzyme loosens up the coil and separates the two mirrored strands of DNA. The open strand of DNA is copied in a process called *transcription* (Figure 6.3). The result is a slightly modified copy of the gene that is called *messenger RNA* (*mRNA*). This mRNA can leave the nucleus and is free to enter the general cell body. (The mRNA is slightly different from DNA in that, instead of thymine as a base, uracil is used; but that uracil bonds to adenine just as thymine does in DNA.) Once exported from the nucleus, it is read by a part of the cell called the *ribosome*. The ribosome is what actually reads the genetic code itself, taking the bases in sequence much the way a string of instructions is read by a computer.

The process of reading the genetic code is called *translation*, and in this process the nucleotides are read by the ribosomes in chunks of three. So a group of three nucleotides (called a *triplet*) are read, and the ribosome attaches an amino acid. The next triplet or nucleotide bases is read, and a second amino acid is incorporated, attached to the first. This process is repeated for the entire gene, until the whole mRNA is read. The string of amino acids produced by this process coils and folds, and the result is a protein (Figure 6.3).

The proteins are the ultimate result of the genetic process, and, in a sense, proteins are what differentiate us from other individuals and species. Humans are roughly 65 percent water, 12 percent fat, 3 percent minerals, and 20 percent protein. The water, minerals, and fats are the same across individuals, but the proteins, in different combinations, are what make us living organisms. Most of what you think of as your body are proteins: the collagen in bones, the hemoglobin in your blood, and the adrenaline flowing through your body under stress. All of those are proteins (composed of amino acids) and all are the result of your genes being copied (transcribed) and read (translation).

This is known as the central dogma of molecular biology: (1) DNA is copied by RNA and (2) RNA makes proteins.

Box 6.2 (cont)

Figure 6.3 DNA transcription and translation. In this image we see the genetic code being copied inside the cell nucleus, then the mRNA passes through the wall of the nucleus out to the cell body, where the ribosome reads the codons. The ribosomes link the amino acids in a chain, which ultimately results in a protein. The DNA is protected by the cell nucleus and so cannot pass through it. mRNA, however, can pass through and bring the genetic code to the ribosomes.

Box 6.3 How to read DNA

Knowing the sequence of nucleotide bases in a gene allows you to predict the sequence of amino acids in the protein. The bases are read in groups of three (triplets), and each triplet makes a specific amino acid. Figure 6.4 shows

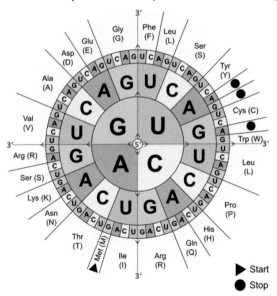

Figure 6.4 Reading the genetic code. Starting from the center, the letters are read moving toward the outside. For example A + A + (G or A) produces the amino acid lysine. Note that, in many cases, a mutation that results in a substitution in the last codon does not result in a different amino acid, although in other cases it does. A substitution earlier in the codon almost always produces a different amino acid. A + U + G starts the reading of the mRNA and U + A + A/G and U + G + A stops the reading of the mRNA.

Box 6.3 (cont)

the patterns, and, as an example, if the mRNA triplet is UCC, then (starting at the center of the circle) the resultant amino acid is serine. CAC would produce histidine. Note that, often, changing the last base in the triplet has no effect on the resulting amino acid. However, changing the first nucleotide virtually always results in a change in the amino acid produced. This has significant implications when talking about mutations.

Box 6.4 The role of mutations

In Boxes 6.2 and 6.3 we see how the sequence of nucleotides determines the amino acids incorporated into a protein. In particular, a change in one of the nucleotides might, or might not, cause a change in the resulting amino acid. A mutation from one nucleotide to another in the last base of the codon often does not cause a change in the amino acid (see Figure 6.4). However, changing the first will almost always produce a different amino acid.

The biggest change, however, can result from the deletion of a nucleotide. Examine the chain of bases below. A change at the sixth position will not produce a different amino acid. However, a deletion of the second base causes the ribosome to read the genetic code entirely differently, from the first codon to the last. This would likely result in a completely defective expression of that gene, and a failure to produce the normal end product. So, for example, this cell might not produce hemoglobin (if this is the gene for hemoglobin).

Normally, an organism has many mutations all the time, with no real ill effect. For example, of your tens of millions of skin cells, if one fails to produce collagen correctly that cell may die. No problem – there are plenty of other skin cells around the defective one to fill in the gap. But there are two examples when mutation can have a much more severe effect.

The most obvious is cancer. Cancer is caused by mutations in genes regulating proliferation and motility, causing cells to grow and reproduce in an out-of-control manner. We tend to think of growth as the main goal of transcription and translation, but as anyone who has ever sprung a leak in their kitchen water line can attest, being able to turn the valve off is as important as being able to turn it on. In a cancerous cell, unregulated growth can lead to crowding out healthy tissues, but more seriously, cancer cells have the ability to migrate, potentially spreading to other tissues or organs. This is known as metastasis. Researchers have documented a wide variety of genes in the

Box 6.4 (cont)

human genome that readily produce cancer if subject to mutation. These are known as oncogenes.

The second type of important mutation is when it occurs in the sperm or egg (the germline). If a skin cell has a mutation, and that cell dies, there is no significant impact on the organisms or its offspring. But if a mutation occurs in an egg, for example, and that egg is fertilized, then that mutation becomes part of the genome of the offspring. If the mutation is unhealthy, it can result in the death of the fetus or defective anatomy in the offspring. This is why lead aprons are frequently applied over the reproductive organs of patients undergoing X-rays.

Box 6.5 The importance of sexual reproduction

Reproduction for single-celled organisms (Archaea) is pretty straightforward: they just make a copy of themselves. Sometimes this is good, because it is rapid and allows the organism to rapidly produce millions of offspring quickly, when the conditions are appropriate, without the difficulties of finding a mate. However, inherent in this strength is a weakness: the organisms are all essentially clones of each other (aside from the occasional mutation). If, for example, the bacteria has a predator that figures out how to eat it, or a host organism acquires a good immune defense to the bacteria, that can spell the end of that kind of bacteria. That is because there is no variation – whatever works on one of the bacterial cells (either predation or immune response) will work on all of them. So a colony of bacteria might rapidly grow but just as rapidly die.

Sexually reproducing organisms, in contrast, are constantly shuffling their genes, both when they produce the gamete cells (egg or sperm), and by the process of mating with others of their own species who also have unique combinations of DNA. Because of this, complex organisms tend not to have excessive genetic uniformity, which gives natural selection something to work with. For example, imagine a species of dog-like animal that is faced with the sudden appearance of an ice age. There will be variation in fur length within that species because of the constant reshuffling of genes. This gives the species a chance to survive, as the furriest are likeliest to survive and pass down their genes. If they were all clones of each other, once one of them was too cold to survive they would all be too cold, and the species would go extinct. None would be furrier, and natural selection would not be able to select for the advantageous genes.

Box 6.6 Ancient DNA and me

Genetic analysis can tell us about relationships among living people, and about relationships between species. This is simply based on the overall similarity of the genome. So, for example, you share a great deal of DNA with your siblings, less with your cousins, and so on. But you also share a great deal of your DNA with chimpanzees (98.7 percent), since that is the most closely related animal to humans, and a surprising amount with other organisms – you share 92 percent of your DNA with the common field mouse, 60 percent with a fruit fly, and 18 percent with a plant such as thale cress. The differences between these other organisms in their DNA also allow us to estimate how long ago our branch of life separated from theirs. In the case of chimpanzees, geneticists have estimated a split between the human and chimpanzee line to be somewhere between 6 and 8 million years, and the fossil record seems to support this (which we discuss in Chapter 10).

Advances in DNA analysis have also allowed us to look directly at the past. We can extract DNA from fossils up to a half a million years old, if the preservation was just right. This has allowed us to determine the genetic

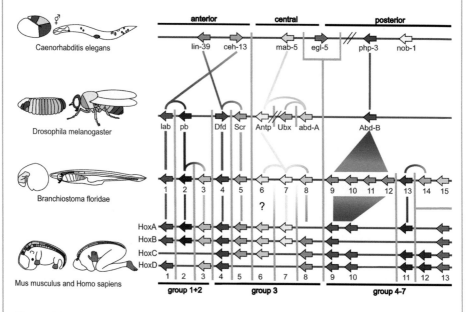

Figure 6.5 Genes are highly conserved across all life-forms. The HOX genes organize the orientation and segmentation of all animals with bilateral symmetry, from houseflies to humans, and are likely to be more than 550 million years old. Trilobites, known from the Cambrian Explosion, at 550 million years, appear to be organized by HOX genes.

Box 6.6 (cont)

relationships among wooly mammoths and elephants, for example, as well as whether Neanderthals and modern humans ever interbred (they did). It is now possible, and not very expensive, to have a genetic analysis done that will tell you if you have any Neanderthal DNA in your genome (I do).

One of the most important families of genes that all complex animals share is the HOX genes. HOX genes are responsible for organizing the development of the body and are seen in animals as diverse as humans, mice, lizards, house flies, and even ancient trilobites (see Figure 6.4). These genes determine which end of the animal is the head, which is the tail, and how the segments in between are organized. The fact that we all share these genes and that their function is conserved in animal development is further evidence for the shared relatedness of all organisms, and that we all share a common ancestor more than 300 million years ago.

Box 6.7 Chimpanzees and humans: close genetic cousins

As I mentioned in Box 6.6, humans and chimpanzees share a great deal of their DNA (98.7 percent of base pairs), and have roughly the same number of genes (20,000), yet if you look at their karyotypes you will immediately notice one big difference. Humans have twenty-three pairs of chromosomes, yet chimpanzees have twenty-four. If we are so closely related, why is the number of chromosomes different?

To answer this question you have to know a bit about the way a string of DNA is constructed. Each time a cell reproduces a little bit of DNA is lost from the end of the strand. After many cycles of cell reproduction, you would expect the genes themselves to start being affected as parts of genes are lost. To counter this problem, a long strand of "dummy" DNA sits at the end of the strand, so that when a bit of DNA is lost at each reproduction cycle the genes are kept intact. These sequences of "dummy" DNA are called *telomeres* and we can readily identify them. Each chromosome has a long sequence of these telomeres and one source of aging is the loss of these later in life, after enough cycles of reproduction have essentially used them up.

If you joined two chromosomes, you would expect to see long sequences of telomeres in the middle. This is exactly what you see in our human chromosome number 2. So, at some time in our deep evolutionary past, two ancestral ape chromosomes joined into a single chromosome. The two halves

---

Box 6.7 (cont)

are nearly identical matches for individual chimpanzee chromosomes. At some point prior to 7 million years ago, we must have shared an ancestor with chimpanzees that had twenty-four chromosome pairs; the chimpanzee lineage retained twenty-four chromosome pairs, while we joined two together and now have twenty-three pairs. While we share a vast majority of nucleotide bases with chimpanzees, we do have an important difference that may have prevented our lineages from interbreeding more than 6 million years ago. This genetic isolation from the chimpanzee line may well be one factor in the appearance of the human lineage.

Box 6.8 The genetics of simple and complex traits

Often, when we think about genetics, we think about relatively simple traits, like the genetics of eye color, or a cleft chin. These are traits that are controlled by a single gene, and they are subject to the laws of Mendelian genetics, in which one version of the gene (the *dominant allele*) is expressed, while the other (the *recessive allele*) is not expressed. So, for example, if you receive a blue allele from your mother and a brown allele from your father, you have brown eyes because the brown allele is dominant.

However, most anatomy is not controlled by a single gene. One well-understood example is the *pax*6 gene in fruit flies (*drosophila*). This is a master gene that turns on all the other genes necessary to make an eye, and when this gene is expressed in the wrong location an eye appears, with all the complex anatomy. So this gene has a "master control over" a suite of other genes that must be activated to make the lenses, nerves, blood vessels, etc., of an eye. This "master control" of one gene over another (or many others) is called "pleiotropy," and in complex organisms there can be complex networks of pleiotropy so that the multiple master genes (regulatory genes) have higher-level regulatory genes controlling them as well.

One implication of pleiotropic control is that one or a few genes might be responsible for dramatic changes in anatomy, so major evolutionary events like speciation might be the result of mutations in regulatory genes controlling pleiotropic networks. The study of these gene networks, how they express themselves in development, and how, in turn, they interact with environmental factors during organismal development is the field of *evolutionary developmental biology* (also known as "evo-devo").

Box 6.8 (cont)

This enormous complexity of networks of pleiotropic genetic control is one of the reasons it is so difficult to predict the anatomy just by looking at the genome of an organism. Right now, scientists are busy trying to determine the relationships within and among pleiotropic networks; the mathematical complexity of this task means that it will be many years before we can really understand how the human genome controls our anatomy.

Chapter 7: The Evolution of Complex Life

As we saw in Chapter 4, the first life forms were simple and single-celled, and probably acquired energy for reproduction the same way plants do today – by converting energy from the Sun into the fuel needed for growth. For most of the Earth's history, life was simply that – single-celled organisms (early cyanobacteria) that converted solar energy to carbohydrates that they could then consume, and converted carbon dioxide into carbon and free oxygen (O_2) in the process. The production of oxygen was the critical element that allowed the evolution of complex life on Earth, but it took more than 2 billion years for sufficient oxygen to accumulate (see Figure 7.1 for geological evidence). Oxygen is a highly reactive element, which is why it can be so dangerous in a pure form, readily binding with other elements. Fire is one way this binding occurs but so is rust, and we see evidence of oxygen build-up in layers of rusted iron as far back as 2.5 billion years ago. In fact, the Earth's iron deposits absorbed this free oxygen and it was not until after a billion years ago that the crust of the Earth was essentially saturated and the gas could accumulate in the atmosphere.

The process of converting solar energy into sugar is known as photosynthesis and it may be the single most important step that led to the evolution of all life on Earth. Photosynthesis is the conversion of the Sun's energy into energy that a plant can use to grow. If a plant has access to water, atmospheric carbon dioxide, and sunlight, those things alone are sufficient for growth. A plant needs no other fuel to achieve this. So when we talk about photosynthesis, we really are talking about the conversion of the Sun's energy into plant matter and the generation of oxygen as waste.

You may have learned about photosynthesis in school, but it is worth repeating how this process works. Sunlight hits the surface of a cell (whether the leaf of a flower, or a mat of cyanobacteria), and a chemical in the cell known as chlorophyll absorbs the sunlight and releases a small amount of electricity. This electrical charge breaks up the bonds in the water (H_2O) and carbon dioxide (CO_2) and reshuffles the atoms in those two chemicals. The carbon and the hydrogen, and some of the oxygen atoms, form a new chemical called glucose ($C_6H_{12}O_6$), and since there are some oxygen atoms left over,

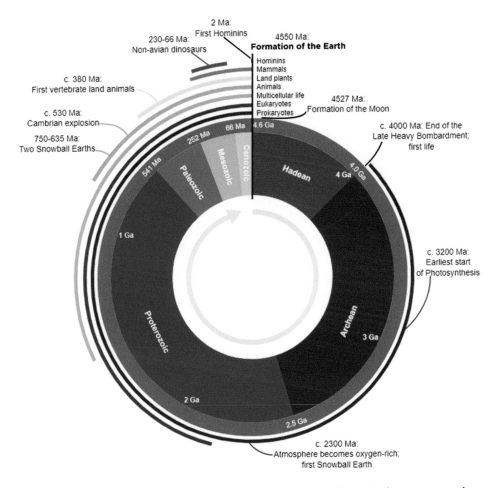

Figure 7.1 The geological timescale shows the events and states leading up to the appearance of complex life. Notice that, for most of the Earth's history, life was only the simple, unicellular cyanobacteria. Complex life probably evolved fairly late, perhaps around a billion years ago.

they form O_2, the familiar form of oxygen you and I breathe. The glucose is a sugar that fuels the cell's growth, and the oxygen is simply released into the atmosphere.

This accumulation of atmospheric oxygen was the critical element that allowed the evolution of multi-celled organisms. These early multi-celled organisms were similar to a modern fungus – the cells were similar, and there was nothing as complex as movement. But there were different cell types, with different roles, to help the organism survive (see Box 7.1). The earliest fossils of multi-celled organisms date to around 2 billion years ago, and we know relatively little about them, but scientists suspect that they formed as symbiotic bacterial colonies, at first as groups of independent bacteria but ultimately merging into one complex multicellular organism (see Figure 7.2).

Figure 7.2 Complex life may have evolved from symbiotic bacterial mats. Here we see an image of just such a mat, from Mushroom Springs at Yellowstone Park. Here, the top layers of algae provides photosynthetic energy, while the lower layers help process the chemicals from the spring. In this image, the top seventeen layers are photosynthetic, out of a total of 300 layers. Such complex life is seen at roughly 2 billion years ago, but preservation of such simple life is very rare.

Multicellularity probably occurred multiple times (perhaps as many as forty) over the evolutionary history of early life. Unlike scientists attempting (unsuccessfully) to generate life under laboratory conditions, scientists have repeatedly been able to take single-celled organisms (such as yeast) and select for groups of cells that appear to cluster. In the example of the yeast, after repeated selection events the yeast shared many of the characteristics of complex life, with a division of labor among specialized cells, reproduction with "adult" and "juvenile" stages, and multicellular offspring. It appears that, if conditions are appropriate for specialization and cooperation among cells, they will readily join into a complex organism, and this multicellular complexity will be passed down to the offspring.

Endosymbiosis

The early photosynthetic bacteria formed the foundation of all later multicellular photosynthetic life, including plants. In fact, cyanobacteria are now part of every organism that undergoes photosynthesis. Inside every photosynthetic cell of a plant is a small structure known as a chloroplast. The chloroplast has the chlorophyll and is responsible for absorbing sunlight and generating the energy for the plant cell. But this small structure has its own DNA and its own cell membrane. This is because, at some time in the deep past, several billion years ago, one bacteria type absorbed a

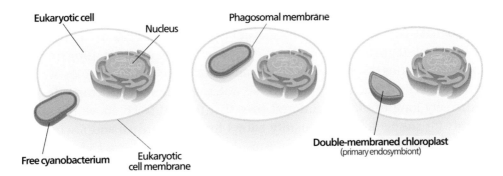

Figure 7.3 One way to become a complex organism is to absorb energy-producing bacteria. This has happened at least twice in the history of life. At some point in the deep past, some simple organism absorbed a cyanobacteria cell and co-opted the energy produced (endosymbiosis). The advantage for the absorbed organisms was the protection of the outer cell wall and reproduction whenever the outer cell reproduced (the chloroplasts have their own DNA independent of the nuclear DNA of the host cell). This happened again when energy-producing bacteria (a relative of *Rickettsia*) was absorbed into an early animal cell, which we now call mitochondria.

cyanobacteria and co-opted its energy-generating capabilities for its own use. That co-opting bacterium had its own DNA, and as time went on that bacterium evolved into more and more complex organisms, ultimately leading to plants. During the growth of this evolutionary lineage, the reproductive capabilities of that cell were also employed to reproduce the chloroplast (or cyanobacteria, if you prefer). And over billions of years, as the lineage evolved more complex organisms, the chloroplast was always reproduced as well. Nowadays, the chloroplast is an essential part of a plant, which cannot survive without it (see Figure 7.3).

Endosymbiosis has appeared at least twice (see Box 7.1) and may be critical for the evolution of more complex life. If a cell, or parts of an organism, can be relieved of the need to acquire energy by absorbing another organism, that frees the host organism to develop specialized cells. For example, the trunk of a tree does not need to have photosynthetic cells because another part of the plant, the leaves, has them and can absorb energy for the tree. Single-celled organisms don't have this luxury because each cell is an organism that must survive on its own. Because of the seemingly infinite variation in anatomical structures available to a multicellular organism, being a *eukaryote* with anatomical variation among different cell types allows complex organisms to adapt to a greater variety of environments. For example, if the ground is covered by photosynthetic bacteria, there would seem to be no way for a plant to acquire enough solar energy to survive. But a plant can extend vertically, growing stalks above the ground and spreading leaves so that it can acquire solar energy. A single-celled organism like cyanobacteria has no ability to do this;

all they can do is reproduce, whereas plants can extend specialized anatomy out to make sure the organism itself acquires enough sunlight to survive.

If we look at plant life, we find innumerable examples of plants using specialized adaptations to survive a wide variety of environments, from water-retaining cells in cactuses, to extensive water-seeking root structures, to the extremely long trunks in rainforest trees that try and reach above the other plants for sunlight. Multicellularity allows for the extensive and fascinating variation we see in complex organisms. In Chapter 8, I discuss some of the variation in early animal anatomy after a billion years.

Box 7.1 Endosymbiosis

Nature likes to be efficient, and one way to simplify growth is to use something that has already evolved. This has happened innumerable times over the course of evolutionary history, and one of the most interesting and important examples is the case of endosymbiosis.

Symbiosis is when two life-forms cooperate, as in the example of the clownfish living within a sea anemone. The clownfish gets protection from predators, because the sea anemone has stinging tentacles that sting any fish other than the clownfish, and the clownfish, in turn, eats small invertebrates that would otherwise harm the sea anemone. They both get something out of the relationship.

In the case of *endosymbiosis*, one life-form literally absorbs the other, and it is advantageous to both (see Figure 7.3). Long ago, between 1.5 and 2 billion years ago, a photosynthetic cyanobacterium was absorbed by another organism, probably a similar, simple prokaryotic bacterium that lacked photosynthetic ability. Both organisms received a substantial benefit from this absorption of one life-form by another. The cyanobacteria received several advantages. The first was additional protection, as it was now within another organism, within its own cell membrane. Viruses and other parasitic organisms would now have to breach two cell membranes to attach to the cyanobacteria. Second, the replication mechanism of the host organisms was now doing much of the work of reproduction, so the cyanobacteria found it easier to reproduce.

The host cell acquired something equally important, if not more important: energy. Whatever the host was doing previously to acquire energy (perhaps digesting dead cyanobacteria floating in the ocean), it now had a power-generating station inside itself. The cyanobacteria needed only sunlight to produce energy in the form of glucose. The host cell now had access to that

Box 7.1 (cont)

energy source. This exact relationship is the foundation of all photosynthetic eukaryotes including algae (single-celled) and plants (multicelled).

These small power generators inside the eukaryote cells are known as *chloroplasts*, and one piece of evidence for their having been independent organisms is the fact that they have their own DNA that is separate from the DNA of the host cell. Not only that, when geneticists looked at the DNA of the chloroplasts it was very similar to cyanobacteria DNA in structure and method of replication.

Mitochondria

Endosymbiosis has occurred several times, including in the animal lineage. Mitochondria are the analogous power generators that exist inside the cells of all animals. Unlike chloroplasts, they do not engage in photosynthesis, as animals do not derive their energy directly from the Sun. Animals survive by consuming other organisms (plants or animals), so the mitochondria assist in converting the bodies of other organisms into energy that the animal cells need.

Rickettsia is the bacteria family responsible for a variety of diseases, including typhus and spotted fever. It is a parasitic bacteria, which means that it feeds off of the resources of the cell it infects. This bacterium has DNA that shows a close relationship with mitochondria, and is likely to be the bacteria that was endosymbiotically absorbed by its host. Because the *Rickettsia* bacteria are already parasitic, and can convert organic material into cellular energy, absorbing them would have given the host cell a substantial evolutionary advantage. Any organism that had such bacteria within their cells could specialize in absorbing the bodies of other organisms; in other words, if they were multicellular these organisms could become animals. Mitochondria are now essential for the generation of energy in animal cells.

Box 7.2 Sexual reproduction

We often take sexual reproduction for granted because it is how all animals and most plants reproduce, and those are the organisms we tend to see on a day-to-day basis. But there are far more organisms that reproduce asexually. These organisms produce clones of themselves. From an evolutionary perspective, this method makes a lot of sense. If the purpose of reproduction is to pass

Box 7.2 (cont)

down your genes, cloning passes down *all* of your genes whereas sexual reproduction only passes down *half*. Also, the DNA replication process is fairly simple and efficient, unlike the complicated process of meiosis in which your genes are shuffled before being put into the sperm/egg. Finally, sexual reproduction requires finding an appropriate mate. Finding a mate is not necessarily a given if the organisms are spread out, and it also means that organisms have to recognize each other as potential mates and invest energy in the mating process. Finally, the mating process can be dangerous, because the first thing any organism thinks about is its next meal, so mating requires being able to discriminate enough to know not to try and eat your mate (or be eaten).

So why is sexual reproduction the dominant form for higher organisms? The main reason is because it generates *variation* and variation is the main driver of natural selection. Take, for example, a single-celled asexually reproducing organism that is adapted to warm climates. If the climate changes and it becomes cold, a cloning organism can only make more versions of itself. If that version is not sufficiently well adapted to the cold (with, for example, fur, or blubber, or a faster metabolism), then not only will that one organism die but the whole lineage will die out. In a sexually reproducing organism, genes are constantly shuffled so that the population will have variation in multiple traits. In the scenario described above, the furrier or more blubbery individuals in the population will be more likely to survive the cold. All of the adaptations you are likely to think of in plants and animals – the wings of birds, the fins of fish, or the brilliantly colored flowers of plants – are the result of novel anatomies that could be selected for because of the variation introduced by sexual reproduction.

This is not to say that cloning organisms such as bacteria don't generate new genetic variation occasionally. They do, but it is typically introduced by mutation, and so is a rarer event. Some bacteria exchange genes (in fact, this is how antibiotic resistance can spread in a bacterial colony), but it is a more haphazard event, and in any event only passes a few genes. But there is no mechanism for consistent genetic recombination as in sexually reproducing organisms. One reason that bacteria do not reproduce sexually is that they are single-celled organisms, and sexual reproduction requires specialized organs, which single-celled organisms could not have.

Finally, it is worth noting that even complex organisms can retain the ability to reproduce asexually if the circumstances are appropriate. Most plants retain the ability to reproduce asexually, and some plants alternate a sexually

Box 7.2 (cont)

reproduced generation with an asexually reproduced generation (ferns, for example, reproduce this way). In animals, the New Zealand mud snail (*Potamopyrgus antipodarum*) reproduces both ways, depending on whether it is in deep or shallow water. In shallow water, it is exposed to a high number of waterborne parasites, so it reproduces sexually, increasing its population variation as a way to evolve resistance to the parasites. In deep water, it reproduces asexually, because there are few parasites in deep water and asexual reproduction is a more efficient way to have a high number of offspring. In some sharks and reptiles, if a mate cannot be found, the female will produce a clone of herself rather than fail to reproduce.

Box 7.3 Specialized anatomy

Imagine a world where all organisms are one-celled and reproduce asexually. They would be able to exist on the surface of the ocean, where photosynthesis could support large algal mats and perhaps to a shallow depth where the sunlight could penetrate. On land, wherever water was available, at least intermittently, bacteria could survive on the surfaces of rocks, much like lichen does today (which is, even now, partially composed of photosynthetic cyanobacteria).

But there would be no swimming, walking, running, burrowing, or flying. All of those activities require bones, muscles, ligaments, and tendons, as well as skin, feathers, claws, fur, and scales. Nor would there be leaves, trunks, flowers, vines, or roots. All of these parts of animal and plant anatomy are specialized types of cells that can only exist in multicellular organisms.

The evolution of multicellularity opened the door to an enormous range of anatomical adaptations, so that today we have organisms that can move and feed deep in the oceans beyond the reach of the Sun, as well as high in the air. What is particularly noteworthy is how rapidly this occurred. Once organisms with complex anatomy arrived on the scene, the combination of sexual reproduction, with its constant reshuffling of genes, and a complex combination of anatomies on which natural selection could work, produced a bewildering array of creatures. After more than 2 billion years of single-celled organisms as the only life on Earth, suddenly there were creatures swimming and crawling through the oceans, as well as new plant variations on which these animals could feed.

Box 7.3 (cont)

The evolution of specialized anatomy is closely linked to the evolution of sexual reproduction. Without specialized anatomy, sexual organs could not exist. And once sexual reproduction takes off, more anatomical variation is possible. So these two elements act together in a feedback loop to dramatically increase the potential anatomical variation in multicelled organisms on Earth, leading to the sudden proliferation of complex life we call the Cambrian Explosion.

Box 7.4 The importance of photosynthesis

You have almost certainly learned about photosynthesis somewhere along the way in your education. It is the process by which plants take the Sun's energy and use it to generate fuel for its own growth. A plant takes in carbon dioxide and water and, using the sunlight, breaks down the carbon dioxide and water into sugar that it uses to store energy in a chemical form. In the process, a waste product, oxygen, is generated.

If you think about it for a minute, you might find that this is a pretty profound transition. Water (H_2O) and carbon dioxide (CO_2) do not contain chemical energy that can be used by the plant, so the amazing part is how photosynthesis rearranges the atoms in each of those molecules in a way that stores chemical energy. One way to think about it would be like taking a piece of thin, strong metal and then bending it slightly. Suddenly you have stored energy (in this case, as a spring). Once that energy is stored, it can be used any number of ways.

So plants use the stored chemical energy to grow. But when animals eat those plants, they are acquiring that same stored energy, which they then convert to growth, or store as fat. If a predator eats an animal, that same energy is passed down to the predator. So, in effect, when a lion eats an impala it is eating an animal that is stored solar energy, only in chemical form.

Humans access this same solar energy when they eat plants or animals, or burn logs for warmth. But what about other forms of energy? When we burn gasoline in our cars, we are using that exact same solar energy. Hundreds of millions to billions of years ago, giant mats of cyanobacteria and other organisms covered the Earth's oceans. When they died and were buried beneath the ocean floor, over time that plant material decayed and slowly converted into what we call crude oil. A similar process occurs during coal

Box 7.4 (cont)

formation, when plant matter is buried and decays over millions of years. In both cases, the product we burn is a stored accumulation of millions of years of solar energy, converted into a chemical form via photosynthesis. The only form of chemical energy on Earth that is not the result of photosynthesis is nuclear energy.

Box 7.5 The accumulation of oxygen

One of the incidental results of 2 billion years of photosynthesis by single-celled cyanobacteria was the accumulation of oxygen in the atmosphere. Oxygen is the byproduct of photosynthesis, and, from the perspective of plants and cyanobacteria, is a waste product. Conversely, animals use this oxygen and the glucose created in part by plants to fuel themselves, releasing carbon dioxide and water as byproducts. Of course, this accumulation has had profound consequences for us, and all other complex multicellular organisms.

$$\textit{Cellular respiration photosynthesis}$$
$$C_6H_{12}O_6 + 6O_2 \leftrightarrow 6CO_2 + 6H_2O + \text{energy}$$
$$\textit{glucose} \quad \textit{oxygen carbon dioxide} \quad \textit{water}$$

Oxygen has some important chemical properties – it makes bonds readily with carbon, so it can form some of the complex organic molecules that exist in all organisms, such as lipids, sugars, and proteins. Since oxygen normally exists as a molecule of two atoms (O_2), it can form chains by bonding with two separate molecules. Oxygen can also readily transfer energy by donating an electron, which is essential for the transfer of energy to cells during cellular respiration.

Complex organisms have large energy requirements and the chemical properties of oxygen make the element essential, which is why we do not see the appearance of complex life until large amounts of atmospheric oxygen had accumulated. Some animals, such as arthropods, are constrained in their ability to absorb oxygen because they do not possess lungs. Instead of using a vascular system, they rely on the diffusion of oxygen through their bodies. This is one of the reasons why modern arthropods are fairly small, and why there is a fossil record of large terrestrial arthropods around 340–280 million years ago, when oxygen levels spiked in the atmosphere.

Chapter 8: The Cambrian Explosion

Once life had evolved to become multicellular, and had acquired sexual reproduction, suddenly natural selection had tools with which it could generate an enormous variety of anatomical forms. Multicellular organisms could acquire organs and tissues with different functions, and sexual reproduction meant that genes were always shuffling, so they could mix and match, enabling nature to experiment with different combinations.

After more than 2.5 billion years of life looking pretty much the same across the entire planet – most of it just green algae – suddenly, in a relatively brief period of time, an enormous variety of plants and animals evolved from a common ancestor. DNA analysis of modern organisms places the divergence of plants and animals at more than a billion years ago, and there are a few multicellular fossils known from that era. (Note, however, that there is no DNA that can be extracted from these early fossils, and we cannot know if the true common ancestor is one of those fossils – relatively few organisms are ever preserved, so it is unlikely we happen to have the right species.) Once these two lineages diverged, they adopted very different adaptive strategies – the plant lineage acquired a symbiotic relationship with cyanobacteria and became dependent on solar radiation as their energy source. Animals became dependent on the energy in other organisms: animals are the branch of life that consumes other organisms (whether plant, animal, algae, etc.).

However, we don't see anything that we would identify, today, as an animal until around half a billion years ago. By this time, oxygen had started to accumulate in the atmosphere in levels comparable to today, and this oxygen enabled the evolution of a wide variety of complex and larger-bodied life. The oldest animal fossil that can be identified as an animal is the sponge. Sponges are the simplest animals and act as filters, sitting on the ocean floor catching whatever passes through their porous tissues. We have fossils that date to half a billion years ago that are virtually identical to modern glass sponges. Sponges were apparently a relatively diverse and successful group, as we have quite a variety of fossil sponges even by half a billion years ago. Plus, sponges are still alive! They are an incredibly successful branch of life by any measure.

Figure 8.1 The best known fossils of the Cambrian Explosion are known from the Burgess Shale. There is a wide variety of animal life known from this time, although most are unknown today. These animals lived in shallow seas, and so were adapted to a well-lit environment. The largest predator is *Anomalocaris*, but there were a variety of bottom feeders, plant-eaters, and other small predators swimming the oceans. Note that they swam by flapping large horizontal plates, rather than moving their bodies side-to-side, as in modern fish.

One of the most well-studied groups of early Cambrian fauna (animals are also called "fauna") are known from the Burgess Shale deposits in British Columbia (see Figure 8.1 and Box 8.2). These organisms show a wide and, in large part, completely unfamiliar set of anatomies that scientists still debate over today. But it does appear that they match some of the niches we see in animals in our modern shallow water environments. For example, it appears that there were wormlike animals in the ocean floor, bottom feeders (much like today's crabs), sponges, and predators. Most of these animals except sponges were arthropods, meaning that they had, at least partially, hard exoskeletons, much the way aquatic arthropods do today (horseshoe crabs, lobsters, crabs, etc.).

There are many other fossil locales that have preserved animals from the Cambrian, and during this time we also see the appearance of some familiar groups. For example, cephalopods, the family that includes nautiluses, squids, and octopuses, originated at this time. Other mollusk groups, such as bivalves (clams, oysters, mussels) and gastropods (snails and slugs on land, conch and limpets in the water), are also known from fossil record from the period around 500 million years ago (0.5 billion years ago). This is also when we see the ancestors of the vertebrates, of which humans are a member. Many of the important anatomical adaptations that we see in animals today appeared during this time, such as body shape (Box 8.5), nerves and brains (Box 8.8), and eyes (Box 8.9).

Because of the good preservation, as well as the early appearance of critical anatomical adaptations, the Cambrian era has long been the subject of intense study. For more than 200 years, paleontologists have been studying the fascinating fauna of this period, and every year new and interesting fossils are discovered. Most modern large evolutionary groups (arthropods, vertebrates, mollusks, plants) made their initial appearance during this time, so the subsequent 400 million years of evolutionary history is, in many ways, simply a playing out of the chess pieces put on the board.

Box 8.1 The oldest animals: sponges

At some point, perhaps 700 million years ago, the earliest animals evolved. These weren't animals as we normally think of them, as they had no legs, eyes, flippers, tentacles, fins, or antennae, nor any of the other myriad physical specializations we see in most modern animals. In fact, these early animals lacked the systems we associate with animals: circulatory, nervous, and digestive systems. However, they *did* digest food, so the first big step in the animal lineage was abandoning the pursuit of solar energy and instead looking to other organisms as sources of energy.

The way they do this is by catching food floating through the water. And sponges do have some specialized cells for such roles as digestion, cell communication, structural maintenance, and reproduction. Sponges are a highly successful group, as they live in deep and shallow oceans, warm and cold waters, and saltwater and freshwater. They digest a variety of marine

---

Box 8.1 (cont)

organisms, primarily small floating plants, but one group has evolved to become carnivorous, trapping and eating small marine crustaceans.

Paleontologists have discovered very old fossil sponges. By 580 million years ago we already have a diversified group that is well preserved in the fossil record. The next major group to evolve was jellyfish, and in this group we see many of the adaptations we associate with later animals: nervous systems, movement, specialized digestion, etc. And from there, the acquisition of specialized anatomy suddenly expanded, as evidenced by the fossil record of animals during the Cambrian Explosion. But all of these later adaptations are dependent on the initial commitment by sponges to live by eating other organisms.

Box 8.2 The Burgess Shale

Much of what we know about the evolution of the earliest animals comes from discoveries in British Columbia made more than a hundred years ago. In 1909, Charles Walcott, a paleontologist from New York State, discovered fossils in shale deposits in the Canadian Rockies. He returned the next year with his wife and children, and, for the next two decades, excavated some of the most interesting and important fossils ever discovered. By the time of his death in 1927, Walcott and his family had excavated more than 60,000 fossils.

Many of these animals completely stumped paleontologists and anatomists because they have anatomy unlike anything we see in modern animals. For example, *Opabinia* (Figure 8.2) and *Anomalocaris* have long, curved, spikey feeding appendages, and *Wiwaxia* appears to be a slug-like creature covered with armored plates and spines. *Hallucigenia* is a wormlike animal with seven or eight pairs of clawed legs and eight long spines protruding along its back.

These animals look nothing like those we see today, so what do we make of them? In all likelihood, these were evolutionary experiments that succeeded for a time, then, as other animals evolved with better adaptations, they became extinct. One thing to keep in mind when discussing evolution is that one of the main drivers of change is competition. Initially, these animals may have had no competition, as they evolved into completely new niches. But, as other animals evolved to become faster swimmers, or more efficient feeders, these strange creatures from the Burgess Shale would have gone extinct.

Box 8.2 (cont)

Figure 8.2 *Opabinia* was a swimming soft-food predator and one of the smaller predators in the ocean at the time. With five eyes and a unique proboscis, taxonomists have disagreed about where to place this creature. It is likely too unusual to make a good ancestor to later animals, and so is likely to be part of a now-extinct branch of the arthropods.

Excavations continue to this day at the Burgess Shale deposits, and new discoveries are made every year. Since we have no modern analogs to these wonderful and lost creatures, the fossils are all we have to go on, and interpretations have changed as new discoveries have been made. Every year, we get another look at the way natural selection shaped anatomy at the very dawn of animal radiation.

Box 8.3 Trilobites

One of the most interesting groups to emerge from the Cambrian Explosion is the trilobites (Figure 8.3). Unlike the Burgess Shale fauna, we know a *lot* about the trilobites. We know so much about them because they were so successful. Trilobites lived for more than 250 million years (from roughly 520 to 250 million years ago) and were geographically widespread – they have been discovered in the Americas, Asia, Africa, Europe, and Australia. To date, scientists have

Box 8.3 (cont)

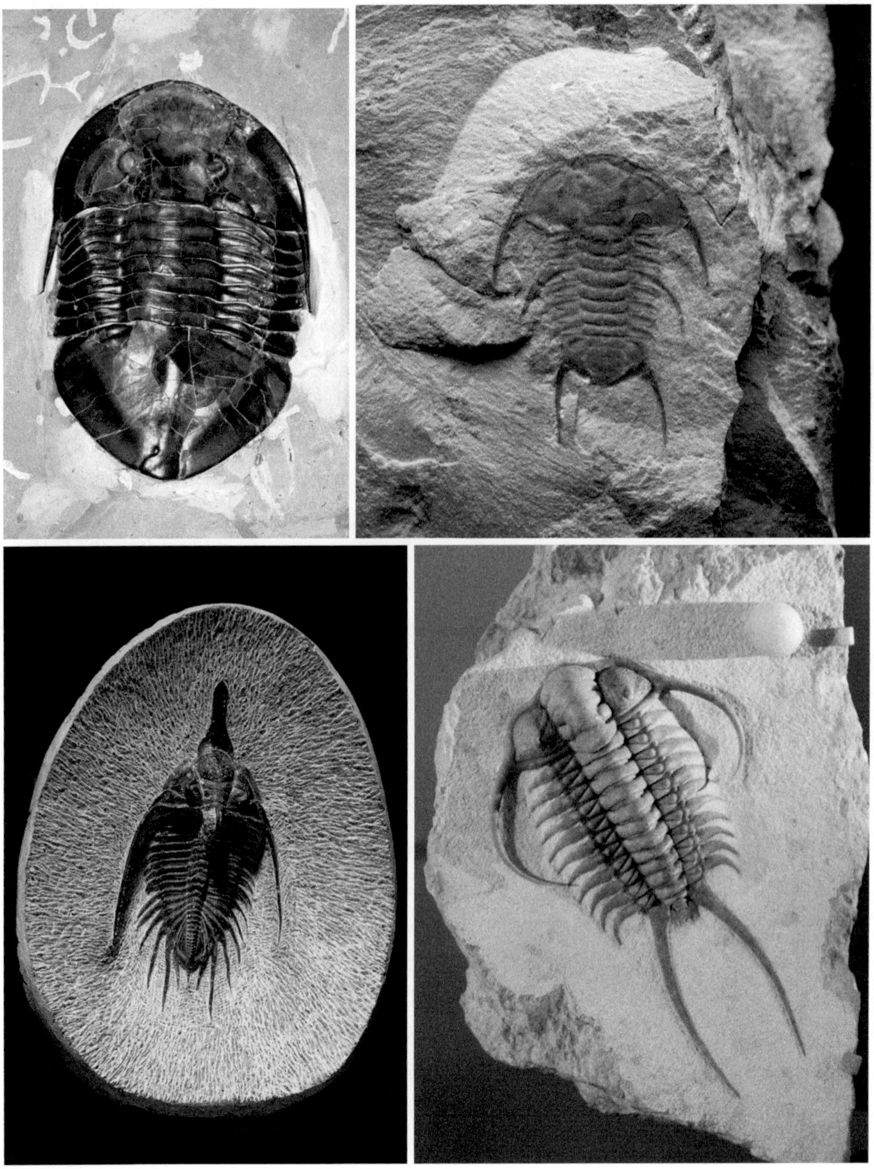

Figure 8.3 The trilobites were the first animals to be geographically widespread and display a significant amount of anatomical variation. Trilobites are known from Cambrian fossil beds around the world, and paleontologists can even find beds with hundreds of trilobites preserved together. They appear to have been highly social, and sexual selection may account for the exotic ornamentation seen in some species.

Box 8.3 (cont)

identified more than 10,000 different species, and because of their diversity we can study all kinds of interesting things about them.

Trilobites were arthropods, which means that they had an exoskeleton, much like crabs and lobsters, and they grew by shedding their hard shell and growing into a softer shell, which then became hard. This process, called molting, is seen in crabs and lobsters, and if you have ever eaten a soft-shell crab you have eaten a crab mid-molt. Much like modern arthropods, they varied in size, and they ranged from a few millimeters (1/4 inch) to 30 centimeters (roughly a foot).

They lived on the bottoms of the oceans. Some species lived in shallow water, and could see one another fairly well with their well-developed eyes. At least some species were social, and there are fossil collections with hundreds of individuals preserved together, as well as interlocking sets of tracks in fossilized mud. This sociality had some implications for their behavior, and the ornate anatomy in some trilobite species has been attributed to sexual selection, in much the way we see ornamentation in peacocks.

As you might expect from a widespread group that lived for more than 250 million years, they had diverse dietary adaptations. Some were predators of seafloor organisms, like worms, and had appendages for extracting animals from the sea mud, whereas some were particle feeders with filter-like mechanisms. Most appeared to live on the bottom of the ocean floor, but some may have been able to swim. Because so many are preserved, we have been lucky enough to get a few with soft tissue, so we know about their nervous and digestive systems, and this helps us understand the ancestry and evolution of these systems. Although rarely preserved, they had many legs, as well as antennae. Their eyes, made of compound calcite lenses, are particularly interesting, and are quite diverse themselves; they can tell us about how deep any particular species lived (deeper water has little light, and no need for complex eyes, whereas some shallow water species actually evolved sun-shades over their eyes).

As long-lived and geographically widespread as this group was, they ultimately all disappeared, in a great event known as the Permian–Triassic Extinction, roughly 250 million years ago, when the majority of life on Earth died off. But before they died, they left a fantastic fossil record that grows with every new discovery. And although paleontologists, such as myself, are sometimes wary of encouraging private fossil collection, trilobites are a great exception. There are so many trilobites that collection has had no negative effect on the scientific research on the group, so there is no reason not to buy yourself a piece of prehistory. Today, you can get yourself a nice specimen of this fun and ancient group for as little as $25.

Box 8.4 The origin of vertebrates

Humans are vertebrates, which means that we have a bony spine, inside of which is the spinal cord, the main stem of the nervous system. This pattern is shared by all vertebrates, ranging from the smallest frog to the largest whale. We lack an *exoskeleton* – our weight is borne by a bony skeleton. There are obvious advantages and disadvantages to having an *endoskeleton*, when compared to the exoskeleton of arthropods. The most obvious advantage of having an exoskeleton is that the shell acts as armor, protecting the inner soft tissue from damage by predators or anything else in the environment. But an exoskeleton imposes constraints. The first is growth. As mentioned in Box 8.3, animals such a trilobites with an exoskeleton must shed their shells to grow in size (molt). When they are still in the soft-shell stage they are very vulnerable, and growth is limited to a relatively brief period of time before the shell hardens (although one advantage arthropods have is that the juvenile stage and adult stage can be radically different, as in butterflies/caterpillars). Further, the outer shell is heavy relative to its strength, which is why you only see large arthropods in the ocean. Finally, the circulatory system of arthropods is inefficient – the blood is simply pumped to large organ areas, then pumped out, whereas in vertebrates each cell is vascularized. This allows animals to get larger, as the circulatory system can be scaled up relatively easily to ensure oxygen is distributed to each cell.

The earliest ancestors to true vertebrates are called *chordates*, which have a central nervous system but have not yet evolved a complete bony spinal structure. A fossil animal, *Pikiai*, has been discovered in the Burgess Shale that looks very similar to what we would expect from the first chordate ancestor. At 530 million years old, this animal has the foundations of later vertebrate anatomy at the same time that arthropods were the dominant animal form. Today, the lancet looks extremely similar to what we see in *Pikiai*, suggesting that, while they may represent the stem form of all later vertebrates, they were a successful enough form that they never died out.

However, it is worth pointing out that, when contrasting vertebrates with arthropods, it is not entirely fair to say that vertebrates are more successful. Vertebrates have evolved to fill many niches, and are found the world over, but there are *many* more species of arthropod (roughly a million species of arthropod compared to 60,000 species of vertebrate), and they have repeatedly survived the Earth's many extinction events, whereas vertebrate groups have been subjected to widespread extinction on several occasions.

> **Box 8.5** Body symmetry, body organization, and HOX genes
>
> When you look at most animals, whether arthropod, chordate, or vertebrate, their body is organized according to a plan: there is a front and back, and a left and right. But very old animal lineages do not show these patterns. A sponge has no particular body plan; it just grows in response to environmental stimuli (temperature, food, water currents, etc.). The oldest animal to show symmetry is a jellyfish. They have what is known as a *radial* body plan. There is no front or back to a jellyfish, although it does show symmetry. But vertebrates, chordates, and arthropods have *bilateral symmetry*. This means they have a left and right side, and typically each side is a mirror of the other. Bilateral symmetry is desirable for a few reasons (e.g., locomotion appears to be easier), but a critical one is that it is genetically and evolutionarily efficient. Let's say there is selection for a larger front limb. Both front limbs are under the genetic control of a few "master genes," so it takes fewer genetic changes to acquire the desired trait. Evolution doesn't need to work on the genes of both front limbs independently – selection on one acts on the other.
>
> The other thing about bilaterally symmetrical animals is that they have a front and a back. Geneticists have identified the genes that determine body layout across all animals with bilateral symmetry; these are known as HOX genes. HOX genes organize the anterior-to-posterior (i.e., front-to-back) layout of houseflies, scorpions, lizards, sharks, grizzly bears, and humans (see Figure 6.4). This is a deeply conserved gene that must be more than 500 million years old, yet is still retained in all these lineages, further demonstrating the common ancestry of all animals.

> **Box 8.6** Plant evolution
>
> In this chapter (and book) we mostly focus on animals, but most of the biomass on Earth is composed of plant matter, and it is the foundation upon which other forms of life are built. Plants evolved from land-based photosynthetic green algae during the Cambrian Explosion somewhere between 450 and 500 million years ago. At some point, probably prior to 500 million years ago, a photosynthetic algae cell was absorbed by some other single-celled form. The photosynthetic algae cell was not absorbed; rather, it became part of the host cell. The host cell became dependent on this source of photosynthetic energy, and was subsequently reproduced within the host.
>
> This differentiation of parts within the host cell required it to become eukaryotic – to acquire a cell membrane to protects its nuclear DNA. As

Box 8.6 (cont)

previously mentioned, to become a multicellular organism appears to be an easy transition for eukaryotes, so over time this organism was able to evolve specialized cells. This differentiation of cells is what allows the variety of plant organ adaptations (e.g., roots, stems, and leaves).

The earliest plants were mosses, which do not possess vascularization (the tiny tubes that transport water throughout the plant). Rather, they are porous, so absorb water through external contact. This means that moss must be close to the ground and therefore close to water. However, soon plants acquired vascular systems – long tubes through which water can travel via capillary action. Fossil plants with roots may be as much as 400 million years old. Once this adaptation was acquired, plants could start becoming more vertical. By 350 million years ago leaves had evolved, which allowed broad flat areas of the plant to come into contact with sunlight (although leaves may have independently evolved as many as four times).

Once plants started to cover the ground, competition for sunlight became fierce. One way to maximize access to sunlight is to grow above other plants. The reason trees are so high is that the tallest tree has access to as much sunlight as it has limbs, branches, and leaves. But being tall requires a trunk, so the evolution of a specialized trunk was a necessary precursor to the evolution of forests. During this early evolution of plants, at 400 million years ago, no plant was taller than 1 meter (3 feet). But once trunks evolved, forests, as we think of them, became commonplace, as ferns and related early trees quickly evolved to be much like modern forests.

Early plants reproduced the way ferns do, with an inefficient spreading of spores, which then grew, on their own, into small organisms called gametophytes, which, in turn, produce sperm or eggs. The sperm must "swim" to find the egg cells. Most modern plants are *gymnosperms*, in which plants produce seeds that are fertilized by pollen from another plant, the plant ultimately producing a fertilized seed. This method of reproduction is more efficient, and gymnosperms became the dominant plant form. Later, angiosperms, or flowering plants, evolved.

The evolution of plants had major impacts on the evolution of animals, since animals are direct consumers of plant material. For example, primates (of which we are one example) are argued to have only appeared once flowering plants evolved. Early primates were small, arboreal insectivores, pursuing the flying insects (such as bees or flies) that were fertilizing the flowering plants. Similarly, many animal groups have acquired anatomical specializations to

---

Box 8.6 (cont)

consume specific types of plants. Horses, for example, have especially long teeth to counter the wear from eating grasses that are high in abrasive silica.

Animals are completely dependent on plant life, so, even as this book focuses on animal evolution and behavior, keep in mind the critical importance that plant life represents, and the wide diversity that evolution has generated.

Box 8.7 Extinctions

Extinctions have been a fact for as long as there has been life on Earth. The first major extinction (the Ordovician–Silurian Extinction) occurred between 450 and 430 million years ago and wiped out many of the major animal groups living at that time. As far as we know, *all* of the Burgess Shale fauna were extinct by the end of this event. However, some animals did survive. Trilobites survived another 200 million years, through the next major extinction (the Late Devonian extinction, at 360 million years ago), only to be wiped out by the next extinction (the Permian–Triassic Extinction) at 250 million years ago. Even more extinction events were to follow. For many of these events, we don't have a complete scientific consensus on the causes. In most cases, large geological events, such glaciation or volcanism, are invoked, with the subsequent Earth-wide changes in climate altering the landscape in catastrophic ways. Other hypotheses invoke astronomic forces, such as a sudden burst of gamma radiation from a nearby dying star, or the impact of a large meteor.

Whatever the cause, extinctions have a way of "wiping the slate clean," so that other organisms can evolve in newly vacated ecological niches. Often organisms with new adaptations cannot gain a toehold in the environment until existing animals are eliminated via extinction. This has occurred multiple times throughout Earth's history, and is why we have subsequent radiations of land animals: amphibians, reptiles, and mammals, as each large groups gets wiped out and is succeeded by another.

But it is important to recognize that, in all of these cases, some organisms do survive the extinctions. For example, horseshoe crabs, which are almost 500 million years old, survived all these extinctions, as did lancets. Similarly, amphibians and birds survived subsequent major extinctions. Extinctions are not only a destructive event (although they clearly are that) but a major enabler of evolutionary diversity.

Box 8.8 The evolution of the nervous system

Although sponges are animals, they are extremely primitive and lack many of the structures associated with later animals. One of those structures is the nervous system. Animals display a wide variety of nervous systems, which are the result of natural selection's attempts to solve various problems encountered by more complex animals.

Sponges do not move, and do not need to coordinate movement in the way that more complex animals do, but cells can communicate with each other via chemical signals. Sometimes a sponge will contract to reduce its size when a predator attacks, or to control water flow through its tissues, and for coordinated movement the cells will communicate using calcium-based chemicals. But sponges contract slowly; once animals were no longer anchored to the seafloor, they needed to be able to coordinate movement more efficiently, and this means cells needed to communicate rapidly.

The cell that is used to communicate information in complex animals is the *neuron*. Scientists are still trying to understand the origins of the neuron, but the advantages are clear. A neuron uses a combination of electricity and chemistry to pass signals, and because of the adoption of electricity the signals move very rapidly. An animal that uses electricity-based communication can respond quickly to the environment, which would give it an enormous advantage over other organisms. Measurements of neuron signals in animals have found the speed to exceed 200 mph.

The simplest nerve systems are found in Cnidarians (jellyfish, coral, and sea anemones). These animals possess a *nerve net*, in which the neurons are distributed around the animal, with no central "lump" of neurons that makes decisions. The cells communicate, but there is nothing analogous to a brain in which decisions are made. These animals tend to respond in very simple ways to a stimulus: the whole body will contract, for example, when in contact with something damaging but will not "know" the direction from which the damage came.

However, animals that show bilateral symmetry have places (typically in the anterior end of the animal) where there is a "lump" of neurons that serves as a central processor for signals coming in from all over the animal. This "lump" is the brain, the center of the *central nervous system*, and it is one of the most critical adaptations found in early animals. In animals with central nervous systems, signals are sent to the brain, where they are interpreted (for example, as either potentially damaging or not) and then a decision is made for a response. If a response is called for, a signal is sent along peripheral nerves to activate the response (for example, moving a limb). In vertebrates, signals are

---

Box 8.8 (cont)

sent along the spinal cord to the rest of the body. In arthropods, the signal is sent along two large nerves that run in parallel down the body from the brain.

Chapter 14 is devoted to brain evolution, so I go into more detail there, but it is important to know that the acquisition of the brain happened very early in animal evolution. Brains were possessed by Cambrian animals such as those found in the Burgess Shale, and some trilobites are preserved enough that parts of their brains have been studied. The importance of the brain for animals responding to complex environments means that the brain has been under constant and strong selection, and many animals have specialized structures in their brains to deal with the different kinds of stimuli available (e.g., vision or smell) or problems to be solved (e.g., navigation or remembering food location).

Box 8.9 Eye evolution

The evolution of the brain (Box 8.8) was a critical first step in the acquisition of problem-solving tools for early animals. However, equally important was the ability to perceive the environment. An animal that can, in some sense, "know" what is around it can avoid danger while pursuing food. Perhaps the most critical organ for knowing what is around is the eye (Figure 8.4).

The eyes of most modern animals are extremely complex: an arrangement of specialized nerve cells a particular distance from a sophisticated focusing lens. However, the earliest eyes were not like this at all, and we can see these ancestral eyes in animals still living today. The simplest form of light perception can be seen in single-celled photosynthetic organisms called *Euglenia*. These organisms are extremely ancient, and actually predate the split between plants and animals. They have simple pigmented tissues on their bodies that perceive the presence and absence of light, probably for synchronizing their metabolisms to the circadian (day/night) rhythm.

Some simple early animals, such as flatworms (planarians, which date to the Cambrian) also have pigmented spots that perceive the presence and absence of light, and in many of these animals the pigmented spot is concave, so that they can actually perceive the direction from which the light is coming. This allows them to move toward and away from light sources. Jellyfish have similar "eyespots" that allow them to rise and fall depending on the presence of sunlight in the ocean.

Once light perception becomes important to an organism, there is reason to protect the sensitive pigmented cells, so many animals have acquired ways to

Box 8.9 (cont)

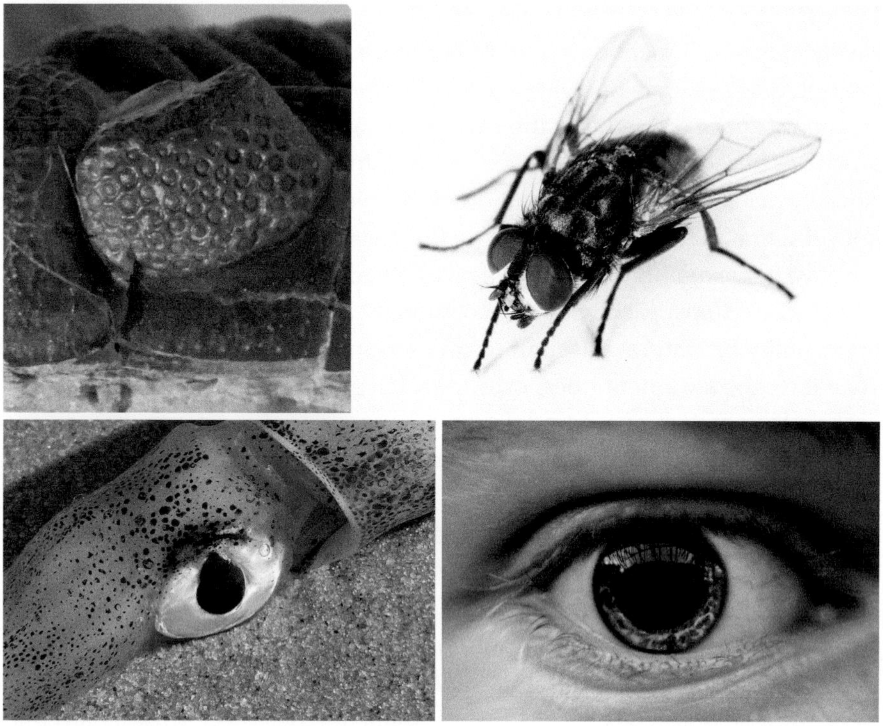

Figure 8.4 The eye has evolved in multiple ways, and has converged on several solutions repeatedly. The multi-lens eye of the trilobite (top left) was independently acquired in the insect lineage, as seen in the housefly (top right). The focusing single-lens eye has independently evolved in cephalopods (squid, lower left) and in vertebrates (human, lower right), although they have a few notable anatomical differences in the organization of the nerve receptor cells.

protect eye tissue. Some animals have recessed the light-sensitive tissue deep inside a protected ring (as in mollusks), whereas others have evolved hardened lenses over the tissue (as in trilobites).

Eye evolution in animals reveals some interesting patterns: for example, several animal groups have independently evolved similar solutions to the problem of the eye. The trilobite and housefly both have compound multi-lens eyes, and these were independent acquisitions by each lineage. Similarly, the "camera eye" of humans and cephalopods (squid and octopuses), in which there is a focusing lens in front of clear fluid, then a nerve membrane (the retina), was also a case of evolutionary convergence. As with many anatomical solutions in nature, it appears that there are only so many "good" solutions, and, given enough time, nature will find those solutions again and again.

Box 8.9 (cont)

The Eye as Evidence for Divine Creation

Finally, it is worth mentioning that the evolution of the eye is sometimes presented as evidence of divine creation, or "intelligent design." Some creationists have argued that the eye could not be the result of evolutionary processes because it is so complex. The argument is thus: there could be no "intermediate" steps in the evolution of the eye, because if any one element is not present and functioning perfectly (for example, the lens, or the light-sensitive nerve membrane), the eye gives no advantage to the animal. Without intermediate steps, each with an evolutionary advantage, there could be no opportunity for natural selection to give advantage to animals with eyes. Therefore, the argument goes, eyes must have "appeared" as completed organs, the product of a divine designer.

The problem with this argument is that we have, in nature, *many* intermediate steps for the evolution of the eye, each of which confers an advantage on the organisms that possess them. Mollusks have been around since the late Cambrian, and they provide all the steps of eye evolution. The pigmented tissues described above are beginning steps, and provide the advantage of knowing if light is present, and the direction of that light in the more advanced concave pigmented spots. The eye of the nautilus provides an ideal example of the next step, in which the tissue is protected by a hard ring, with a small hole for light to pass through. This hole provides protection but can also be made smaller or larger to allow focusing of the light rays, much as a pinhole camera focuses light. The evolution of a clear lens is present in ocean snails, isolating the nerve tissue from potential contaminants in the seawater. The most complex stage, with the focusing lens and the "camera eye," is seen in octopuses and squids. So, although removing any element of a fully functioning "camera eye" would result in its failure, that does not mean that it did not evolve in a stepwise fashion from more primitive stages.

The argument probably stems from a failure to understand that humans are animals, so comparisons to other animals are seen as invalid by creationists. In fact, understanding that humans are part of the animal kingdom, rather than apart or above it, is one of the most important messages that you should draw from the first half of this textbook.

Chapter 9: Fish and Land Animals

The Cambrian saw the appearance of most of our current large and complex organism groups, including arthropods, vertebrates, mollusks, and plants. Many of the mollusks and arthropods would be recognizable today, and some have persisted since that time, but the vertebrates had just gotten started. And since we are vertebrates, we are especially interested in this group.

The earliest vertebrates evolved from animals with spinal cords (*chordates*), like the lancet. Since these animals had no protective exoskeleton, such as seen in trilobites and other contemporaneous arthropods, their bodies were much more vulnerable to damage. The vertebral column was an evolutionary response to this, and it protected the most important parts of the animal – the nerves and blood supply. The candidates for the oldest true vertebrate date from between 450 and 500 million years ago, toward the end of the Cambrian (some are considered controversial because preservation is poor for this group, since these animals lacked hard tissues).

But by around 400 million years ago, animals that are clearly vertebrates start to appear in the fossil record. These animals were limbless and jawless and still exist today in the modern lampreys and hagfish. They ate food from the ocean floor, and it is from this group that fish evolved.

The Devonian and the Evolution of Fishes

Fish are the first large vertebrate group to *radiate* (evolve and diversify) throughout the oceans, and are perhaps the most successful vertebrate group of all. They first appear at the end of the Cambrian with groups such as the Ostrecaderms. These were bottom-feeding fish with a heavy armored head for protection from marine arthropod predators. During the periods following the Cambrian (the Ordovician and Silurian), vertebrates, especially jawless fish, expanded throughout the oceans, and we have many fossils from the period of their adaptive radiation. But these early fish were primitive in many ways, and subsequent evolutionary transitions and acquisitions found them unable to compete with later fish.

Within vertebrates, probably the most important adaptation was the acquisition of the moving lower jaw (see Box 9.1). We tend to take the *mandible* for granted, which moves against our upper jaw to allow us to crush food, but early vertebrates had to use muscular power, with hard denticles embedded in the tissue, to try and grind food. The moving bony jaw permits animals to generate enormous crushing power, and has been critical in allowing vertebrates to eat a much wider variety of foods. The earliest jawed fishes known are the Placoderms, which were large, heavily armored, predatory fish that date to 420 million years ago, during the transition between the Silurian and the Devonian.

The Devonian saw a sudden radiation of fish types, and for this reason is known as the "Age of Fishes." Once fish acquired the moveable bony jaw, their adaptive niches expanded dramatically. It is during this time that we see the evolution of many of the modern fish lineages, including the sharks and rays (with cartilaginous skeletons) and well as the bony fish (such as tuna, trout, and perch). For our purposes (as terrestrial vertebrates), perhaps the most important transition was the appearance of the *lobe-finned fishes*.

Most modern fish are known as *ray-finned fish*. These are the fish that you are probably most familiar with, and are the ones you might see on your dinner table: tuna, salmon, halibut, etc. If you look at the fins they are thin, with membranes between the bones; the bones themselves are long and slender, radiate out from the base of the fin, and are described as rays (like rays of sunlight). However, back in the Devonian and Silurian there was a group of common fish with another fin pattern known as the lobe-finned fishes. The fins of this group are not thin, with long, slender bones. Rather, the fins are heavier and fleshier, and the bones themselves are thicker and articulate (they have joints). The fin bones strongly resemble the limb bones of later land animals because it is from this lineage that all later land animals (with their load-bearing limbs) evolved.

Sometime around 360 million years ago, a lobe-finned fish used its strong, bony limbs to haul itself out of the water. We don't really know why it did this; they were predators, so perhaps they wanted to pursue fish that lived in very shallow water, and could only do so by being able to carry themselves across mud flats or flooded grasses. But once out of the water, they acquired a group of adaptations that allowed them to stay on land. For example, their strong limbs, with their thick bones, acquired a bony attachment to the spine. This became the shoulder girdle (for the forelimb) and the pelvis (for the hind limb). However, they retained some aquatic adaptations as well, and so were able to traverse between land and water.

Rise of the Amphibians

Today we know this group as the *amphibians*, which are represented in the modern world by frogs, salamanders, and caecilians (limbless amphibians). They lay eggs in water, and have a juvenile aquatic stage (as in tadpoles), but as adults become air-breathing terrestrial quadrupeds. Modern amphibians tend to be smaller than many of the larger vertebrates, but when they first emerged from the oceans some 360 million years ago there were no other large terrestrial animals against which they had to compete. They quickly evolved to fill many of the ecological niches we today associate with modern vertebrate groups. Some of these amphibians were large and carnivorous; the largest was *Eryops*, which could be 3 meters (10 feet) long and had rows of sharp, curved teeth in a large crocodile-like mouth.

Amphibians were the predominant terrestrial vertebrate form for almost 100 million years. Other groups, such as reptiles, with their complete commitment to living and reproducing on land, had appeared but only in small numbers. But there is a reason why there are no large, carnivorous amphibians roaming the Earth today. Roughly 250 million years ago the largest mass extinction event in Earth's history occurred. This event, known as the *Permian–Triassic Extinction*, wiped out 70 percent of all land animals and more than 90 percent of aquatic animals (including, sadly, trilobites).

The causes for this mass extinction are still under debate, although there are many candidates (meteor strike, volcanism, microbes, celestial events, releases of trapped poisonous underground gases). But what is known is that this extinction cleared the path for the rise of the reptiles.

Rise of the Reptiles

The period following this extinction found early ancestral reptiles able to outcompete the terrestrial amphibians. One key adaptation was the amniotic egg, which could be laid on land or retained in the mother but, critically, did not have to be laid in water (as in fish and amphibians). This key adaptation liberated the reptile lineage from dependence on large bodies of water for reproduction and opened up a broader terrestrial adaptation.

One group, the Archosaurs, was particularly successful, as their descendants included the crocodiles and alligators, as well as the dinosaurs and birds. The period following the extinction, known as the Triassic, and the period after that, the Jurassic, are well known as the age of dinosaurs, as reptiles filled the ecological niches that had been formerly filled by amphibians. By some 30–50 million years after the extinction, the land was filled was a remarkable array of terrestrial reptiles. In fact, reptiles were so successful that some (mososaurs, pliosaurs, plesiosaurs, etc.) were able to evolve to become aquatic (although breathing air), outcompeting their aquatic ancestors.

The most well-known fossil reptile group is called the dinosaurs (although this is a popular, rather than scientific term – see Box 9.3). We know a great deal about this group because their bones are so well preserved (due to their large size), they lived on every continent, and they were successful for so long. They had an enormous range of dietary adaptations (herbivores, carnivores, scavengers, even insectivores). They had a similarly large range in size, which is unequaled in modern terrestrial animals. And some of them appear to have been social, which we know from their footprints as well as from anatomy that appears to be related to social or sexual competition.

Cretaceous–Tertiary Extinction

This group was so diverse and so successful that they would likely still be the dominant terrestrial animal group were it not for the *Cretaceous–Tertiary (K–T) Extinction*. Roughly 65 million years ago, a large meteor hit the Earth in what is now the Gulf of Mexico. This enormous meteor (see Box 9.4) caused a large flood in North America, and threw enough silt into the air to block photosynthesis for roughly a hundred years. The lack of photosynthesis caused the food web to collapse, and all of the large terrestrial animals on Earth went extinct.

Age of the Mammals

Many of these animal groups ultimately did recover, although, in many cases, not in their original forms. For example, one branch of the dinosaurs, the *saurischians*, are represented in today's birds. Some groups persisted largely as they were, as in the crocodile and turtle groups. And some simply went extinct, as in the large ocean-going reptiles (mosasaurs, ichthyosaurs, and pliosaurs), and the *ornithischian* dinosaurs. But one group that had formerly been relatively marginal suddenly exploded. The mammals, with their high reproductive rate and their maternal care of offspring, were able to rebound from the extinction event and fill the niches formerly filled by the dinosaurs. Large herbivores, large carnivores, scavengers, insectivores, aquatic fish-eaters – these were all niches that the mammals were rapidly able to evolve into very successfully.

Mammals evolved in stages, and, remarkably, there are three kinds of mammal alive today, representing ancestral and more evolved forms (see Box 9.5). We still live in the age of mammals, and, barring some catastrophic extinction event, we expect that to continue for the foreseeable future. But evolution works on very large timescales, and it is essentially impossible to predict what life will look like in another 1 million, 10 million, or 100 million years.

Box 9.1 The evolution of the skull and mandible

Early vertebrates inherited a very simple body plan from their chordate ancestors. Since early chordates were boneless, they had none of the bony adaptations that we associate with vertebrates. As mentioned before, the vertebra protects the spine, but the evolution of the bony skull was necessary to protect the brains, eyes, and olfactory sense organs of these early animals. Large predatory arthropods, such as the sea scorpions, probably acted as a selective pressure on ossification of the cranium. Early jawless fishes, such as the ostracoderms, lacked other features seen in vertebrates, such as limbs and jaws, but they had a strong bony skull.

The evolution of the skull was probably relatively straightforward, as hardened cartilaginous cells formed around the neurological tissues in the head. Early skulls were composed of many independent hardened plates which, in later species, coalesced into the complete cranial vault (Figure 9.1).

However, the evolution of the bony mandible was a more remarkable evolutionary transition. Unlike the cranial vault, which is simply a group of hardened plates, the mandible acts as a hinged joint, articulating against the upper jaw. Where did the jaw come from and how was it acquired?

Surprisingly, the answer does not come from fossils, but rather from examining the embryos of existing animals. Evolution often works by changing the organisms through development, and, because the ancestor of all existing jawed vertebrates was a fish, we all go through a fish stage in our development. If you look at the embryos of fish, reptiles, rats, and humans, we all look essentially similar early in our developmental stages. This reflects our fish ancestry.

Because we have a "fish stage," we also have gills. During the various developmental stages, the bones supporting these "gills" (the gill *arches*) shift into the head to become our upper and lower jaw. In jawless animals, such as lampreys and hagfish, this transition never occurs; the gills remain gills throughout development, and they continue like that through their adult stages. In jawed vertebrates, the top two sets of gill arches shift to become the jaws. This transition can be directly observed in developing organisms, and modern geneticists can actually examine the genes controlling this developmental shift. When these genes become altered through mutation, pathologies, such as a cleft palate (in humans), can result.

The earliest known animals with jaws (*gnathostomes*) are the Placoderms, and these animals were large-bodied predators. They used their jaws to crush food that would have been otherwise unavailable to them, and this acquisition of jaws allowed them to expand into many predatory niches – eating animals of all sizes with exoskeletons or shells.

Box 9.1 (cont)

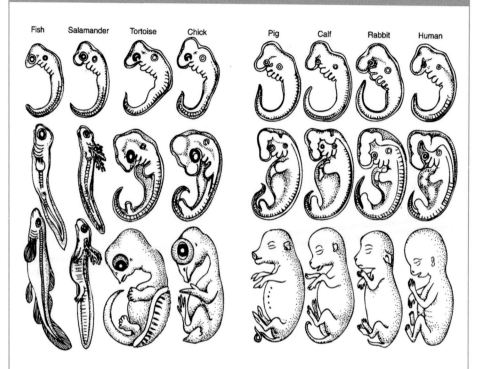

Fish   Salamander   Tortoise   Chick   Pig   Calf   Rabbit   Human

Figure 9.1 Evolution of the jaw. The jaws of modern vertebrates evolved from the gill arches of primitive, jawless fish. Although there is no fossil record of the transition between jawless and jawed fishes, there are multiple lines of evidence from modern animals. One famous piece of evidence is the shared developmental stages we see across all vertebrates. This image comes from nineteenth-century anatomists Ernst Haeckel and George Romanes. In the sequence, you can see that the earliest embryological stages of animals as diverse as fish, salamanders, birds, and mammals all lack jaws in the early stages of their development. They possess gill arches, the first of which (the top arch) then develop into upper and lower jaws. Fish retain the rest of the gill arches, since they have gills, but in tetrapods the gill arches become other parts of the anatomy, including the hyoid bone (which supports the voice box/larynx), the bones of the inner ear, and the other parts of the anatomy of the neck.

The Recurrent Laryngeal Nerve

The fact that the gill arches have become the upper and lower jaws has an interesting implication for the anatomies of later animals. In animals with necks (mammals, for example), the nerve controlling the voice box (larynx) is known as the recurrent laryngeal nerve (see Figure 9.2). This nerve passes from the brainstem, down through the chest, under the aorta (one of the large blood vessels of the heart), and then back up into the larynx in the neck. In animals with particularly long necks, such as giraffes, the nerve must descend five feet, then return another five feet, even though the larynx is only a few inches below

Box 9.1 (cont)

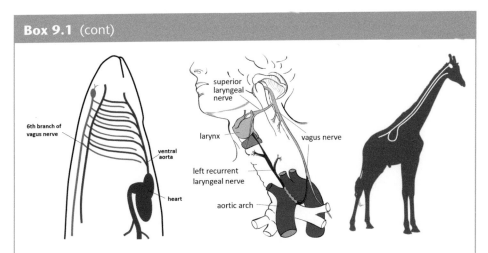

Figure 9.2 The recurrent laryngeal nerve. One piece of evidence that the bony elements of the fish gill structure evolved into the elements of the neck can be seen in the route of the recurrent laryngeal nerve. The recurrent laryngeal nerve was, ancestrally (in ancient fish), the nerve that controlled the musculature of the sixth gill set. This nerve passed below the vessel that provided blood for that gill (left image). When the bones and muscles of the fish gill structure were co-opted by vertebrates for the anatomy of the neck, the nerves had to accommodate the changes. The lowest blood vessel in the gill arch became the aorta in tetrapods, whereas the bones of that gill arch became the larynx (middle image). The nerve controlling the larynx could not simply detach and pass over the aorta, so, to retain nerve control of the larynx, it grew in length as the necks of tetrapods developed.

In animals with long necks, that meant that the nerve would have to pass down the neck, into the chest cavity, and then back up the neck and into the throat. In giraffes (right), the recurrent laryngeal nerve can be over 15 feet. This is despite the fact that the larynx only sits a few inches from the brainstem. This is strong evidence against "intelligent design," since the design is so clearly not "intelligent," but it is also strong evidence that humans evolved from a fish ancestor.

the skull. Why would the recurrent laryngeal nerve have such a long path, when it would be much more efficient to have the laryngeal nerve pass a few inches from the base of the skull into the larynx?

The answer has to do with the evolution of the gill arches into the jaws. In fish, the nerves and blood vessels for the gills pass around the bony gill arches. One of the gills evolved into the larynx, but the corresponding blood vessel remained in the torso to become the aorta. As the bony arches ascended into the skull, to maintain the nerve connection to the larynx the nerves had to follow the bones up but also remain passing around the blood vessel – they couldn't simply detach and find a new path. As the neck became longer, the nerve simply elongated.

This is strong evidence for the mandible evolving from the gill arches. And, perhaps more broadly, it is strong evidence for natural selection and against "intelligent design." This is because the design is not actually a "good"

> **Box 9.1** (cont)
>
> design – the nerves must go far away from the most efficient path. Yet, this is the only way natural selection could do it – by tweaking an existing design, while still largely constrained by existing anatomical relationships.

> **Box 9.2** Limb evolution
>
> The earliest vertebrates were, like modern lampreys and hagfish, limbless. We don't know much about the earliest evolution of the limbs, but we do know that, fairly early on, fish divided into two groups. One is the type of fish you are familiar with – the ray-finned fish. These fish are the ones you have probably have eaten, like trout, perch, and tuna. They have fins with long, slender bones radiating out from the base in a "ray" pattern. The other type of fish had stouter fins, which had thicker bones.
>
> It is from these fish that we evolved. In this early group of fish, the fins had bones that we can actually recognize today as the same bones we have in our limbs (Figure 9.3). We have one upper arm bone (the humerus) and two lower arm bones (the radius and ulna) and then a collection of wrist and hand bones. We see essentially this same pattern more than 400 million years ago in the lobe-finned fish. Their limb/fin bones are different shapes than ours, but they are recognizably the same bones.
>
> If you look at the bones of your arms and legs, you may notice that they share a pattern. There is one upper bone (the humerus in the arms and the femur in the leg), then two lower bones (radius/ulna in the arm, and the tibia/fibula in the leg), then the same number of wrist/ankle bones and the same number of hand/foot bones. This is not an accident. At some point around 400 million years, lobe-finned fish had only forelimbs, then the forelimb was essentially copied to make the hind limb, and the genetic controls were copied when the limb itself was copied. The group of organisms with four limbs are known as *tetrapods* (from the Greek for the number four, plus the word for limb). Interestingly, this replication occurred only once in evolutionary history, to the best of our knowledge, because there are no six-limbed vertebrates.
>
> These heavily boned fins were easily adapted to bearing weight on land, and this lobe-finned fish lineage became the ancestor of all land vertebrates. We have good fossils of early land animals, and the early adaptations of fins to bearing weight are apparent in the anatomy. For example, the fossil species *Tiktaalik*, which dates to around 375 million years ago, has fins that are adapted to swimming, yet the bones are also connected to the spine via the shoulder girdle so that the fins can bear weight. *Acanthostega*, at 360 million

Box 9.2 (cont)

Figure 9.3 Tetrapod evolution. This image depicts the evolution of tetrapods from aquatic ancestors. At the bottom, Eusthenopteron was a lobe-finned fish from roughly 385 million years ago. At the top is a terrestrial amphibian, *Pederpes*, the earliest known terrestrial animal, and dates to 340–350 million years ago. Between these two animals we see several animals with intermediate adaptations to land (*Ichthyostega, Acanthostega, Tiktaalik, Panderichthys*, bottom to top). In the lower left, the evolution of the limbs, from lobe-finned fish to tetrapods, shows the changes in the bones necessary to become terrestrial.

years ago, has recognizable limbs, with digits (fingers) and a pelvis attaching the lower limbs to the spine, even though it retains some swimming adaptations – notably the long tail to help propel it in water.

The transition from water to land was obviously one of the most critical changes in our evolutionary history, since we evolved from these early land animals, and we are lucky to have a good fossil record of this early transition. But keep in mind that this transition is only important to us because we are land animals. From the perspective of a fish, everything has been fine since the Devonian, and the rest is just noise.

Box 9.3 Extinctions

During the time when all of these organisms were evolving, there were multiple large-scale extinction events. Probably the most dramatic was the *Permian–Triassic Extinction* at roughly 250 million years ago when between 75 percent and 90 percent of all plants and animals were wiped out. The one people are likely to be most familiar with is the *K–T Extinction*, which, at 65 million years ago, wiped out the majority of dinosaur and aquatic reptile species. But there were many others over the many hundreds of millions of years since life evolved on Earth. More than twenty large-scale extinctions have been documented in the fossil record (see Figure 9.4).

The causes of extinctions are under study by paleontologists and geologists, as we only know the cause of a few: we know the K–T Extinction was caused by a meteor strike, and we know the Great Oxygenation Event was caused by photosynthetic bacteria (good for us, but deadly to previous anaerobic organisms). The other extinction events are being examined using new techniques and technologies to try and figure out the causes, and there are many candidates – nearby astronomic events such as volcanism, gamma ray bursts from supernovas, changes in the Earth's magnetic field, global climate change, and others. But one thing is certain: before the extinction there were many organisms that became extinct afterwards, so *something* happened to kill those lineages.

At first consideration, an extinction seems like a purely destructive event. There are a lot of people who would love to have dinosaurs around to see today, and lament the fact that they were all killed off by the meteor strike 65 million years ago. And I would love it if we still had big populations of the many different species of trilobites in the oceans. But extinctions are also essential to increasing biological diversity. Each time there is a big extinction event some of the existing lineages persist, but the ecological space gets wiped clean which allows other lineages to evolve. And these lineages would often not get a foothold without having the slate cleared for them.

For example, before 65 million years ago, dinosaurs and other reptiles filled every mammalian ecological niche on Earth, from grass-grazing and tall-tree-browsing herbivores to carnivores of all sizes. There were even small insect-eating dinosaurs. Mammals had long evolved – in fact, mammals had been around for more than 100 million years when the meteor hit, but during that 100 million years the dinosaurs filled almost every ecological niche available, so the only mammals were small rodent-like creatures. If a mammal lineage evolved to become larger and tried to compete, say as a small predator, it would not have had a chance, because there were already multiple existing dinosaur lineages in that niche. (One analogy would be business: imagine

Box 9.3 (cont)

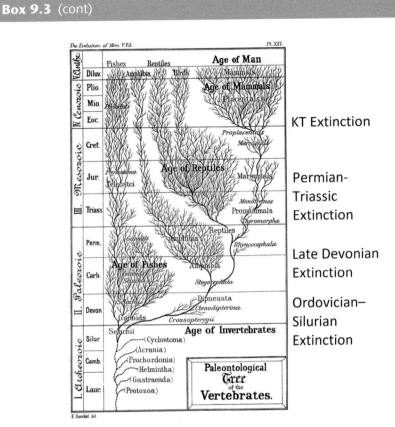

Figure 9 4 Extinctions. This classic early twentieth-century (1910) illustration, from Ernst Haeckel, does a good job of showing the importance of extinctions for understanding our current animal diversity. On the right margin I have added the names we now attach to the extinction events. As is visible in this illustration, the radiation of large animal groups, such as the reptiles and mammals, was only possible once an extinction event had cleared the path. Not all groups have thrived just because another group was reduced by an extinction event, but those that had an adaptive advantage (such as the reproductive advantage in mammals) were able to thrive once the dinosaurs were no longer filling all the available ecological niches.

trying to start a book-selling website today – you would never get started because everybody already goes to Amazon.) So they never had a chance to expand beyond that rodent-like niche.

But the meteor strike at 65 million years ago killed off the majority of dinosaur lineages. They weren't all wiped out – we still have birds today – but all the large land animals died, presumably because the dust thrown in the air by the meteor strike so reduced photosynthesis that there was not enough food to support large animals, possibly for a hundred years. And once the skies cleared, the mammals were competing on a more even footing with the

Box 9.3 (cont)

remaining small dinosaurs. It turns out that mammals have a few reproductive advantages, so they quickly became the dominant land animal group. But without that meteor strike there would be no large mammal groups, and humans would certainly not exist.

A similar pattern is seen at other extinction events – the Permian–Triassic Extinction at 250 million years ago cleared off the large terrestrial amphibians and allowed reptiles to compete, which is why the dinosaurs themselves arrived. Life on Earth is a zero-sum game. Each extinction event throughout our prehistory is both destructive and creative, and has been essential for the adaptive radiation of the large animal groups.

Box 9.4 The story of the coelacanth

The importance of the lobe-finned fish in the ancestry of tetrapods has been known since the nineteenth century. Fossil discoveries of lobe-finned fish were not uncommon, and actually predate Darwin. The coelacanth (Figure 9.5) was described by anatomist Louis Agassiz in 1839. Once Darwin published *Origin of Species*, the significance of the bones in the lobe-finned fossil fish was well understood, and it was clear that it was, in essence, our great uncle, if not our great grandparent.

Fossil coelacanths date to around 400 million years ago and, with their heavily boned lobed fins, are the ideal anatomical precursor to the later species that acquired more terrestrial adaptations. In the broad explanation of how animals evolved from fish to tetrapods, the role of the coelacanth was clear: it predated land animals but had the right bones to be the ancestor.

For about a hundred years this was the state of our knowledge on the coelacanth. Until in 1938, in a fish market in East London, South Africa, a local museum curator, Marjorie Courtenay-Latimer, identified a coelacanth among the day's catch. It was a big fish (almost 1.5 meters, or 5 feet, long), but she took it in a taxi to her museum and preserved it, where scientists from around the world could come and study it. Now we know that coelacanths, long thought to be extinct for many hundreds of millions of years, have actually persisted through many worldwide extinctions, and have two species in populations off the east coast of Africa, in the Indian Ocean. They are deep water fishes, and have so far been able to survive relatively unmolested by

Box 9.4 (cont)

Figure 9.5 The coelacanth. The coelacanth is likely to represent the lineage from which all subsequent tetrapods (land vertebrates) evolved. This fish has been known since the eighteenth century, and was thought to be extinct by at least 400 million years ago. However, in 1938 a specimen was discovered in a fish market in South Africa. Since then, populations of this fish have been discovered in deep waters in the Indian Ocean, and scientists have been studying this animal for clues to our deep ancestry. Notice the heavy, fleshy fin (top right). This fin possesses bones that are the precursors to our own limb bones, and the pattern in those limbs is essentially our own: a single bone for the upper part of the limb, and two bones for the lower, with a larger number in the wrist and hands. Scientists are currently working on the DNA of this animal to identify the genetics of our ancient limb structures.

events on the surface of the Earth. With luck, this group may persist another 400 million years, because clearly they have been a highly successful and resilient group.

Box 9.5 The two kinds of dinosaurs

Often, when you watch movies about bringing back Tyrannosaurus from ancient DNA, or read about the Triassic in popular science books and magazines, you will see the term "dinosaur" used. And we scientist use this term as well, waving our arms in the lab, trying to make some point about

Box 9.5 (cont)

evolutionary theory. But it is actually a bad term. It is bad because it doesn't actually reflect the true evolutionary history of the animals of this time period.

There are actually two branches of dinosaurs, and they are pretty distantly related. About 20 million years after the Permian–Triassic Extinction (250 million years ago), fossils appear that make good ancestors for the later reptile families (including dinosaurs and crocodiles). But the two dinosaur branches diverged by 210 million years ago into two groups, which means that, at the moment of the K–T Extinction, 65 million years ago, the two groups had been separated for 150 million years.

The two groups are known as the lizard-hipped group (*saurischians*) and the bird-hipped group (*ornithischians*). And, much like placental mammals and marsupial mammals (see Box 9.6), they converged on various anatomical solutions. In both groups there are quadrupeds and bipeds, and both groups had wide size ranges. However, there were some specializations seen only in one group or the other. All the meat-eating dinosaurs (*Tyrannosaurus*, and *Allosaurus*, and the various raptors, for example) are Saurischians. The armored dinosaurs (*Triceratops*, *Stegosaurus*, and *Ankylosaurus*, for example) are ornithischians.

You have probably heard that dinosaurs survived, in the form of modern birds. This is true. Modern birds are saurischians, which means that, ironically, birds evolved from lizard-hipped dinosaurs. Saurischians included *Archyopteryx*, which was one of the stems of this lineage that led to modern birds. Unfortunately, no ornithischians survived the K–T Extinction. But this is why scientists don't really like the term "dinosaur" – because using it to describe a bunch of giant lizardy things that lived over 65 million years ago obscures the fact that it refers to two distantly related groups, one of which still survives in the present. When someone says "dinosaur" you don't think of a sparrow or pigeon. But they are dinosaurs as much as *Tyrannosaurus* or *Triceratops*, and they are certainly not going extinct any time soon.

Box 9.6 What is a mammal?

When you think of a mammal, what probably comes to mind is something furry, maybe a bear, or a rabbit, or a monkey. And these are, obviously, mammals, but they are a specific type of mammal – the *placental mammals*. But what makes a mammal? And are these the only kinds of mammal?

Box 9.6 (cont)

Technically, what makes a mammal is the presence of mammary glands to nurse young, the presence of fur/hair, and some specializations of the brain and inner ear. But that actually covers a wide array of animals that are quite distantly related.

There are three kinds of mammal. The most "primitive" mammals, those that retain the most characteristics from their reptile ancestors, are the *monotremes*. These likely represent the earliest form of mammal. They lay eggs, as reptiles do, and also lack some of the brain adaptations we see in placental mammals. They don't nurse their young because they lack nipples. Rather, they secrete milk onto their skin, where the young lick it off. However, they do care for their young. Monotremes also have some interesting adaptations – for example, they have a venomous spur on their hind legs, and modern monotremes lack teeth. Today, the only surviving monotremes are platypuses and echidnas, but they were probably far more diverse in the deeper past.

Marsupials are a very successful and diverse group of mammals that is characterized by the presence of a pouch, in which mothers keep their offspring while they are still young enough to nurse. Today, marsupials are largely constrained to Australia and New Guinea, but they were once widespread across Antarctica and South America. Continental drift pushed South America up against North America, with the contact occurring some 1–3 million years ago. At that point, marsupials moved north, competing against placental mammals. Ultimately, the placental mammals were able to outcompete all the marsupials of South America, with the exception of the opossum. All animals went extinct in Antarctica, as it became the south polar continent covered with ice, but Australia (connected to New Guinea) was geographically isolated, and marsupials have persisted there to this day.

Today we think of marsupials as kangaroos, koalas, and wombats, but this group was once far more diverse. There were once large-bodied predators the size of modern lions. There were also large grazers, much like African antelope. However, the arrival of modern humans in Australia some 50,000 years ago, along with changes to the climate of Australia, drove these amazing marsupials to extinction. As recently as the 1930s a wolf-sized marsupial predator roamed the Australian island of Tasmania, but sheep farmers killed them all off as they often preyed on young sheep. Today, the largest marsupial predator is the fox-sized Tasmanian devil.

When marsupials and placental mammals compete, the placental mammals seem to be more successful. Scientists are not completely clear why this is the case. There may be more behavioral flexibility in placental mammals – they

Box 9.6 (cont)

have a large set of brain circuits (the corpus collosum) that allows rapid communication between the right and left hemispheres of the brain, and this is absent in marsupials (and monotremes). There may be other anatomical constraints imposed on marsupials. When young, marsupials must have strong claws to climb up to the pouch (after birth), and to climb from the bottom of the pouch to the lactating teats. This is probably why you do not see hooved marsupials – they must all have claws. However, the most important difference may lie in the name of placental mammals – the placenta. The placenta appears to provide more nutrients, and allows for a longer period of in utero development. This makes the young more likely to survive gestation, birth, and the early juvenile period.

The isolation of the marsupials in Australia for tens of millions of years has provided us with an excellent natural experiment. We can look at the marsupial mammals and placental mammals and examine how they adapted to similar environmental circumstances. What is apparent is how much these two groups have converged on similar anatomical solutions to the same problems. The most famous example is that of the thylacine, which was the large-bodied predator on Tasmania. It looked remarkably like a wolf: it had a long snout, with long legs, and similar position of the eyes and ears. This is because they both are/were olfactory (smell)-focused predators who pursue(d) their prey over long distances.

For evolutionary biologists, this is the classic example of evolutionary convergence (*homoplasy*), since the last common ancestor of marsupials and placentals looked like a tree shrew. Over tens of millions of years, those two independent lineages converged on a similar anatomical solution. It turns out that, when you compare marsupials and placentals, there are many examples of convergence. There are marsupial equivalents (in size, shape, and behavior) to moles, rats, anteaters, lorises, flying foxes, hippopotami, deer, tapir, and pigs. There was even a saber-toothed predator very similar to placental saber-tooth lions. Many of these marsupials are extinct, but more fossils are discovered all the time. This pattern of marsupial–placental convergence is perhaps the best example of the power and flexibility of the evolutionary process.

Chapter 10: Protohumans

Once the competition from dinosaurs ended with the K–T Extinction, mammals established themselves as the dominant terrestrial group. They quickly diversified into the various lineages we see today (see Box 10.1). One of these mammalian lineages, the *Primates*, was a group of small, arboreal (tree-living) animals who survived by eating insects. They probably lived in flowering trees that attracted insects trying to pollinate the trees. We know that they were insectivores because they were so small – large animals have a hard time tracking down enough insects to maintain a large body (unless they are eating social insects, as anteaters do). Also, these small creatures had sharp, spikey teeth for cutting through the hard insect shells. Their eyes faced forward for good binocular vision, much as we see in carnivores today (compare an owl's eyes to a pigeon's). They had grasping fingers, much as we do today, for holding on to thin tree branches as they pursued insects. We can see what these animals looked like by studying modern animals that don't seem to have changed much in 50 million years. One of these groups is the galago/loris group of primates, and they look very much like the fossils of early primates. They are small, relatively solitary creatures who survive by eating insects, often at night, using their excellent vision (see Figure 10.1).

But the primate group evolved another lineage, in which the animals became large-bodied and social. This is the group of primates that consists of monkeys and apes. You are probably familiar with this group – they have spread all over the world and are one of the most successful mammalian lineages. By becoming large-bodied and social, this group (sometimes known as the Anthropoids) became able to resist predators much better than the small nocturnal galagos, who have to worry about owls and other predators in the trees. But they retained the arboreal adaptation, which allows them to avoid the larger terrestrial predators, like leopards, hyenas, and lions.

Along with their large bodies, the monkeys and apes evolved larger brains. In general, primates have much larger brains, with many more neurons, than the more primitive lorises and galagos. Scientists are still studying primate brains for clues as to how these more complex brains evolved, but it is a clear characteristic of monkeys and apes that they have greater problem-solving

Figure 10.1 Two primates – a galago and a chimpanzee. The galago (left) represents the more ancestral branch of primates. Galagos are small (less than 5 lb/2 kg), solitary, nocturnal insectivores (this galago is eating a mealworm) with brain sizes between 5 and 15 cc (cubic centimeters). On the right we see common chimpanzees, who are large-bodied (up to 150 lb/70kg), very social, large-brained (300–350cc), diurnal (active during the day), and with a diverse diet that is focused on fruit, but includes fishing for termites (seen here), and occasional hunting of small mammals.

Box 10.1 The Primates

After the K–T Extinction and the death of all large terrestrial animals, this absence presented mammals with an opportunity to compete with reptiles for the now-open ecological niches. Mammals responded by rapidly evolving a wide array of adaptive responses that have led to the mammalian branches we see today. Within 5 million years, we have the appearance of the ancestral branches of ungulates (hooved mammals), rodents, carnivores, bats, elephants, moles, rabbits, tree shrews, and primates. This was a time of incredibly rapid evolutionary change, and reinforces the importance of extinctions for evolutionary diversification.

Our branch is known as the primates. Primates are one of the most successful and diverse mammalian lineages, as primates can live in almost any climate (the driest deserts seems to be the exception, as primates need to drink every day). The very first primates probably looked something like tree shrews, and there are some primitive primates that still have small bodies and insectivorous appetites. In general, we divide primates into two groups: the Prosimians (also known as the Strepsirrhines) and the Anthropoids (also known as the Haplorrhines). The Prosimians fall into two groups: the most primitive group is the Loris/Galago/Potto group, and they retain many primitive (or ancestral) traits that we associate with the mammals that existed right after the K–T Extinction: they are small, nocturnal insectivores with wet noses and eyes that reflect light (like dogs and cats do – this is an adaptation to seeing well in the dark), are socially solitary, only coming together with others of their species to mate. Lemurs are the other Prosimian group, and because of their isolation

Box 10.1 (cont)

on Madagascar, where there are no monkeys or apes, they evolved into many of the monkey niches – they are large, social, and diurnal (meaning active during the day). But they still retain the primitive anatomy: long snouts, wet noses, and eye shine (the anatomical term is tapetum lucidum).

The other main primate group is the Anthropoids. Anthropoids are the monkeys and apes, and they lost many of the primitive traits seen in the Prosimians. They lost the long snouts, their noses are dry, and they do not have the nocturnal eye shine from a tapetum lucidum. They are also very social, and are widespread across the world. Only Australia and North America lack Anthropoids, and in both cases it was simply the difficulty of getting there (in the case of Australia, crossing the Timor Sea, and for North America, crossing the Great Southwestern Desert). But once established, they are successful and adaptable. Monkeys are obviously successful in forests, but they can also live in dry grasslands or otherwise inhospitable environments – baboons are big, terrestrial monkeys who are just as happy on the grasslands as they are in the deserts of Ethiopia. Baboons are even successful living in the suburbs of towns and cities across Africa.

For us, the most important Anthropoids are the great apes. Multiple species were once widespread across Africa, Europe, and Asia, even though today only a handful of species exist. This group has acquired large brains and sophisticated social and tool-making behaviors. This group also has one unusual anatomical trait that you have – they lack tails. This is one sure way you know that you are an ape!

skills than most other mammal lineages. This trend continued with the evolution of the apes more than 25 million years ago.

This epoch, the Miocene (roughly 25–5 million years ago), is known to paleontologists as the Age of Apes, as apes from this period are found all over the Old World, including places where apes no longer live: Greece, Italy, Spain, China, India, Turkey, and the Arabian Peninsula. There may have been as many as a hundred species of ape that lived during this period. Forests covered these parts of the Earth, and where the forests were, apes were.

Some apes became even more large-bodied, and this large body size allowed them to descend from the trees, as few predators other than the largest cats (lions and leopards) care to mix it up with a full-grown ape. African apes, particularly, are largely terrestrial, and some extremely large-bodied relatives of modern orangutans were terrestrial; *Sivapithecus* was a large terrestrial ancestor to orangutans, and *Gigantopithecus* may have been twice the size of a modern gorilla, and would have subsisted on terrestrial vegetation. This terrestrial adaptation provided new feeding opportunities, even as those new feeding opportunities would have imposed new requirements.

To access these new foods on the ground, apes needed to evolve new strategies. For example, chimpanzees fish for termites from termite mounds. This requires the chimpanzee to find an appropriate "termite fishing stick," which must be relatively straight and must be modified by stripping away the leaves. This is tool manufacture and use, and it requires a relatively large brain. Chimpanzees are known to use other tools, including a hammer-stone and anvil-stone for cracking nuts. This behavior is not seen in monkeys, nor in the galago/loris lineage.

It is from this ape stock that the earliest humans evolved. As the Miocene ended, the climate cooled and dried, and forests started to recede. All over the Old World, ape species simply went extinct as their habitats disappeared. However, in Africa, rather than go extinct, one ape group figured out how to survive the drier, cooler, and less forested environment in East Africa. According to DNA analysis, this occurred sometime between 8 and 6 million years ago. We share a common ancestor with chimpanzees, which is reflected in the fact that we share 98.6 percent of our genetic material with them. Our common ancestor with the chimps probably looked quite a bit like a modern chimpanzee, and might have been a knuckle-walker like modern chimps but also with the long powerful arms chimpanzees possess, to allow them to scamper into a tree if a lion happened to stroll by.

Since the DNA places the split between the chimpanzee lineage and the human lineage at 6–8 million years ago, paleontologists have focused on that time period to look for the "first human." We don't really have a good human fossil record for that time period, but there are a few fossils that have been put forward as candidates for the first humans. It isn't until around 5 million years ago that we have fossils we can reliably place on the human lineage. This genus, *Ardipithecus*, known from deposits in the north of modern Ethiopia that date to 5.2–4.5 million years ago, was very ape-like, with long arms, a small brain, opposable big toes, and small thumbs, all traits we see in modern chimps. However, it appears to have started the transition to bipedalism, and bipedalism is one of the key traits that characterizes our lineage. By about 4 million years ago we have a good fossil record of a truly bipedal human ancestor, *Australopithecus*. This was a widespread genus, with multiple species in eastern and southern Africa. You may have heard of Lucy, a small female specimen of *Australopithecus afarensis*, a species that is known from Ethiopia, Kenya, and Tanzania that persisted from 3.8 to 3 million years ago.

Figure 10.2 The foot and hand of modern chimpanzees. Chimps are partially arboreal, so the hand and foot are specially adapted to climbing in trees. The big toes of great apes are opposable, so they can grasp limbs as we do with our hands (left image). However, their hands have small thumbs (middle image); instead of using their thumbs for climbing, they hang from their strong fingers (right image). One major change seen in humans between 3 and 4 million years ago is the development of the strong thumb, which is critical for holding stone tools. Another critical change is the shift in the big toe to align with the rest of the toes. The big toe allows for efficient human walking and running.

Australopithecus was, in many ways, intermediate between apes and later humans (Figure 10.2). It was still short, like chimpanzees, and had long arms for climbing, but it was truly bipedal, with a big toe like ours that was perfect for walking and running and definitely not opposable. The thumb of *Australopithecus* was also large and powerful, like ours, and the fingers were shorter, so that they could make a precision grip (think about how you hold a pencil), or grab a stone tool the same way you hold a baseball. The skull was also intermediate. They no longer had the big canine teeth we see in monkeys and apes, even though their faces protruded like an ape's face, and their brain was about 50 percent larger than a chimp's (Figure 10.3).

If you had seen *Australopithecus*, they would have seemed very ape-like, yet very familiar. Imagine an ape – about 4 feet tall, covered with hair, with long arms, yet striding bipedally, with arms swinging, the same way we do. This bipedalism was a critical adaptation – bipeds can cover long distances much more efficiently than knuckle-walkers, and as the climate dried out and the forest started disappearing, food sources were likely to be much more spread out. A biped could have crossed cross miles of open country to get to forest patches for food, whereas a knuckle-walker might have struggled, because bipeds are about seven times more efficient at terrestrial locomotion (see Box 10.2).

But this adaptation to a terrestrial lifestyle would have imposed even more pressure on brain size, because living as a biped means a real commitment to finding food on the ground. Although their preferred food was likely fruit, much as in modern apes, the drier climate would have meant that this was more and more scarce, and they would have had to look for other food sources. Whatever they were eating must have been much more demanding to chew

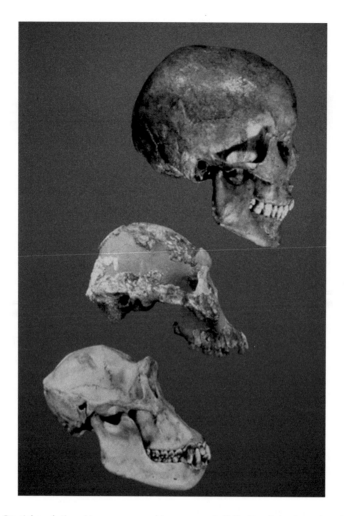

Figure 10.3 Cranial evolution. Here we see a chimpanzee skull (bottom) representing the likely ancestral condition for humans. In the middle is *Australopithecus afarensis*, which dates to roughly 3–4 million years ago. It has a slightly smaller face and a slightly larger brain (~25–30 percent) than chimpanzees. But this species had modern feet and hands, and walked much as we do. The genus *Australopithecus* clearly represents an intermediate stage in human evolution and was very successful, with multiple species across South, East, and North Africa for more than 2 million years. Later evolution in the human lineage continues this trend (top), with larger brains and smaller faces evolving over the next few million years.

than fruit, because their jaws and teeth are much thicker and more heavily built than we see in chimpanzees. Currently, we do not know what this food would have been, but perhaps they were pulling tubers from the ground that were very tough to chew (think of uncooked potatoes). This food would have been gritty and worn down their teeth quickly, because the enamel on their teeth is much thicker than we see in chimpanzees and *Ardipithecus*, yet is almost always worn down in adult fossil specimens.

Box 10.2 Human and ape anatomical differences

Although we often think of brain size as the critical difference between apes and humans, in fact the earliest humans had brains no larger than modern chimps. We see two critical adaptations appear when our brains jump a bit in size, as it is likely that these two changes enabled the expansion of our brains. And these two differences are not in the skull at all – they are the hands and feet.

Ardipithecus, which is known from deposits 5.2–4.5 million years old, has an opposable big toe, and an ape-like hand with a thin, short thumb. But by about 3.9 million years ago, with the appearance of *Australopithecus afarensis*, we see a brain that is about 30 percent larger than a chimp's. This brain increase is accompanied by a change in the hands and feet. It is worth exploring the details of these differences because they may well have been the key changes that enabled us to become the big-brained apes we are today.

Hands

The chimpanzee hand is extremely powerful, and they have long fingers that allow them to climb or hang from trees for long periods of time. But they only use the first four fingers to hang, not their thumbs. In fact, the thumbs are thin and weak relative to their fingers because they just don't use them very much. Humans have a hand that is, by comparison, smaller, because our fingers have shortened, and we can't hang in a tree very long before tiring. But, in contrast, our thumb is more powerful than a chimp's, and we actually have a muscle that chimps don't, known as the *flexor pollicus longus*, which runs from the tip of the thumb, through the palm of the hand, back to the forearm. This muscle allows humans to grip with the thumb in opposition to the rest of the fingers. When you hold a baseball or pick up a suitcase by the handle, your thumb is in opposition to the fingers, and is really doing as much work as the rest of the fingers.

So we have good hand strength, thanks to our thumbs, but we also have precision. When you thread a needle, your thumb and forefinger are in precise opposition. Humans can make very fine movements with their fingers and thumb. Writing with a pencil is another example. None of these things are possible for chimps. With their long fingers and short thumbs, they cannot hold anything small with precision and strength. This may well have been one of our critical adaptations, because making tools is what we do well. While chimps can make a few tools, our hands allow us a great deal more precision. Some bonobos (a type of chimpanzee) have been taught to make stone tools, and they are smart enough, but they struggle to hold the small pieces of stone

Box 10.2 (cont)

to make sharp edges. Our seemingly weaker hands give us a huge advantage in making stone tools.

Feet

Our other major change is in our feet. All apes have opposable big toes, and this is a critical adaptation to climbing in trees. If you have ever tried to climb a tree, you know that even if your hands can easily pull you up, you will tire quickly if you can't find some way to help yourself with your feet. Apes can simply grab a limb with their feet. The human foot has a big toe that is in line with the other toes and is essentially useless for climbing.

The advantage of having a big toe in line with the other toes is that it can be used for terrestrial locomotion – walking. Humans use their big toes every time they take a step but especially when running.

Humans are very efficient walkers and runners. Quadrupeds are faster, even small quadrupeds are fast. Anyone who has seen someone trying to catch their small dog in the park knows that humans are slow and clumsy in comparison to quadrupeds. However, we are very, very efficient. In comparison to a chimpanzee knuckle-walking, a human uses roughly 1/7th the energy to cross open ground. In a climate where forests were drying out and becoming patchy, the ability to efficiently cross a grassland to get to another forested patch to find food would have been a critical adaptation. And humans are able to do this steadily. Well-conditioned humans today are able to walk or run 50 or even 100 miles (160 km) essentially without stopping.

The shape of our feet is essential for this, as the big toe allows the foot to act as a lever and shock absorber while walking and running. The big toe generates most of the propulsive force, and is essential to any movement except the slowest stroll. Mountain climbers are famous for losing toes to the cold, but they are always the most worried about their big toe because they know that they can climb more mountains without any of the smaller toes, but not without that one big, powerful toe.

Often, fossil species, especially dinosaurs, are known from only a single fossil. But when we examine living species we want to know all kinds of things that can only be known by looking at the variation in anatomy and behavior across the species. For example: How big or small can they get? Are males larger than females? How long are the juveniles dependent on the parents? How large are the social groups? What kind of social group does the animal have? We can rarely access this kind of information for fossil species.

However, several species of early humans have such good fossil records that we can examine exactly those important parameters we normally only study in extant species. For *Australopithecus africanus*, from South Africa, and *Australopithecus afarensis*, from East Africa, we have adults and juveniles, and males and females, in good numbers. We know that the children grew quickly, like apes, and would have been able to reproduce by seven or eight. We also know that they would have had a gorilla-like social structure, in which there was one male and multiple females (see Box 10.3). We also have collections of footprints that tell us how they moved across the landscape, and how the social groups were composed. Because of this, paleontologists who study human fossils often think of themselves as biologists of the past, since they are getting at much of the information field biologists use to study populations.

Box 10.3 Early human societies

Chimpanzees live in mixed multi-male, multi-female social groups. And their mating systems are complex as well. They do not form long-term *monogamous* (one male and one female) bonds, as we see in gibbons (the lesser apes), nor do they have the system seen in gorillas, where one dominant male mates with multiple females (*polygyny*). Rather, it is a fairly open and fluid system in which females often make decisions about with whom to mate, and males must often negotiate for the right to mate. These decisions are complex and females will often use their fertility as a bargaining chip, making decisions based on the health of the male, whether the male provides something in exchange (if a male has made a recent kill he may exchange meat), or based on other social obligations (perhaps the male helps her raise her offspring). Larger, more powerful males get more opportunities, but the females are typically spread out across the landscape, and smaller males can also find plenty of opportunities to mate; females often mate with multiple males when they are in estrus (ovulating).

Because modern humans have complex social and mating systems, it is natural to assume that our early ancestors also had some sort of complex multi-male, multi-female social system. However, this appears not to be the case. It is obviously difficult to figure out details of social systems from the fossil record, but if you have enough fossils it is possible to at least understand the overall structure. The way to do this is to look at the size of males relative to females. In mammals, when the social system is polygyny the male is usually much larger than the female. This is the case in herding animals (deer, elk, cattle), carnivores (lions), and primates. Gorillas and orangutans, who engage

Box 10.3 (cont)

in polygyny, have much larger males – in fact males are roughly twice as large as females. This is because they must fight for the right to mate with the females, and large size is important in physical contests between males. This system is only present in primates where the females tend to congregate around a food source and a large male can drive away the other males. In chimpanzees, because the females are widely distributed across the landscape, no male can monopolize multiple females. This means that there is no evolutionary advantage to being much larger (large size imposes costs – if there is a drought and food is scarce, large animals, who need more food, are the first to die of starvation).

We have a few fossil species that are well-enough preserved that we can determine the relative sizes of males and females. In *Australopithecus afarensis* and *Australopithecus africanus*, for example, the males appear to be roughly twice as big as the females (Figure 10.4). We have evidence from both the fossil record, in which we can examine the sizes of individuals directly, as well as in footprints. At the famous footprint site at Laetoli, Tanzania, fossil human

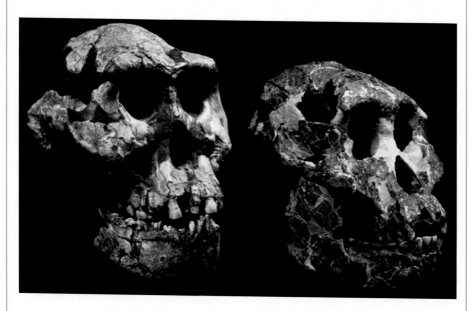

Figure 10.4 The sexual dimorphism of early humans. The female of *Australopithecus afarensis* was roughly half the size of the male. This is the pattern we see in modern gorillas and orangutans, as well as in other mammalian species with a single-male, multiple-female mating group. This pattern is seen in multiple species of early humans, and persisted until the appearance of our genus, *Homo*, between 2 and 3 million years ago. We currently do not know *why* we transitioned from the polygynous social group to pair-bonding, and this transition is the subject of considerable scientific interest.

Box 10.3 (cont)

footprints from 3.6 to 3.8 million years ago display a size difference between the largest and smallest footprints consistent with very large males and relatively small females. This pattern of sexual dimorphism is the case in every early human species for which we have a good fossil record. So it appears that ancestrally, we were polygynous. When, after own genus, *Homo*, appears, we transition away from polygyny (somewhere between 2.0 and 3.0 million years ago) and the fossils show the differences we see in humans today.

We don't yet know why we changed. Normally, the mating system is determined by the females and how they are distributed across the landscape, so we assume that sometime around 2 million years ago females switched to a different food source that was more widely distributed. We see in Chapter 11 that the appearance of evidence of meat eating roughly corresponds to this change, so as we became meat-eaters we lost polygyny. This transition had significant consequences for the future of our lineage.

The Radiation of Humans

You will sometimes see a T-shirt or bumper sticker with the evolution of humanity portrayed as a linear transition from a monkey to an ape to a "cave man" to a modern human (sometimes the last modern human is portrayed as having devolved into an overweight office worker, or something similar). This linear representation of the evolution of humans has a long history. In the nineteenth century, without a good fossil record, early evolutionary biologists assumed that the evolution of the human branch was linear, and that there would be intermediate "ape-like" humans in our past. Into the twentieth century, as a few fossils were uncovered, this idea stuck, and it created a conundrum: Why were there so few human species, when across nature we see multiple carnivores and herbivores living in close proximity? In the USA, for example, foxes, coyotes, and wolves all live in roughly the same areas at the same time, so why weren't there multiple human species found in the fossil record? Some paleontologists came up with the idea of "competitive exclusion," in which there was only one "human niche" to explain the lack of human species.

But, as the twentieth century passed, more and more fossil species came to light. Soon it was impossible to argue that there was only one human niche, as multiple contemporaneous species were discovered. In fact, at roughly 2 million years ago, there were as many as seven human species all living at the same time across Africa. We now think of human evolution as an adaptive radiation, just as we see in other animal groups. Once humans evolved, they adapted to multiple ecological niches across the landscape, and they appear to have been doing very different things to survive (see Box 10.4). We now have almost thirty different species identified in the fossil record and there is no reason to think that more will not appear as researchers probe unexplored caves and remote deserts.

Box 10.4 The human radiation

Humans are the only species on Earth to study its own evolutionary history. This makes understanding our place in nature a bit tricky because we are naturally tempted to treat ourselves as something special. You see this impulse in studies of history as well: Americans tend to write histories of American victories in World War II, Russians tend to write histories of Russian victories, the British do the same. This is a natural impulse. But it does not mean that those events are especially unique, and often reflects more about the writers than the events.

This is the case with evolutionary history as well, and was especially notable in early studies of human evolution. Darwin published *On the Origin of Species* in 1859 and biological scientists immediately understood the implications for humans. However, there was no fossil record, and of course there was only a single species of human. Since there are multiple species of canid (dog), felid (cat), rodent, butterfly, worm, etc., the fact that there was only one species of human required some sort of explanation. Our "uniqueness" stood out as a likely explanation, and this was supported by cultural biases that emphasized how special humans were in relation to nature. For example, in Genesis 1:26 of the Old Testament, "man" is the only creature created in God's image, and has dominion over the rest of the natural world: "Then God said, 'Let Us make man in Our image, according to Our likeness; let them have dominion over the fish of the sea, over the birds of the air, and over the cattle, over all the earth and over every creeping thing that creeps on the earth.'"

The attitudes toward the uniqueness of humans are apparent in early descriptions of human evolution, in which the human lineage is portrayed (unlike every other plant or animal) as a more or less direct line, with an ape-like ancestor that slowly changed in intermediate stages, finally arriving at modern humans. This model held into the early twentieth century, but slowly the fossil record was filled in, as scientists made discoveries across Africa of early human fossils that were dramatically different, even though they were discovered in the same temporal horizon – often at the same locality. There was a significant debate within human evolutionary studies, and some researchers proposed that there was a single human evolutionary niche, a

Box 10.4 (cont)

"cultural" niche, and it could not accommodate multiple contemporaneous species, applying the model of "competitive exclusion" to humans.

This was an assertion essentially without evidence, and showed a lack of imagination of what human adaptation could be. There was no particular reason that there could not be humans doing different things for a living. As the fossil record grew, it became clear that there were many different *hominin* (fossil human) species, and they appeared to have many different anatomical adaptations. Over the last hundred plus years of paleontological exploration,

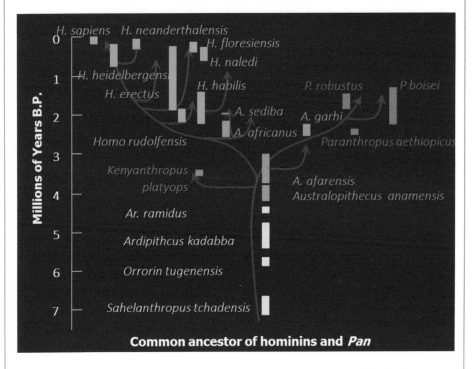

Figure 10.5 The human radiation. Humans evolved into a new adaptive position in the environment and radiated out to fill the many niches available to adaptable, large-brained, large-bodied, bipedal primates. This is consistent with what we see in many other organismal groups, where an initial successful adaptation opened the door to many related species and genera evolving into suddenly available niches. Humans originated with a single species (yet unknown) some 6–8 million years ago, but by 2 million years ago there were likely seven species of human living across Africa, Asia, and Europe. Note that there were two genera of human at 2 million years ago – *Homo* and *Paranthropus*. *Homo* ultimately culminated in our species, *Homo* sapiens, but the other genus, *Paranthropus*, persisted for more than a million years across South and East Asia (see discussion in Chapter 11). Figure adapted from Fig. 1 in Wood, B, Boye, E. 2016. Hominin taxic diversity: fact or fantasy? Yrbk. Phys. Anth. 159:S37–S78.

Box 10.4 (cont)

scientists have found many different human species, and differences can be stark: some humans possess small brains and large teeth and chewing muscles, whereas other human ancestors, living at the same time, in the same geographic locality, have large brains and small teeth. Some are tall (as tall as modern humans), whereas some were shorter than four feet. Brain sizes range widely. As of the publication of this book, scientists have discovered almost thirty species across Africa, Asia, and Europe (Figure 10.5).

Humans have radiated out into diverse environments, and, especially for earlier species, their anatomies reflect their biological (rather than cultural) adaptations. This means that humans evolved in ways that follow patterns broadly similar to those seen in the evolutionary histories of other organisms. Once a successful adaptation is acquired, organisms radiate out into multiple environments. The broader implication is that humans are animals, and have responded to environmental demands as other animals do, and that our anatomy and behavior is the result of millions of years of natural selection.

In Chapter 11 we explore the evolution of our genus, *Homo*, our diverse adaptations, and the ways in which we were able to spread around the world and survive in virtually any environment. We will also examine the reasons why there is only one species of human that survives today.

Chapter 11: The Genus *Homo*

The earliest humans lived a very ape-like lifestyle for roughly the first 5 million years of human evolution. At first, they appear to have been relatively sparse, but for at least 2 million years, humans, in the form of the genus *Australopithecus*, were quite successful in some sort of partially arboreal, partially terrestrial adaptation. With their long arms and powerful shoulders, they were better climbers than any human is today, but with their hips, legs, and feet adapted to bipedalism, they were better walkers than any ape and able to cross open country to find food in ways apes never could.

But this anatomy was partially an adaptation to a forested environment, and about 3 million years ago, we see changes in the environment. Animals that are associated with the forest (like colobus monkeys and the leaf-eating antelopes) start to decrease in the fossil record, and drier climate animals (such as baboons, grass-grazing antelopes, and zebras) appear in greater numbers. This shows that the climate was drying out, and this would have meant that the preferred habitat of *Australopithecus* was disappearing. When the climate changes and habitat disappears there are two responses by organisms: go extinct or adapt. Most organisms simply go extinct (more than 99.9 percent of all species that have ever existed are extinct), but obviously some do persist. We are fortunate that humans found a way to adapt.

They did this by diverging into two distinct lineages with very different adaptations. One lineage is *Homo*, which we will discuss in detail here, but there was another unique and unexpected lineage, with an unusual adaptation, that was successful for more than 1 million years across South and East Africa (see Box 11.4).

Our genus, *Homo*, adapted to this drier climate by altering critical aspects of behavior and anatomy. In the skeleton below the skull (the *postcrania*), the earliest members of *Homo* were similar to *Australopithecus*. They were small, with long arms and short, bipedal legs. However, they had distinctly larger brains and smaller teeth and jaws. These changes also appeared roughly contemporaneously with the very earliest stone tools used for cutting meat. The adaptation to meat was probably critical, because brain growth requires enormous resources and meat is highly dense in calories, fat, protein, and amino acids. In fact, becoming more reliant on meat may have enabled the brains of early species of *Homo* to evolve to above 600 cc (see Box 11.2).

This shift to meat dependence may have been the key to our later success. Without claws or fangs, and with an arboreal-adapted body, early *Homo* was ill-equipped to find their preferred forest foods in the dry savannahs opening up across East and South Africa. And with a relatively slow bipedal gait, it was unlikely that these humans could catch prey directly. But the savannah ecosystem has a large niche for meat scavengers – after a kill, hyenas, jackals, vultures, crows, and a variety of other animals feast on what is left behind after a lion kill. The remains of animals eaten by early *Homo* show cut marks that are on the types of body parts ignored by lions (who get first pick), but eaten by hyenas. So humans were in the scavenger niche, much as hyenas and jackals are today.

There is ample evidence for adaptation to meat eating. This is the time period when we first see evidence of stone tool manufacture and cut-marked bones; further, the DNA of human parasites are closely related to the DNA of carnivore parasites, and the parasite lineages appear to have diverged between 2 and 3 million years ago.

Suddenly we could feed our evolving brains. And once the brains evolved further, that of course meant that ever more complex problem-solving abilities were available. This became a feedback loop, as more complex behaviors enabled greater acquisition of meat, and the additional meat enabled more brain development. Repeated over and over, the brains of the genus *Homo* more than doubled from 2 million years ago to today.

Box 11.1 The anatomy of walking, running, and throwing

Walking

As you can see in Figures 11.1 and 11.2, there are big differences between the skeletons of human and apes. Much of our skeleton reflects the fact that our anatomy was once ape-like, and was modified over millions of years by evolutionary forces. There are obviously big differences in the limb lengths, and the fact that the chimp relies so heavily on its forelimbs (arms) for support in trees and on the ground is reflected in their length and robustness. The humerus (the upper arm bone) is longer than the femur (the upper leg bone) in apes, whereas in modern humans the pattern is reversed. However, on the skeleton of Lucy (*Australopithecus afarensis*), those two bones are roughly the same length. That reflects the fact that this early human had an intermediate adaptation – it was able to walk bipedally but was also adept at climbing in the trees. In other words, it had not made the full commitment to a terrestrial lifestyle. This "intermediate" adaptation, which was successful for several

Box 11.1 (cont)

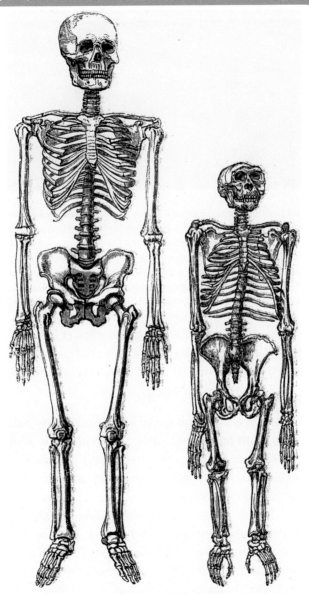

111.—Skeleton of Man.   112.—Skeleton of Chimpanzee.

Figure 11.1 Comparison of human and chimpanzee skeletons. Both have the exact same number of bones, and the bones perform the same functions, indicating their shared ancestry. The differences are in the shapes of some of the bones. For example, the human has short arms but long legs, and in the chimp this is reversed, reflecting their different locomotor modes. The human, in its long legs, short arms, and voluminous lung capacity, is particularly well adapted to long-distance terrestrial locomotion, even as it lacks the powerful shoulders and arms of the chimp.

Box 11.1 (cont)

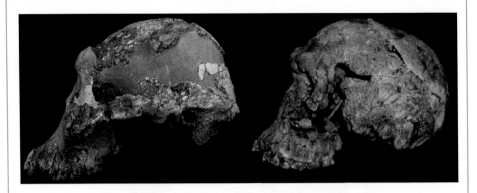

Figure 11.2 A comparison of *Australopithecus* (left) with early *Homo* (right). Note the continued trajectory of human evolution – early *Homo* has a smaller face and larger braincase than *Australopithecus*, which is the pattern established in our lineage by about 4 million years ago. The subsequent species *Homo erectus* has an even bigger braincase and smaller face, as do we (*Homo sapiens*), with the largest braincase relative to the face. This pattern of change reflects the decreased use of our faces as tools (for fighting, or tearing open hard foods), and the increased dependence on our brains to generate, design, and build tools out of materials found in nature.

million years, is also apparent in the fact that the lower half of the body, including the pelvis and feet, look very similar to ours, whereas their upper bodies, including the long arms but also the shoulders, were still ape-like.

Running

The complete anatomical commitment to a terrestrial lifestyle did not come until *Homo erectus*. With this change came a suite of major alterations to the body. We saw that the upper limbs became smaller and the lower limbs longer, and this allowed a significant increase in bipedal efficiency. But there were other changes that reflect the fact that we have become efficient long-distance runners. The rib cage, which in apes is shaped like an inverted funnel to allow for the heavy shoulder musculature, changed to a "barrel" shape to accommodate much larger lungs. There are also soft tissue changes – for example, modern humans have Achilles tendons, whereas in apes the calf muscle directly attaches to the back of the foot. Our Achilles tendons give us a spring-like step for running on the balls of our feet.

Although humans are slower and clumsier than almost every mammal (even small quadrupeds are faster), we can run longer and more efficiently than any of those other animals. There are modern human groups (e.g., the San of southern Africa and the Tarahumara of the Sierra Madre Mountains in Mexico) known for their ability to pursue prey, such as antelope or deer, until it is too exhausted to continue running (this is known as *persistence*

Box 11.1 (cont)

hunting). Humans are the only animal on Earth that can (if properly trained) run continuously for 100 miles (160 km), as is regularly done in super-marathons such as the Leadville 100 (which is run at an elevation of up to 12,000 feet/3,500 meters).

Throwing

Equally important to changes in the lower body are important changes to the arms and shoulders. *Australopithecus* had ape-like arms and shoulders, adapted to pulling their bodies through the trees using the arms alone. Even a small chimpanzee has stronger arms than the strongest power-lifting human. In the 1920s, at Bronx Zoo, an agitated female chimpanzee (at roughly 90 pounds/40 kg) performed a 1,200 pound (540 kg) deadlift with one arm. No human, even modern weightlifters, has ever approached that level of strength. The muscles and bones of apes are built for raw strength, to suspend their bodies and catapult them through the trees.

That kind of strength seems useful, so why would we lose it? Humans have acquired something else by sacrificing raw strength: speed. A trained human adult can throw a baseball between 60 and 90 mph (100–145 kph). Our arms and shoulders have reorganized the anatomy to give us the ability to project force at a distance. This is a relatively rare phenomenon in nature – only a few animals, such as spitting cobras, can defend themselves or strike prey without exposing themselves to danger from physical contact. This ability would have given us a significant advantage over other animals, whether hunting or defending ourselves against large predators.

Homo Erectus

These major transitions first appeared with *Homo erectus*. Combining the overall increase in body size to that of modern humans with the locomotor efficiency of the lower body, and alterations to the shoulder that make throwing more efficient, *Homo erectus* was a formidable predator. This was the first time that humans were able to hunt prey and to defend themselves against other predators on the open savannahs of Africa. This shift in physical abilities compatible with open country survival and locomotion, along with a 50–100 percent increase in brain size, made *Homo erectus* able to adapt to a much wider variety of climates, and in a relatively short time after it evolved its anatomy we suddenly see it appear in the fossil record of Asia and Europe. Because of these reasons, for many researchers *Homo erectus* is the first hominin that we recognize as "us."

Box 11.2 Demography and technology

Between 250,000 and 150,000 years ago, we see a change in the archaeological and fossil record. Over a period of some 100,000 years there is a shift toward technological sophistication that was not present previously. This change, called the Middle Stone Age transition is a departure from the relatively unchanging archaeological record of the previous 1–1.5 million years. It is during this time that our species, *Homo sapiens*, appears in the fossil record, first in Africa, then, over some 50,000 years, around the world.

Because of the importance we place on our own species, scientists have focused on this transition to a considerable extent. For many decades the model of a "revolution" held sway. The fossil record is relatively low-resolution, and it seemed to many researchers that humans appeared relatively suddenly, and that some radical neurological transformation must have occurred in their brains to produce such sophisticated tools, along with art and, possibly, religion. However, more recent research suggests that the appearance of new technologies in the fossil record has more to do with the demographic changes of our species.

Demography

Humans, who had ready access to meat, and who were under pressure to solve complex resource-related problems, as well as store important social information, developed larger and larger brains. In terms of time intervals, the changes occurred over tens and hundreds of thousands of years, but at some point humans developed skills that allowed them to survive almost anywhere on Earth. Once populations of humans were distributed across the Old World (Africa, Asia, and Europe), subsequent population growth increased population densities rather than causing migrations. It was probably at this point that demography (population density) became a major factor in the spread of technology.

Imagine you are a young modern human (*Homo sapiens*) living comfortably somewhere in East Africa, with very low population densities. Your social group is probably your own family: your parents and siblings, and potentially some uncles and aunts. If you want to learn how to make stone tools, or spears, or fishhooks, you only have a few adults who can teach you. However, if the population increases, such that your area includes another five or ten families within walking distance, you might be able to look around to see who makes the best fishhooks, for example, and learn from that person. Suddenly, you have access to a more sources of knowledge. And, as you become an adult, you in turn become a source of information for younger humans in nearby families who want to know how to make the best fishhooks. Repeated

Box 11.2 (cont)

over thousands of generations over millions of square miles, the increased population densities of modern humans pushed technology steadily forward. With relatively high population densities, innovative information could literally travel around the world.

This pattern actually repeats over and over throughout human history, and is why cities become centers of innovation. And when civilizations collapse and population densities decrease (as in the fall of ancient Rome), technologies can become lost.

Box 11.3 Language, art, and religion

At roughly the same time that we see the appearance of innovative Neolithic technologies, we also see the representation of abstract ideas. This is most visible in the art of early modern humans across the globe. This art takes several forms: cave paintings, petroglyphs pecked into cliff walls, and figurines. In all of these art forms, there are clearly ideas being expressed, much as we see in art today (see Figure 11.3). This ability to take an idea or thing and portray it visually represents an important advance that probably allowed humans to outcompete all other human species across the globe. In all likelihood it is associated with the same intellectual advancements that give us language. Language is, after all, an abstract representation of ideas. Other animals can communicate using sounds, but as far as we know only *Homo sapiens* has the ability to create a true language, in which sounds represent any idea even though they sound nothing like the idea itself.

When a child calls a train a "choo choo," that child is representing the object using something that "sounds" like the idea. But when, as an adult, you call a train a "train," there is nothing in that sound that is particularly train-like. It is an abstract representation of an idea. That is a particularly powerful ability, because it means all manner of ideas can be represented and communicated to other people. In fact, one of the characteristics of language is the ability to shuffle and reorganize words in ways that communicate new ideas not necessarily related to what is in front of the speaker.

One thing that appears contemporaneously with art is religion. Many of the early expressions of art appear to show objects that do not appear in nature, and are more like shamanistic religious expression. For example, one of the most well-known pieces of early European cave art in the Trois-Frères cave of

Box 11.3 (cont)

Figure 11.3 Rock art. Here we see ancient (25,000–50,000 years old) rock art from Australia (top), Africa (middle), and Europe (bottom). The appearance of rock art is one of the surest indicators that modern *Homo sapiens* has evolved, and that it is something fairly distinct in its behavior from previous species. No other human species makes clear, representative art. We tend to associate the appearance of rock art with the appearance of language, since they are both ways of representing specific ideas. Increases in brain size are likely responsible for this ability to "abstract" ideas and communicate them effectively.

Box 11.3 (cont)

southwestern France is "The Sorcerer." This figure is a hybrid, with the basic body of a bipedal human but also antlers, animal paws, a fox's tail, and other animal anatomy. We will never know the specifics of what this represented to the people who made it (although that doesn't stop people from offering their opinions). But it is clearly not a simple story of hunting elk, so we are forced to conclude that something else was being communicated. Among modern foraging societies, "shamans" often have hybrid identities, and there is no reason to expect that early humans would not have had people with such roles in their societies.

Such early "shamanistic" art appears around the world, and suggests that early humans had some form of religion tens, if not hundreds of thousands of years ago. Currently, the earliest evidence that supports an early religious belief is the burial, 90,000 years ago, of a woman, with her body coated in red ochre. Again, we cannot know the significance of this special burial, but the body must have been prepared in some way after death, and this is commonly seen as a religious act in societies today.

It may be that the same neurological abilities that gave humans the ability to abstract ideas into language provided the opportunity for the development of religion, since all religions represent something that is not directly observable and "concrete." Religion appears to be a universal phenomenon across the Earth, as a society has never been discovered that does not have some form of religious thought and expression (more than 5,000 distinct religions have been documented to date). I discuss religion in more detail in Chapter 18, but it is worth noting the universality of this practice among human societies, and that it first appears when technological and behavioral sophistication is visible in the archaeological and fossil record.

Box 11.4 *Paranthropus*, the mystery lineage

When Charles Darwin published *On the Origin of Species* in 1859, there was no known human fossil record. But he understood that humans shared a common ancestor with apes, and if you could bring him forward 150 years and show him *Ardipithecus*, *Australopithecus*, *Homo habilis*, and *Homo erectus*, he would not be surprised. Over some 4–5 million years there was a relatively steady progression from ape-like to human-like, with changes in the brain, teeth,

Box 11.4 (cont)

arms, legs, pelvis, hands, and feet that reflect the transition to what we see in the bodies of modern humans.

However, there is another lineage, discovered in the 1950s, that came as a complete surprise, largely because it does *not* reflect this pattern of change over time. In this lineage, the genus *Paranthropus*, early humans acquired a new and unique adaptation, and remained in that niche, largely unchanged, for more than a million years. *Paranthropus* is known from fossils in East and South Africa, and dates from roughly 2.6 to 1.4 million years ago. There are three species that have been discovered so far, and we have enough fossils that we know quite a bit about this genus.

Probably the most notable thing to know about *Paranthropus* is that, unlike the *Homo* linage, in which the teeth and jaws get smaller as the brain increases, there is no increase in *Paranthropus* brain size relative to its ancestor, *Australopithecus*. Whatever *Paranthropus* was eating it must have imposed enormous demands on the teeth, because the molars and premolars are extremely large and heavily worn, even in relatively young individuals. The muscle attachments and bones of the jaw and skull are similarly overdeveloped, suggesting extremely heavy and repetitive loading of the jaws and teeth.

We don't know what *Paranthropus* was eating, but it might have been something tough and fibrous, possibly with a lot of grit in it. Perhaps it was some sort of tuber or plant root that required heavy chewing. Whatever it was, it must not have imposed any particularly difficult problem-solving skills, because the brains in this genus are not particularly large. This was probably a relatively low-quality diet, in which a great deal of food had to be eaten to provide enough calories for survival. And this diet probably would not have provided enough calories, protein, and fat for the development of large brains.

They are found at the same time and place as fossils of early *Homo*, so they might have coexisted with early members of our genus. But because their dietary niche was so different from that of early *Homo*, they could have been present without presenting competition.

So, when the climate changed around 3–2.8 million years ago, and the forests receded, there were at least two ways to survive on the new, drier landscape. One was by eating (at least part of the time) the animals found on the savannah, and that lineage led to us. The other was to eat low-quality vegetation in bulk. To me, one of the most interesting conclusions from this is that there were at least two ways to be a human between 3 and 2 million years ago.

> **Box 11.4** (cont)
>
> Ultimately, *Paranthropus* went extinct. We don't know why, but it is likely that further climate change eliminated their niche. Since they had not evolved large brains, and the broad problem-solving abilities that accompany them, they may have been faced with the disappearance of their preferred foods and no way to compensate. But they stand as a good example of why we need to think of human evolution in much the way we think of organismal evolution generally – animals and humans alike will evolve unpredictable adaptations, and into unpredictable niches, and we need to keep an open mind when we think about what it *is* to be human.

Advanced *Homo*

By 2 million years ago another major transition appears in the human fossil record. This is when we see *Homo erectus*, arguably the first "modern" fossil human. *Australopithecus* and earliest *Homo* were small, with long arms and short legs, and they would have looked very much like apes. However, *Homo erectus* was tall with modern limb proportions – long legs and reduced arms. This was a human well adapted for long-distance travel in hot climates and was likely to have been the first human to be an organized hunter. Major changes appear in the skeleton consistent with a long-distance running adaptation, including larger lungs and longer, more efficient legs. Further, the shoulder and arm changed – not just becoming smaller (for more efficient running), but acquiring a shoulder especially well adapted to throwing. Running and throwing are two adaptations that would have allowed them to hunt animals rather than just scavenge.

Homo erectus was an incredibly versatile animal, and as far as we know was the first human able to adapt to a broad variety of climates. This is when we see humans push out of Africa into Asia and Europe, so that by 1.8 million years ago, *Homo erectus* is well-established in Southeast Asia, thousands of miles from Africa. *Homo erectus* was also relatively long-lived, and existed from roughly 2 million years ago to fewer than half a million years ago. Over that 1.5-million-year run, brain size steadily increased, from under 900 cc to 1,300 cc, which is in the low range of humans today.

Several technological innovations are associated with *Homo erectus*. Perhaps the most well known is the "Acheulean hand-ax." The interesting difference between Acheulean tools and the earlier "Olduwan" tools is not so much their function but their appearance. Early *Homo* made tools that had very sharp edges, but there was no underlying design. A few rocks were knocked together until a sharp edge was achieved, and then it was used. Acheulean hand axes, in comparison, all have a similar design: they are teardrop-shaped, and rather than a few whacks necessary to make one, it must have taken considerable skill and a longer time. This tool technology reflects an increase in mental sophistication in two ways: first, the tool-maker is clearly making the tool to match a mental template; and second, there is a significant increase in skill necessary to make this tool. All of this speaks to the increased intellectual power of *Homo erectus*.

Although the evidence is somewhat controversial, and research continues on the topic, *Homo erectus* has been attributed with the first use of fire by a human species. The importance of fire for later humans is difficult to overstate. Fire allows many things. The first is obviously warmth – we see later humans, such as Neanderthals and *Homo sapiens*, succeeding in close proximity to glaciated environments in Europe and Asia, where it must have regularly been well below freezing for long periods of time. It is difficult to imagine that this would have been possible without the mastery of fire. Fire is also important for diet. Cooking meat kills parasites, and would have therefore been important for the health of the population. Cooking also enables us to eat foods that are otherwise very difficult to chew and digest (imagine eating a raw potato), because the heat breaks down the tougher, more fibrous material. Finally, fire has the potential to change animal behavior. A brushfire might be used to flush game from bushes for hunting or trapping, or to keep predators, such as lions, from a campsite.

This period of time is also when humans made the transition from a gorilla-like polygynous social system, with significant sexual dimorphism between males and females, to a more modern human pattern. We see the disappearance of strong sexual dimorphism with the appearance of *Homo erectus*, in which males and females are roughly the difference in size we see today. We do not know the reason for this change, but since social systems are, in essence, the product of female dietary patterns on the landscape, it must have had something to do with a shift in the foods eaten by these human populations – possibly as a result of the transition to dependence on more meat in the diet.

Because *Homo erectus* was so successful for so long, it is, in a sense, the "source" population for several human species. There are multiple species of later *Homo* that all sprung from this worldwide pool of successful, large-brained humans. One of these populations is today known as the Neanderthals, and they lived in Europe from roughly 250,000 years ago to just 25,000 years ago, when they went extinct. This was the fate of several species of later *Homo*. There has been considerable debate over the reasons why these populations went extinct whereas we did not. And, although there is not a consensus on the cause, it is likely that they were, in some way, not able to compete with *Homo sapiens*.

The Appearance of *Homo Sapiens*

Somewhere, perhaps 500,000–700,000 years ago, while populations of *Homo erectus* were still present in various places around the world, one branch split off. This branch acquired even larger brains and populated parts of Africa and Europe. These populations lived successfully, but separately, for 200,000–300,000 years, as they were kept apart by the Mediterranean Sea. This separation, and apparent lack of genetic flow, allowed differences to accumulate. The European population became the Neanderthals, whereas the African population became us, *Homo sapiens*.

By roughly 150,000 years ago, the modern humans in Africa had acquired significant technological and behavioral advantages over all the other species of humans across the world. Over time, these modern humans migrated out of Africa and spread out across the continents of Asia, Europe, and Australia (North and South America were not populated by humans until much, much later), ultimately replacing the local populations (derived from *Homo erectus*, as in the case in Asia, or the Neanderthals in Europe). We don't know the ways in which this replacement occurred – for example, we don't have any direct evidence of interpersonal violence from the fossil record, and the resolution of the fossil record makes it difficult to know how long the replacement took in any one place. But it is clear that multiple species went extinct upon the arrival of modern humans (*Homo sapiens*), wherever they were.

And although we don't know the details of the replacement, we do think we have a pretty good idea of why *Homo sapiens* was able to outcompete local populations. Modern humans have a suite of formidable capabilities that we don't see in previous human species. The most obvious is probably our large problem-solving brain. There is a group of technologies that appears only with the evolution of modern humans. Tools like fishhooks (made from bone or shell), bone needles (used for sewing hide clothes), and more sophisticated stone blade technologies first appear in the African fossil record when modern humans arrive, and then appear elsewhere only with the arrival of modern human populations migrating out of Africa.

But perhaps the most important shift has to be inferred, as it cannot be directly observed. The appearance of symbolic art first appears with modern humans (see Box 11.3). But while the art is, itself, important, what is more important is what it implies. Symbolic art is a method for communicating complex ideas, and it is the ability to communicate so effectively that may have given humans their ultimate advantage. Language is the natural human method of communication, and language permits the transmission of extremely complex ideas. You can imagine the advantage this would give the early modern humans over any other human species. The ability to explain to others of your group where the game herds were, and provide a geographic description of an area, for example, would allow you to exploit the game animals much more efficiently than earlier human species. Methods for innovative technologies could be explained verbally, and readily passed from group to group. Dangers could be explained in detail, and plans made for dealing with them.

Of course, there is no way to directly observe language in the fossil record, but the ability to make representative art is likely to have directly correlated with the evolution of the ability to represent concrete ideas symbolically. And the art is still effective – when we examine rock art from caves in Africa, Europe, Asia, or Australia, we can see that it is making direct representations, and often we can interpret the events even today.

Chapter 12: Human Variation

The Science and History

The biological study of modern humans starts with their evolution, in Africa, some 200,000–300,000 years ago. Over the next quarter of a million years, modern humans, much as other animals have done before them, migrated out into diverse environments across the globe (Figure 12.1).

However, when scientists have examined humans scientifically in the past, they have had trouble treating humans as they have other animals. Instead, in many cases, they have presented perspectives that reflected more about their political motives or personal biases than scientific reality. Because this is a book on humans, and science has failed so catastrophically when looking at humans scientifically, it is important to present the history of this work in light of what we now know about human biological variation.

The Question of "Race"

Much of the focus of biological anthropology, from the late nineteenth century through World War II, was on the question of "race." This was the subject of considerable research, almost all of which is now debunked. However, the role that race (as a concept) played in the nineteenth and twentieth centuries, and the centrality of its role in some of the most destructive acts in human history, means that it is extremely important that we understand the meaning of the "race" concept, as well as how it was created and studied.

The Biology of "Race"

There are three premises on which the concept of "race" is built. The first is that race is, in some sense, a real thing – that labeling someone according to a particular race reflects an underlying biological "reality." In modern terms, we might phrase this in a way that reflects ancestry, or shared genetics. For example, when modern Americans call someone "white" or "black" they are making what amounts to a biological statement: they believe that there is a biologically discrete group of people that can be labeled with the shorthand "black," and there is a biologically discrete group of people that can be labeled

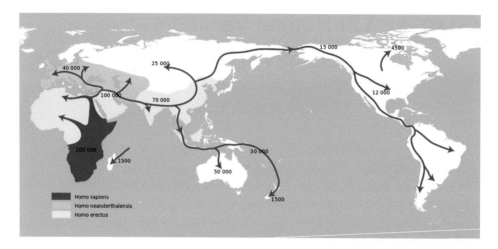

Figure 12.1 Humans migrated out of Africa starting roughly 100,000 years ago, outcompeting local hominin populations over time, so that today there is only one human species on Earth. The genetics of modern humans largely reflects this migration pattern, as people nearer geographically are more closely related. Genetic variation is partly a result of differentiation accumulated over evolutionary time, and because humans originated in Africa genetic variation is greatest there; the Americas, which were populated relatively recently, have much less genetic variation.

with the shorthand "white." Generally, these groupings are held to be in some way 'natural' and to have some important underlying biological significance.

The second premise is that the groups identified by the "race" are truly discrete; that is, they are different things. These "black" and "white" groups are, under this model, meaningfully different from each other. This is quite unlike like color difference in, say, gibbons, where the colors are not held to be meaningful in terms of grouping.

The third premise is that the groups are intrinsically (i.e., biologically) different in some important and meaningful way. Normally, this is held to reflect something about the intellect of the groups: intelligence, morality, work ethic, etc. So a "white" person has a different intellect or character than a "black" person (or by whatever racial categorization used).

However, *all* of these premises are incorrect.

Skin Color

The first thing to understand about human biological variation is that it is real. Humans have been around long enough for selection to have acted on some relatively simple traits of anatomy. For many of us, the most obvious example is skin color. Human groups do have different skin colors, and this reflects the fact that solar radiation is a source of genetic mutations leading to cancer. When you map human skin color onto the globe, you can see that humans (prior to relatively recent global migrations) at latitudes closer to the equator

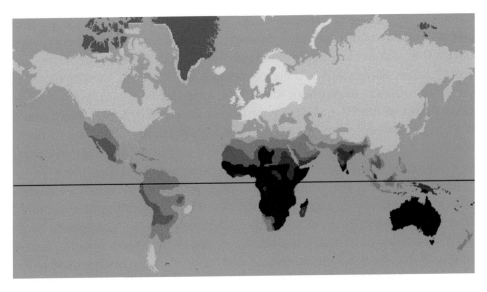

Figure 12.2 The broad pattern of human skin color (prior to the global migrations of the last few hundred years) maps onto latitude because melanin affords protection from solar radiation. Darker skin tends to be found closer to the equator, but does not necessarily indicate close genetic affinity. It has regularly been used to categorize humans into "races," but it simply reflects an adaptation to the strong solar radiation closer to the equator.

Australia and Africa have populations with extremely dark skin, but they are about as distantly related, genetically, as any two populations on Earth. Other genetic markers do not map onto skin color. For example, blood-type variation does not match skin color, nor do other traits such as height, body build, and body hair, because they are not adaptations to solar radiation.

tend to have darker skin (Figure 12.2). Skin is colored by a protein called *melanin*, and having melanin in the skin dramatically reduces the solar radiation passing through the outer skin into the rest of the body. Over the last several tens of thousands years, some groups have acquired (or maintained) dark skin as a way of protecting themselves against skin cancer. However, in groups that migrated away from equatorial regions, the issue of too much solar radiation became less of a factor. Instead, in those areas, the need to allow solar radiation into the skin to help process vitamin D became important. In these groups, selection acted against the presence of melanin.

One implication of this fact is that similarity in skin color is no particular indication of deeply shared ancestry. The darkest skin colors are found in Australia and Africa, and the lightest in northern Europe. Yet Africans are more closely related to northern Europeans than they are to Australians.

There are many other body traits that follow this model. People from equatorial regions tend to be slender, a trait that facilitates the radiation of body heat, whereas heavier and more muscular groups tend to come from colder areas, where compact body forms are more thermodynamically advantageous. People living at high elevations (Tibet, the Andes, the Ethiopian Highlands) all tend to have larger lungs to help cope with the thinner air at high elevation.

So human biological variation is real, and it does reflect something – the adaptive history of our species to different environments. But are these groups discrete? A look at the map of skin color shows that variation is continuous. For example, within Africa there is considerable variation in skin color, as populations native to southern Africa have relatively light skin, as do populations native to northern Africa. Populations in southern Europe have much darker skin than populations native to northern Europe. This pattern holds across all the continents – there are continuous gradations of traits that reflect, to a large extent, the adaptive pressures on that group.

The fact that, in the United States, we have people who can be labeled "black" and "white" is simply a reflection of the migrations to the United States. The two largest populations that migrated to the USA happen to come from very different latitudes, but between those two latitudes populations are in shades of color between the "black" and "white." Those intermediate populations just did not happen to form major parts of our population, due to our particular history ("brown" people, who are part of the native populations of the Americas, are another group as well, but I am simplifying to clarify).

Finally, Africa represents a special case. Since modern humans evolved in Africa, and existed in Africa for perhaps 150,000 years before migrating out, there is far greater genetic variation within that continent than any other. The differences between African populations can be quite large – the genetic distance between the San of southern Africa and, say, the Ethiopian Highland peoples is greater than between the Irish and the San. The non-African peoples of the world descended from African groups, so, in a genetic sense, they are a subset of African variation. Because of this, it does not make genetic sense to have a category of people labeled "Africans" – they are not a coherent and separate genetic group.

> **Box 12.1** Slavery and race
>
> The question of race is particularly important in the United States. Between roughly 1513 and 1865 around 600,000 people were violently removed from their homes and taken to the United States. They were primarily used for agricultural labor in the American south, and in several states formed the majority of the population.
>
> Slavery itself, as an institution, is probably universal throughout human history. A recent study found that no society up to the mid-nineteenth century had existed without slavery at some point in its past, across all continents, for as long as recorded documents were kept. Slavery was present in ancient Mediterranean societies such as Persia, Mesopotamia, and Egypt, in Asian societies in China and Mongolia, in African societies such as the empires of Aksum, Ghana, and Mali, and in Native American societies like the Maya, Aztec, and Inca.

Box 12.1 (cont)

Throughout history (and prehistory), there has been resistance to slavery – slave revolts date back into antiquity. At no time prior to the seventeenth century did any society question the morality of imposing slavery on *other* people in any systematic way. However, during the *Enlightenment*, in Europe, philosophers began to raise questions that challenged the very morality of slavery. Influential writers (see Chapter 19) presented the argument that people were individuals with a core set of natural rights. Immediately the question of slavery presented itself: If people were individuals, with the right to make their own decisions, how could we morally justify slavery?

The trend toward increased literacy and publishing during the Enlightenment may have played a role, as the narratives of escaped slaves became a well-known literary form. Escaped slaves would also go on speaking tours. Both of these may have preventing the benefactors of the slave economies from being insulated from the ethical implications of the institution.

Slavery was one of the central issues that the founders of the US government faced. In the American South, there were strong economic interests in keeping slavery legal, as enormous wealth was generated by large plantations using a slave-based labor force. The founders were well aware of the arguments against slavery, even as they allowed it to continue after the formation of the new US government. The conflict can be seen in the behavior of two of the "Founding Fathers." George Washington wrote against slavery and freed his slaves in his will, but he did this on his deathbed having benefited from the chattel labor his entire life. Thomas Jefferson also wrote against slavery, even as he kept a large plantation staffed with slaves to maintain his standard of living (he also fathered children with one of his slaves, Sally Hemmings).

They were aware of the contradictions in founding a nation on Enlightenment principles while keeping slaves (no Roman or Persian leader ever expressed such compunction). Washington and Jefferson understood that slavery's end was inevitable, given the trajectory of history. In effect, they knew it was immoral but expected later generations to solve the problem, largely because they put their own economic self-interest ahead of the moral issues.

In any event, they turned out to be wrong. The economics of the American South meant that slavery did not simply wither and disappear through the natural course of history. At the outset of the US Civil War, it is estimated that *half* of all equity (cash, material assets, real estate, etc.) was the value of the slaves in the southern states. So there were enormous economic incentives to keep slavery legal, even if much of the world had decided that slavery was immoral (e.g., England and France outlawed slavery in 1818, Spain in 1817, Mexico in 1830, and Ottoman Turkey in 1847).

Box 12.1 (cont)

So how could advocates of slavery defend the institution? One way was to argue that people of African descent were in some way inferior, and that they could not look after their own interests. This argument, absurd as it is, became the central platform for the defense of slavery. Alexander Stephens, the vice president of the Confederacy, in 1861 made his "Cornerstone Speech," in which he argued that slavery was the "cornerstone" of the Confederacy, and was morally justified for racial reasons:

> The new (Confederate) Constitution has put at rest forever all the agitating questions relating to our peculiar institutions – African slavery as it exists among us – the proper status of the negro in our form of civilization. This was the immediate cause of the late rupture and present revolution . . .
>
> The prevailing ideas entertained by (Jefferson) and most of the leading statesmen at the time of the formation of the old constitution, were that the enslavement of the African was in violation of the laws of nature; that it was wrong in principle, socially, morally, and politically. It was an evil they knew not well how to deal with, but the general opinion of the men of that day was that, somehow or other in the order of Providence, the institution would be evanescent and pass away.
>
> Our new Government is founded upon exactly the opposite ideas; its foundations are laid, its cornerstone rests, upon the great truth that the Negro is not equal to the white man; that slavery, subordination to the superior race, is his natural and normal condition.

This was a fairly honest presentation of the Confederacy's position. They wanted to secede so that they could continue slavery, and they maintained that slavery was a natural institution for people of African descent. For the next 150 years, long after slavery was abolished, this argument was repeatedly made to justify segregation, Jim Crow laws, and a variety of prejudicial behaviors, large and small. Of course, this argument is a lie, but it is not the first lie made to protect strong economic interests.

Today, we are still dealing with the aftermath of this lie. Unlike in some ancient societies, where slaves often had the same skin color as the slaveholders, in the United States it was easy to know who had been a former slave, and to subject them to persecution (Figure 12.3). The somewhat accidental fact that the two founding populations happen to come from geographic areas separated by significant differences in latitude, and that descendants of the two groups are so readily distinguished by superficial anatomical features, are the reasons why race has become one of the central antagonizing factors in American social life.

Box 12.1 (cont)

VIRGINIA

HEALTH BULLETIN

| Vol. XVI. | MARCH, 1924. | Extra No. 2 |

The New Virginia Law

To Preserve Racial Integrity

W. A. PLECKER, M. D., *State Registrar of Vital Statistics, Richmond, Va.*

Senate Bill 219, To preserve racial integrity, passed the House March 8, 1924, and is now a law of the State.

This bill aims at correcting a condition which only the more thoughtful people of Virginia know the existence of.

It is estimated that there are in the State from 10,000 to 20,000, possibly more, near white people, who are known to possess an intermixture of colored blood, in some cases to a slight extent it is true, but still enough to prevent them from being white.

In the past it has been possible for these people to declare themselves as white, or even to have the Court so declare them. Then they have demanded the admittance of their children into the white schools, and in not a few cases have intermarried with white people.

In many counties they exist as distinct colonies holding themselves aloof from negroes, but not being admitted by the white people as of their race.

In any large gathering or school of colored people, especially in the cities, many will be observed who are scarcely distinguishable as colored.

These persons, however, are not white in reality, nor by the new definition of this law, that a white person is one with no trace of the blood of another race, except that a person with one-sixteenth of the American Indian, if there is no other race mixture, may be classed as white.

Their children are likely to revert to the distinctly negro type even when all apparent evidence of mixture has disappeared.

The Virginia Bureau of Vital Statistics has been called upon within one month for evidence by two lawyers employed to assist people of this type to force their children into the white public schools, and by another employed by the school trustees of a district to prevent this action.

Entered as second class matter July 28, 1908, at the Postoffice at Richmond, Va., under the Act of July 16, 1894.

Figure 12.3 In 1924 Virginia passed a law outlawing the marriage of people identified by law as 'white' from marrying anyone with any 'trace of the blood of another race'. The law was in effect until the State of Virginia prosecuted and convicted Mildred and Richard Loving, a mixed-race couple married in Washington DC, but who lived in Virginia. The couple appealed the conviction, and in 1967 the US Supreme Court struck down the law (and all similar race-related marriage laws in the US).

Box 12.2 The social and economic context of racism

The earliest documented use of human skin color as a category comes from the ancient Egyptian *Book of Gates*, which dates between 1550 and 1200 BCE. Egypt was one of the trading centers of the world, as well as an imperial power, and so would have had people from around the world pass through. Other writers, such as Linnaeus (the inventor of taxonomy), also used "race" (in his case assigned by continent) as a way to categorize humans. The first *scientific* discussion of skin color comes from the German scientist Johann Friederich Blumenbach, around the end of the eighteenth century. He studied a collection of human skulls and discovered that, by measuring the shapes of the skulls, one could determine where they had probably come from (geographically) and what color their skin would have been. Today, forensic scientists still use his principles when they attempt to identify the ethnic origin of a deceased person from the measurements of their skull (you may have seen this applied in one of the many forensic law-enforcement television shows).

Blumenbach has sometimes been considered the first "scientific racist," but, like in the *Book of Gates*, there was no mention of anything other than visible difference. In fact, given the time, when slaves were commonly being taken from Africa, he went out of his way to denounce racism. But scientific racism soon found a foothold. Why?

One of the most important things to understand about the history of racism is that it *always* takes place in a social context in which someone has something to gain by espousing it. This has been shown repeatedly, and I will offer a few historical examples. One of the most obvious is the European racism that was prevalent up until the end of World War II. In this case, many of the European powers were engaged in a practice known as *colonialism*.

Colonialism, as a way to conquer a people and take what they have, is a strange practice. Under the ancient conquerors like Genghis Khan, or the Persians, or Romans, there was no need to justify conquest – it was understood that the strong take from the weak. But, by the time European colonialism took hold, quite recently in our world history, we had already had the Enlightenment, in which the rights of peoples to self-determine was clear. But the Europeans still had the desire to conquer. So, how did they justify it?

The simplest way, for the colonialists, was to argue that they were doing the conquered peoples a *favor* by conquering them. As inherently inferior peoples they could not govern themselves, so the reasoning went. This was phrased in various ways, but probably the most famous was the "white man's burden" defense. Under this idea, controlling and ruling places like Africa, Asia, and Australia was a moral obligation. Of course, it was always a lie, and the colonial

Box 12.2 (cont)

powers were always, as in times of old, only really interested in extracting as much as they could from the conquered powers.

Modern Racism

In the twentieth century we saw the peak of racial thought. This was, of course, during the 1930s and 1940s, up to and during World War II. Adolf Hitler and the Nazis (partly inspired by eugenics practices in the USA and UK) saw one "race" – the "Nordic" race – as the natural conquerors of all other people, who were to serve as slaves or be exterminated altogether. Hitler's concepts of race were completely absurd (in a scientific context; in a moral context they were abhorrent), but it is worth noting that Europe was not the only place where racism was taking a brutal toll. The Japanese Empire, in justifying the conquest of parts of mainland China, Malaysia, and Indonesia, presented ideas similar to those of the Europeans – arguing that the Japanese were an inherently superior "race" to other Asians. To this day, Japan struggles with racism, and modern politicians exploit ideas of race for political purposes.

When examining racism in the world today, the main question is: Who has something to gain? Racism is always found in cases where one group has something to gain or lose and sees racism as a way to gain. Are people afraid of losing jobs or status? Who gains economically or politically by making the racist claims? From Alexander Stephens' "Cornerstone Speech," to the colonialism of the European powers, to the Imperial Japanese Empire, racism has always been about justifying the right to subjugate another people for one's own economic gain.

Box 12.3 Absurdities: how many races?

One of the curious things about racism, as well as a reason to reject it, is its logical inconsistency. If you examine the historical literature, you will find a wide array of groups that have been called "races" by one writer but then lumped in with other groups by another writer. In the United States today, race is often used to distinguish between people of European and West African descent, but this has not always been the case.

Various groups from within Europe have been considered races: for example, the Irish were called an "inferior race" by many British writers during the nineteenth century (this was, coincidentally, when Ireland was a colony of the

Box 12.3 (cont)

British). Famously, the Jews were considered a race by Europeans, as a justification for economic and social persecution and genocide. Historically, East Asians have sometimes been grouped into one "race," but to the Japanese they themselves have historically been a very distinct "race" and superior to other Asians.

Within Africa, there is more genetic diversity than among all the other peoples of the world. This reflects the fact that humans originated in Africa, and that modern humans have been in Africa far longer than any other place on Earth. European racial thinkers showed no consistency here, either – they broke Africa up into as many as eight "races," or as few as one.

Sometimes racial groups would be subdivided into "subraces." William Ripley, in 1899, divided the "Caucasian" peoples (the peoples of Europe) into three groups: Teutonic, Alpine, and Mediterranean. The same year, Joseph Deniker broke Europeans into ten groups. Later, in 1939, Carleton Coon organized the "Caucasoid race" into nine subraces. These groups often enjoyed esoteric names, such as the Dinaric race, Iranid race, Osterdal race, and Vistulian race.

Hopefully, the complete inconsistency of this work demonstrates why it was bound to fail. These writers were determined to break up a continuous distribution of superficial soft tissue traits, such as skin, hair, nose shape, and eye color, as well as body build and skull shape, into discrete groups (this exercise in artificially lumping things together is known as *typology*). Of course, if the groups are not truly discrete they cannot be logically broken into categories. Imagine how many ways there are to name colors, but it is always just our categorization of what is actually a continuous distribution of light-wave frequencies.

Sometimes, in nature, there are discrete groups. The obvious example is species. Foxes and coyotes are truly and naturally distinct, and they will not normally interbreed. But, within species there is simply continuous variation. Since, to our knowledge, there is no human group that cannot readily (and happily) interbreed with members of other groups, there is no evidence of a natural break. Foxes and coyotes separated more than 7 million years ago, whereas humans evolved very recently by evolutionary standards. There is no biological reason to expect discrete groups in such a young species as *Homo sapiens*.

Ultimately, the various racial categories (typologies) reflected more about the writer (and their social prejudices) than any underlying biological reality. If it served political ends to call the Irish an "inferior race," then they would be called that by British politicians. As with any form of racism, racial typology is a means to a political or economic end rather than any reflection of underlying biological reality.

Box 12.4 Eugenics: the "science" of racism

One of the darkest chapters in the history of science is the discipline now known as "eugenics." Eugenics was the use of seemingly scientific or mathematical techniques to identify "races," and in some cases to determine which groups were superior or inferior in some capacity. The techniques were always highly biased and in actuality unscientific, but their use of seemingly scientific methods lent them an air of legitimacy that was used to justify the prejudices of those societies. Today, eugenics stands as a warning to scientists to examine their biases and to be very aware of the social implications of their work.

The concept of race actually predates science itself. The first written expression of races as categories comes from the Egyptian *Book of Gates*, which dates to sometime between 1200 and 1550 BCE. By the time modern science started to appear, in the late Renaissance and early Enlightenment (the late fifteenth and early sixteenth centuries), the concept of race was already part of the European consciousness. Explorers traveling around the world documented the wide array of human variation, and these descriptions formed the foundation of racial categorization.

The history of eugenics and "scientific racism" in some ways parallels the development of science: as people started to look around the world systematically, they naturally applied some of those principles to understanding humans. This was the case even before Darwin and Wallace had given us a framework for understanding the driver of biological variation. More than 200 years before *Origin of Species*, the Swedish naturalist Carl Linnaeus had organized all known organisms into what we now call the Linnaean Hierarchy (*Systema Naturae*, 1735), which we still use today. In this system, he grouped humans in with the apes, clearly making a statement about humanity being a part of nature that could be studied, much like any other organism. His application of taxonomy grouped humans according to geography (by continent) and has ascribed somewhat absurd (and assuredly racist) character traits to the various races. These traits spanned appearance (e.g., "flowing blond hair"), to personality ("Strict, haughty, greedy"), to styles of dress ("covered by loose garments").

About forty years later (1779), Frederich Blumenbach, a German scientist, organized humanity into five groups based on their geographical origin. He argued that the variation in people's skin color, eye color, hair color, and other soft tissue characteristics showed a pattern that could be mapped onto their geographic distributions. He was also able to demonstrate that he could determine someone's geographic origin by measuring their skull, based on his

Box 12.4 (cont)

examination of some sixty skulls from regions around the world. He was correct in both of these things, and although he is sometimes credited with being the father of scientific racism, he actually went out of his way to refute the racism prevalent at the time. He was simply documenting anatomical differences.

But it was the publication, in 1859, of *On the Origin of Species* that really gave scientific racism a mechanism by which it could explain the superiority of one race over another. At this time, the Europeans had reached their worldwide ascendancy in terms of political and military power, and were engaged in the control of vast regions of the globe through their colonial enterprises. In their minds, there must be some biological reason why Europeans were more powerful than other regions of the globe, and they often invoked natural selection and what they saw as the inherent biological superiority of Europeans as the reason why.

It should be noted, at this point, that Darwin specifically argued against such ideas. He explored human evolution in depth, and in *The Descent of Man* argued against the idea that there were races and significant biological differences between them. However, this did not stop other scientists from employing his model for their own ends.

Natural Selection and the "Cold Climate Hypothesis"

If one race were to be intellectually superior to another, there must be a cause. Natural selection provided the mechanism that scientific racists could invoke. One such theory was the "cold climate hypothesis," in which a colder climate imposed greater selective forces on problem solving, therefore driving natural selection for greater intelligence in cold climate populations. Even a brief examination of this theory shows its weaknesses. The first refutation comes from the fossil record; Neanderthals were a specifically cold-adapted population, and evolved in partially glaciated Europe in one of the coldest periods in Earth's history. Much of their anatomy was specifically adapted to the cold. Yet, when modern humans, who had evolved in warmer Africa, appeared in Europe they rapidly outcompeted the Neanderthals.

The second refutation comes from the historical record. The scientific racists were trying to explain why Europeans were in a position of greater political power. But that was a relatively recent phenomenon. If cold weather populations were inherently superior, we would expect them to have driven technological innovation throughout history and prehistory. This is manifestly *not* the case. There are multiple locations throughout the globe where

Box 12.4 (cont)

agriculture, probably the most important technological advancement since the control of fire, appeared throughout prehistory, often during a window 10,000–5,000 years before the present. In not a single case did agriculture appear first in cold weather populations – it appeared first in Turkey, China, Central Africa, and South/Central America. The cold weather populations of northern Europe were among the last to acquire agriculture. This is not because the cold weather populations are inferior; rather, climatic conditions in warmer areas are more conducive to agriculture.

European technological, political, and military dominance are a recent occurrence. Before roughly 1400 CE, Europe was a relative backwater. Technology appeared first in warmer areas, China and the Mediterranean being the two main hubs of technological innovation from roughly 5000 BCE until 1400 CE. Europe only acquired world power in the last 10 percent of recorded history.

Finally, there is no biological evidence that cold weather imposes greater demands on intelligence than warmer climates. In other animals we do not see this pattern. For example, northern gray wolves are not known to be smarter than desert wolves; polar bears are not smarter than grizzlies. Yet, we know that other elements of cold selection that apply across the natural kingdom *do* apply to humans: both humans and wolves from hot climates are more slender that cold weather humans and wolves and have longer limbs. If cold weather were a driver of intelligence, we would expect to see this pattern across nature, and we do not.

For these reasons, we reject the cold weather hypothesis (which occasionally crops up again, and has to be refuted, again) and others like it. In science, we call this a "just-so story," in which a hypothesis is tossed out that the writer thinks might be true, and that is consistent with their biases, without presenting it as a hypothesis that has actually been tested. For these reasons, we reject it.

The Eugenics Movement

During the late nineteenth century and early twentieth century, scientific racism had found itself in a position of authority, and politicians used it as a justification for a wide array of political acts that warranted persecution, segregation, and ultimately genocide. It is far too deep a subject to give much more than a perfunctory treatment here, but it is important to present the outlines.

During the nineteenth and early twentieth century, there were a variety of migrations, many into the United States, that were caused by climatic,

Box 12.4 (cont)

economic, and political upheavals around the globe. Whenever there is trouble, people flee, as they have for millennia, and during the nineteenth and twentieth centuries large numbers of migrants from China, Ireland, and Italy migrated to the USA. These groups were often seen as economic threats to established groups, and "racial inferiority" was often used as a justification for keeping these groups out of the country. For example, the Chinese Exclusion Act of 1882 specifically restricted Chinese immigrants. The Immigration Act of 1924 set specific quotas, strictly limiting immigration from certain countries where the citizens were deemed undesirable because of racist reasons.

Policies varied around the world, but the underlying premise of eugenics was to control the genetics of humanity by increasing the populations of some groups while restricting it in others. This was expressed in ways other than simply regulating immigration. In various US states before 1967, it was illegal for "whites" to marry any other race for fear of "polluting" the white gene pool (these are known as anti-miscegenation laws - see Figure 21.3). In Nazi Germany, Aryans were forbidden to marry Jews or other non-Aryan peoples. In some cases, individuals deemed especially genetically or socially unfit (the poor, uneducated, disabled, or criminals) were forcibly sterilized or, in Nazi Germany, killed.

These policies (and others like it, both in the USA and in other countries) were often sold to the public under the guise of science. Using techniques such as measuring skull shape or appearing to measure intelligence, justification was offered for various repressive policies. In all of these cases, the material was presented as a science, with all the "authority" that carried. Although eugenics is typically associated with the UK, the USA, and Germany, eugenics policies were found worldwide, including in countries such as Brazil, Australia, Canada, Sweden, Norway, Japan, and Korea.

The Rise and Fall of Eugenics

Eugenics reached its peak with the Nazis in the 1930s and 1940s. Using rationalizations from some of the policies enacted in the USA and Britain, and generating their own racial categories and quantitative techniques, the Nazis found justification for the persecution and genocide of Jewish and eastern European populations. They employed pseudoscientific and mathematical techniques to attempt to quantify the degree to which someone was "Aryan," and millions of people were murdered simply because they fell outside the range of what the Nazis found acceptable (Figure 12.4).

In the decades after World War II, when the world looked around to see all that the Nazis had done, people realized the full implications of eugenics.

Box 12.4 (cont)

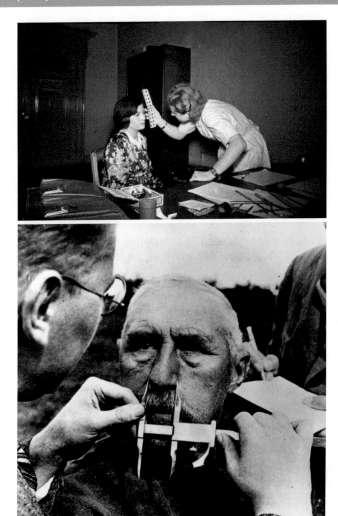

Figure 12.4 Eugenics has a very sordid history, and was used as a tool for persecution across Europe and the USA particularly. Most often it was used to justify policies designed to prevent some groups from acquiring economic or political power, or from creating future generations. Probably the most frightening use was by the Nazis during World War II, when the quantitative techniques derived from eugenics were literally used to decide who would live or die. Measurements of the shape and color of the skull and face were used to distinguish so-called Aryans from Jewish, Slavic, or Romani peoples. A person who was thus deemed "subhuman" would typically be sent to a concentration camp for eventual execution.

> **Box 12.4** (cont)
>
> Within biology, especially, scientists realized that the use of categories, and the typological thinking that was typical of prewar Europe and the USA, was simply untrue. As genetics became more advanced, it became ever clearer that there were not "types" of people, and that the "scientific methods" employed by eugenicists and scientific racists were not just incorrect but immoral, as they were really just ad hoc arguments generated to support racially biased and repressive social policies.
>
> By the 1960s racial typology had been rejected by most scientists, especially within academic disciplines like biological anthropology. However, some of it remains today in medicine, where there are still contexts where people are categorized by "race." Although these uses are not typically framed in terms of "racial superiority," they can have detrimental effects because of the essential inaccuracy of using these outmoded categories that do not reflect the underlying pattern of biological variation.

Race and Intellect

For much of the late nineteenth century and early twentieth century, the question of intellect and character was central to the debate over race. Various groups that were deemed "races" (African Americans, Chinese, Irish, etc.) were held to be unintelligent and immoral. In the case of American slavery, it was often justified by the argument that Africans were somehow more suited to manual labor because of their race.

One critically important thing to understand is that these claims were *always* made in a political context. For example, one justification made by the English for closing the border to Irish fleeing the great potato famine of 1845–1849 was that they were less intelligent, and were poor workers with bad habits. Similar arguments were made to justify restricting Chinese immigration by the US government (for example, the Chinese Exclusion Act of 1882). In many parts of the USA, this type of argument served as the justification for segregation, as well as restricting African Americans from voting in southern states. Even today, this prejudice remains, although politicians nowadays take pains to hide it.

However, there has never been any demonstration (biological or historical) that any human population has a different level of biological intelligence from any other. Further, there is considerable evidence that this is not the case. For example, in virtually all studies of student success the greatest predictor of academic success for students in the Unites States is their parents' education and income level, irrespective of race. IQ closely tracks educational level, and has risen among minority groups and people of color as their access to education has increased, which is the opposite of what we would expect if IQ were an immutable biological quality.

And there is no biological reason why we should expect any one population to be more intelligent than another. In all environments, there are considerable environmental challenges that humans, over the last 100,000 years, had to face. There was never a time when a *lack* of intelligence would have afforded some advantage to survival, and it was always an advantage to have more.

In the United States, the issue of race and intelligence has, in particular, been used to justify excluding African Americans and other groups from various educational opportunities. The prevention of education, of course, has the effect of keeping poor groups poor. There are significant economic reasons for this, as having a poor underclass to whom educational opportunities are denied creates a group for whom the only available economic opportunities are generally undesirable, underpaid, and often dangerous.

For virtually any modern biologist, the question of racial inferiority or superiority is no longer an issue. It is well understood in the sciences that "race" is a political issue, and trying to establish that one group or another is in any way "better" simply reflects a cynical attempt by politicians to exploit fear and ignorance. However, the idea that "races" represent discrete groups lingers – in medical technology, for example, there are sometimes "adjustments" to measures of various anatomical characteristics (lung capacity being one example) that reflect a fundamental misunderstanding of the variation in human anatomy. These misunderstandings, whether intentional or not, can have significant implications for the health of the patients.

Chapter 13: Evolution and Human Behavior

One of the most important conclusions that Darwin forced people to face is that humans are animals and are subject to the same natural forces that shape all living things. While evolution itself is controversial in some circles, in general most people (and virtually all scientists) understand that human anatomy is the product of evolution. We have hair because we are mammals; we have opposable thumbs because we are primates; we lack tails because we are apes. So *our* anatomy is the anatomy of a large-brained ape. However, much as behavior is shaped by evolutionary forces in animals and humans, behavior is similarly shaped. This means that we can interpret broad patterns of human behavior using animal models. Many of the human behaviors that are common across all populations are the result of evolutionary forces driving human decision-making, although they can express themselves in different ways.

Using animal models to describe human behavior has been controversial, and there are reasons to be cautious about how we apply the principles of evolutionary biology to behavior, but, as in anatomy, the application of evolutionary biology has enormous explanatory power. Furthermore, using evolutionary principles has the potential to help us understand, and potentially control, destructive human decision-making.

> When we talk about the evolution of behavior, we must keep in mind that the brain is a complex organ, and we may not know *why* we feel the way we do at any given moment. In fact, we often feel impulses that have an evolutionary origin that we know we must suppress or control. Take hunger for example – we have a natural drive to seek out fat, sugar, and salt, because in the environment in which we evolved those foods were scarce but important. Nowadays, when we are choosing what to eat, we often have to suppress the instinct to eat potato chips in favor of vegetables, even though the salty/sweet/fatty foods are more satisfying. This same principle applies when we talk about other emotions – sexual desire, anger, jealousy, rage, love, etc. These emotions are the product of evolutionary forces, and we may not be able to simply "turn them off," even if we know, intellectually, that we need to control them in a modern society.

Evolutionary Models 1: Reproductive Sexuality

One of the most important aspects of animal behavior is sexuality, since reproduction is critical for the continuation of a lineage. Natural selection is extremely strong in shaping animal sexual behavior, and specific environmental circumstances drive mating anatomy and behavior. Take two extreme examples. In gorillas, red deer, elephant seals, and some other mammals males compete for access to a large number of females. The females congregate because of the spatial distribution of resources, as well as protection from predators, and a single strong male can monopolize the reproduction of all the females. In this case, a single large male can provide much of the genetic material for local gorillas for decades. A contrast is provided by species of polyandrous (single female, multiple male mating group) birds that live in particularly rich environments. Here, the female birds have sufficient food resources to produce multiple clutches of eggs in different nests, each fertilized and cared for by a separate male. In these species (often shorebirds) the female guards her territory to keep other females out, away from her nests and males. So while the principle of sexual reproduction is present in both of these animal groups, the external environment has determined how these animals meet, mate, and reproduce.

Humans are mammals, so we will focus on mammalian mating behavior. In mammals generally, there are three factors that drive patterns of mating behavior:

1.   The cost of reproduction is high for females but low for males. Females bear the metabolic burdens of pregnancy and lactation, as well as the time demands of raising young to independence. In some species males contribute to raising offspring, but they cannot appreciably assist in pregnancy or lactation. For a most male animals, the cost of reproduction is minimal: simply the time and energy required for copulation.

2.   There is much larger potential variation in the reproductive success in males when compared to females. A female tiger, for example, can produce roughly one to two cubs every twelve months. Over her lifetime, she might be able to produce ten to fifteen surviving offspring. In contrast, a large, dominant male tiger that mates with several females might produce six to ten cubs per year, and have a lifetime potential of more than a hundred cubs. However, a smaller male, who cannot compete with the dominant male, might *never* get a chance to mate, nor produce any cubs. While healthy females can always produce offspring, no female could produce as many as a dominant male, but will always produce more than the nondominant males. So there is potentially far more *variation* in the reproductive potential of males than in females.

3.   Females always know that their offspring are theirs, whereas males can never truly know. Even in species that pair bond (such as humans and gibbons), there is always the possibility that another male impregnated the female while the male was away (which is common in animals).

These differences produce quite different mating behaviors in males and females. For males, the main driver of behavior is acquiring access to mates, since this is the biggest constraint on male reproductive success. For a female, finding a male with whom to reproduce is trivial, since there are equal numbers of males and females born. For a female, the two most important factors are ensuring that the mate is genetically fit. In some social animals, a female may be able to select a male who can assist reproduction by providing resources to support her offspring during gestation as well as helping to rear the offspring to adulthood. This pattern is common in birds, for example, but is seen in mammals as well (Figure 13.1).

One thing that must be kept in mind when discussing human evolution (and in particular, behavior) is that most of us no longer live in the kind of environment that imposes strong evolutionary pressures. Before the advent of modern societies and agriculture, populations of people regularly starved to death, so the ability to acquire food, and to have a body that used food efficiently, was always a major evolutionary pressure. Furthermore, preagricultural peoples were highly mobile, and carried their possessions on their backs, so physical health was very important. As you think about how natural selection might have affected our ancestors, keep in mind that our behavior and anatomy evolved in very different circumstances than we enjoy today, and in the past simple decisions might well have made the difference between life and death for yourself or your offspring.

As far as we know, *Homo sapiens* have always lived in mixed male–female social groups, and in these groups there are not normally dominant males driving other males away. This kind of social group imposes specific kinds of selective forces on human mating behaviors. In a mixed-sex social group, both men and women are selecting among potential mates and engaging in behaviors that will enhance their reproductive success. For women, with a relatively finite number of offspring that can be produced, a woman must be quite careful about reproductive opportunities, and make sure that those opportunities are maximized. This typically means two specific strategies:

1.   Choosing a mate with a healthy set of genes (displaying physical health and vigor).

2.   Finding a mate who will contribute resources to the development of the child.

Figure 13.1 Two ape mating systems. Gorillas (top) and gibbons (bottom) provide a useful contrast when looking at how evolution shapes mating systems. For gorillas, food is often concentrated in a single (or a few) trees, and the females, whose priority is to acquire enough resources to feed themselves, as well as gestating or nursing offspring, will go where the food is. For males, who do not have to feed gestating or growing young, food is not the priority – mating opportunities are. So the males map themselves onto the females. If a male is large enough he can drive away competing males. Gibbons, on the other hand, have widely distributed food and females never congregate. Males must choose one female, and they tend to form long-term pair bonds.

The determinant of the mating systems is the evolutionary pressure for reproductive success, as reflected in the distribution of food. Both systems are the product of evolutionarily driven rational choices by the apes

Both of these strategies make evolutionary sense, since she has a finite number of opportunities and they must be maximized. To mate with a genetically unhealthy male who is unlikely to provide resource assistance is, from an evolutionary perspective, potentially costly – a female might use a large fraction of her reproductive potential giving birth to, and raising, a child that has little chance of survival to adulthood (whether because of poor health of the child, or because the male does not help provide resources for the raising of the child). So there is, potentially, a substantial cost to mating without careful consideration of the male.

Males also employ two reproductive strategies:

1. Mate with multiple partners.

2. Invest heavily in a single (or a few) reproductive partner(s).

For males, there is almost no cost to reproduction – they don't carry the offspring during gestation, nor can they nurse their young. Males have a virtually infinite potential for repeated mating with almost no expense (beyond the effort required to mate). So males have less incentive to be choosy when assessing females for a mating opportunity. A male can engage in a reproductive bout at the cost a just a few calories. From an evolutionary perspective, if it produces offspring it was successful but if not, the cost was modest. However, humans tend to pair bond, so males can also engage in a second strategy in which they invest heavily in one or a few mates, to help ensure that the offspring of those mates survive to adulthood. Because of the pair bond, when these pairs produce offspring the males typically help with resources for the children. In these cases, the males can be extremely choosy about the genetic and reproductive health of the female, as, in this case, his investment is considerable.

The Role of Competition

Since natural selection predicts that individuals will act in their genetic self-interest, they will often be in conflict with other individuals. We are well aware that males compete against other males; however, males are also competing, in a sense, against their own mates. For example, infidelity is a way of maximizing self-interest while "hedging one's bets." Consider a male that invests heavily in one woman and her offspring. If other low-cost reproductive opportunities present themselves, it may makes evolutionary sense for him to mate outside the pair bond.

For a woman, it may make sense to engage in infidelity if she comes across a man with better genetic health than her mate. So her mate may help raise a child that is not his (genetically). For the woman, she is getting both the superior genes *and* the resources. Consider that a woman always knows that her children are hers, but a man cannot ever really know (unless she has been kept in a tower away from all other men). Male sexual jealousy is a way of attempting to deal with this evolutionary problem (for the male).

Human females are unlike some other primates, in that female humans do not advertise when they are ovulating (in estrus). Female chimpanzees advertise with swellings around their genital area, which alerts the nearby males to the female's fertility and provokes males to compete for access. Female humans do not do this, but this gives human females an advantage, because males cannot know *when* to compete for access to the female. This is likely to contribute to pair-bonding in humans – since men cannot know when women are in estrus, they must remain a more or less permanent presence in the woman's life to ensure that any offspring are actually their own. In some (or most) complex human societies, males have created institutions to help ensure their paternity, generally by limiting the reproductive autonomy of women.

Scientists have examined human mating preferences, and many of them fall in line with the predictions of evolutionary biology. For example, the economic assets of a potential mate are generally more important to women than men when choosing mates, which makes sense since these assets are what help ensure the success of her children (although this varies considerably across societies, and is strongly mitigated by external circumstances – see Box 13.4). Males prefer women in the optimal age range for reproduction (roughly 18–35) because these women are more likely to have successful pregnancies. The age of the man is less important for the woman, because men can reproduce successfully much later in life.

Sexual jealousy also falls in line with the predictions of evolutionary biology. Males tend to be much more jealous of *sexual* infidelity, because it introduces the possibility that they will unknowingly raise another man's child. Women tend to be more jealous of *emotional* infidelity, because of the possibility that the resources of the man will go to another woman's family. It is important to emphasize that these patterns are tendencies, not absolutes. In modern, more egalitarian societies especially, many women find sexual infidelity in their mates to be completely unacceptable, as do many men of emotional infidelity.

Evolutionary Models 2: Aggression

Violence, especially murder, rape, genocide, and warfare, were once seen as more or less uniquely human. However, over the last fifty years, scientists studying primate behavior have found examples of all these acts in various species of primate (Figure 13.2). Much as in reproductive behavior, violence in animals follows patterns that would be predicted by evolutionary models. For example, in chimpanzees, males will occasionally engage in a killing of another male in a way that, in humans, would clearly be called "murder." Typically, the killed male was in competition with the killer for political power in the social group. Sometimes chimpanzee males will gang up on a single powerful male in something analogous to the overthrow of a human leader.

One common animal behavior is infanticide. If a male gorilla or a male lion overthrows the alpha male, one of his first acts is to kill all the young. This does two things for the male: it immediately puts the females into estrus (so he may

Figure 13.2 Ape aggression. Humans are well known to be violent, and at various times in history it has been argued that humans are uniquely aggressive. However, other animals clearly show aggression, and our aggressive instincts are likely to be, at least partially, inherited from our ape ancestors. Ape aggression can be quite sophisticated.

In this image, taken in 2013 at Fongoli National Park, in Senegal, several adult males coordinated to kill the alpha male of the group. The discovery of the pervasiveness of ape violence has led to a reassessment of potential evolutionary origins of the human tendencies toward aggression and violence. A study of violent aggression across all mammals has revealed that primates, in general, are a violent lineage, although canids (cats), rodents, and tree shrews are also violent. Interestingly, tree shrews (*scandentia*) are the next most closely related group to primates, so the evolutionary origins of violence in our lineage might extend back 60 million years or more.

impregnate them) and ensures that any of his protective behaviors for the social group work toward the furtherance of his genetic line. In humans, infanticide is less common, but according to crime statistics stepfathers are forty times (not 40 percent) more likely to kill the mother's infant than a genetic father. (Although infanticide is relatively uncommon in humans, and there are many, many loving stepfathers.)

Sexual aggression has been studied in primates, and is found in cases where males cannot otherwise acquire reproductive access. For example, in orangutans, females will not willingly mate with any other than the dominant male in the region. But the forest does have smaller, nondominant males, and they will, given the opportunity, force themselves onto the female orangutans. The females do resist, but a male is typically much larger. This behavior falls in with evolutionary predictions – nondominant males take the opportunities when they can in order to have a chance to reproduce.

Coalitional Violence

Groups of chimpanzees also engage in organized violence against other groups. These chimpanzee troops are composed of males, who patrol the edge of their territory looking for members of other troops. If they find a male alone, or in a much smaller group, they will attack and attempt to kill him. If they succeed in killing all the males in the nearby troop, they take over their territory and absorb the females into their troop. Exterminating the male competition and acquiring additional females is very much an evolutionarily predicted behavior, and not dissimilar to that seen in human groups, ranging from the armies of Alexander the Great to the Yanomami hunter-gatherers of the Amazon rainforest.

Commonly, in human societies, preliminary to violence is the identification of who is part of the group and who is outside the group. A member of the *in-group* is normally expected to join in the violence against members of the *out-group*. This dividing up into in-group and out-group can be by language, culture, religion, skin color, citizenship, or even which soccer team you support. But once an individual has been identified as a member of the out-group, violence is often justified simply by membership. Wartime propaganda often focuses on the in-group/out-group affiliation.

Evolutionary Models 3: Altruism

Social mammals can have very sophisticated relationships. They often trade resources or favors, or, between males and females, reproductive access. Some exchanges appear to have no current value for the animals but are in the form of a favor that can be "cashed in" later. Social animals tend to have excellent memories and know exactly who owes what to whom. If, for example, a monkey in a troop continually takes favors but never reciprocates, the other monkeys will remember and stop granting favors to him.

The granting of favors appears to be more common among more closely related animals. Siblings, for example, readily exchange favors, even favors that may not be cashed in for long periods of time. This makes sense from an evolutionary perspective, since helping your brother or sister is, in effect, helping your own gene pool, at least partially. Researchers studying animal altruism have found that patterns of altruism follow the patterns of genetic relatedness: you are most likely to help your sibling, somewhat less likely to help your cousin, and least likely to help someone unrelated.

People may still help those unrelated to them but at a reduced likelihood. Evolutionary models predict that when they do so, it is with the expectation of assistance later in return. This is known as *reciprocal altruism* and it forms the basis of politics among humans. When humans receive favors or resources but do not reciprocate they are known as *free riders*, and can be punished by the social group.

Box 13.1 The conflict over the evolution of human behavior

Among biologists, evolution by natural selection is, essentially, universally accepted as the explanatory model for anatomy and behavior in the natural world. And it is also accepted as the explanation of human anatomy. However, some scientists have resisted applying evolutionary explanations to human behavior. The reasons for this have as much to do with history as they do with scientific studies of human behavior, and it has deep roots in the philosophical studies of humanity.

Writers as far back as Plato have tried to understand how much of our behavior is innate and how much is learned. Plato, among many others, argued that a great deal of our knowledge is inherent, whereas writers in the Enlightenment (see Chapter 19) argued that humans were, in essence, blank slates (this was in some ways a revolutionary idea, since one implication was that a poor peasant was not *inherently* inferior to a bishop or king; the difference was in education and the environment).

As discussed in Chapter 12, the eugenicists of the late nineteenth and early twentieth centuries argued that certain human groups were genetically inferior, intellectually and behaviorally. This movement, and the fact that it was so thoroughly discredited, meant that people were leery of making statements about the biology of behavior. But to any biologist it is clear that some elements of behavior, in humans and animals, are the results of their genetic ancestry. Sea turtles never know their parents, so any behavior that a newly hatched sea turtle has (such as crawling into the sea) *must* be a genetically predetermined behavior. Among humans, however, it has been less clear what role biology plays in behavior.

The most important study that established the biological foundation of behavior was based on studies of ants. In 1975 E.O. Wilson, a Harvard professor of entomology, published *Sociobiology*, a 450-page study of the behavior of ants. Ants engage in very complex behaviors, such as building complex structures, raising offspring in a collaborative environment, engaging in organized aggression, and capturing other insect species as "slaves." Wilson clearly demonstrated the biological bases of behavior in a wide variety of ants and other animals, but when, in the final section of the book, he proposed that human behavior is subject to the same evolutionary forces as seen in other animals, a controversy exploded.

Well-known scientists such as Stephen Jay Gould and Richard Lewontin aired their positions in publications such as *Time Magazine*, the *New York Times*, and the *New York Review of Books*. Leery of the return of eugenics, these popular and important evolutionary biologists pushed back against evolutionary

Box 13.1 (cont)

models, arguing that human behavior was too complex to tease out the biological from cultural influences. The idea of biologically determined differences among people, in their view, contradicted ideas of egalitarianism. Proponents of the sociobiology model such as Richard Dawkins and Wilson himself pushed back, denying that humans present any kind of a special case.

A recent and informative entry into this debate came from cultural anthropologist Napoleon Chagnon, who studied the Yanomami peoples of the Amazon Basin. These people live deep in the rainforest and survive as hunter-gatherers, which is the ancestral lifestyle of all humans. In a series of long-term studies of these remote peoples, Chagnon found that Yanomami behavior followed models predicted by evolutionary biology, and by patterns seen in other primates: notably that they were violent, male-dominated, and engaged in aggression patterns against other groups in ways that were seen in chimpanzees. More importantly, he identified the way that natural selection might favor male aggression: more aggressive Yanomami males had more wives (they are polygamist) and offspring than less aggressive males. This model contradicted the idea that culture was the primary determinant of male aggression (although no one rejects the overall idea that culture can have an effect on behaviors like violence). Because Chagnon was a cultural anthropologist, he found strong resistance to his conclusions, and some tried to have him formally sanctioned and even made false accusations to discredit him.

The Debate

So what were the main objections? One problem with any biological interpretation of human behavior or anatomy is that differences can be used to justify social policy. For example, historically almost all warriors have been men. The justifications offered for this are partially biological: men are physically larger than women, and they tend to be more aggressive. This difference has been used to justify keeping women out of combat roles in the military (this logic is flawed: women have been very effective warriors when given the opportunity, and in a modern army the difference in raw physical strength has become less important – see Box 13.5). So some academics are concerned that identifying biological bases for behavior, and determining that there may be differences between groups, may be used to justify repressive, sexist, or racist social policies. Because of the history of eugenics and other forms of group-based oppression, this is not necessarily an invalid concern.

However, the fact that one group may, *in general*, have a different behavioral tendency from another is never a legitimate justification for preventing or

Box 13.1 (cont)

discouraging them from specific goals. This is for several reasons – both philosophical and mathematical.

First, although we may notice differences that give some people advantages in specific fields, we have no justification for preventing them from a attempting to succeed. For example: tall people tend to be better at basketball, because the hoop is 10 feet from the ground. However, there have been several great players who were quite short: Mugsy Bogues was only 5'4" but played in the NBA for more than ten years, as did Spud Webb (5'6"). Although being taller may have made it easier for them to succeed, the most important fact was that they succeeded. It would have been prejudicial to have prevented them from playing. Some studies have determined that women are, on average, better verbal communicators, while men may be, on average, slightly better at math. But the question is how well any individual does in math or English – you could never justify preventing someone from studying one field or another because of the tendency of a group. That *would* be prejudicial.

Second, all of these behaviors or instincts fall on a bell curve, just as other biological parameters do. Men are, on average, taller than women, but the bell curve of male height significantly overlaps the bell curve of female height. So there are many women who are taller, sometimes much taller, than most men. Evolutionarily inherited behavior follows the same pattern. Males may, for example, be more aggressive on average, but there are many aggressive women and many who are much more aggressive than most men. So, to think of a "male" behavior or a "female" behavior is to miss the enormous range of overlapping variation in the behavior patterns of both sexes.

One problem for early sociobiologists is that they overstated the ability of the evolutionary model to explain specific behaviors in individual people. For example, some early advocates in the 1970s argued that sociobiological explanations of behavior would replace (partially or completely) various other disciplines, including psychology, economics, justice studies, and others. This means that they believed it was possible to reduce virtually all the behaviors of any individual to their biological origins. However, it is impossible to attribute any specific biological origin to any but the simplest behaviors or impulses (hunger is obviously biological) for any one person. People are clearly the result of a combination of biological *and* environmental factors, and it is not really possible to point to the causes of any one specific behavior.

Rather, done properly, sociobiologists attempt to explain large-scale patterns across multiple cultures, to determine how evolution shapes our behaviors and to sift the cultural factors from the biological. So, for example, while it is not

Box 13.1 (cont)

possible to explain, in evolutionary terms, why Monet was an Impressionist and Picasso was a Cubist, it is possible to explain why people use artistic expression to communicate, and why males might compete using art as a medium.

The Debate Continues

Today there are still academics who resist Wilson's ideas, although they tend not to be in the field of biology. Behavioral scientists have repeatedly demonstrated that there is a common pattern of behavior shared across all humans (and, indeed, other animals) that is likely the result of our evolutionary past. Humans are animals, and it is always reasonable to be suspicious of proposals that humans are so special that biology does not apply to them – we have seen again and again that humans are very much animals.

Box 13.2 Morality

One of the things people find so confounding about studying the evolution of behavior is that, unlike anatomy, we tend to think of behavior in moral terms. Having red hair, for example, is not immoral, but having multiple spouses is seen as immoral (and illegal) in many cultures. This is to be expected – if we are to have laws that regulate behavior, and punish or prevent such things as infanticide or rape, we need to be able to frame behavior in moral terms.

Yet it is clear that many of the instincts for acts we see as quite immoral do have evolutionary origins. Chimpanzees engage in killing that we, in a human context, would consider murder. So, when a human murders another, the instinct likely comes from our deep evolutionary past. Does the fact that the violent instinct has an evolutionary origin offer any kind of a defense? Would we give the murderer a pass?

The answer is a resounding "no." The English philosopher David Hume observed that nature only tells us what "is," never what "ought to be." The fact that nature occasionally rains hot balls of lava on us does not mean we can pour molten rock on our fellow citizens. Hyenas tear meat out of each other's mouths – we cannot use that as justification for robbing one another.

Nature is rife with examples of organisms helping one another – symbiotic organisms are extremely common (our own intestinal tract is full of helpful bacteria that survive by breaking down our food). Parasitic organisms are just as common, as are predatory animals. Some animals seem unusually cruel – *Ampulex* wasps

Box 13.2 (cont)

(jewel wasps) reproduce by planting an embryo in their hosts (often a beetle), and those juvenile wasps eat their way out of the paralyzed, living hosts. Despite our feelings about these insects, we don't really regard them as "immoral," and none can offer any guide to how humans must behave toward one another.

Ultimately, humans must make rules about behavior that reflect how they wish to be treated by other humans, or that reflect their moral judgments. We outlaw murder (and judge it to be morally wrong), but that law and moral judgment is entirely human; it in no way reflects some "natural law" because there is no deeper natural law. Some laws seem to be relatively universal across cultures (murder and theft are pretty much universally outlawed). But some are culturally specific: homosexuality was common and legal in ancient Greece and Rome (among other cultures) but was outlawed in many countries after Christianity and Islam spread throughout Africa and Europe. But homosexuality is common in nature, so outlawing it because it is 'unnatural' is both incorrect and philosophically flawed.

Box 13.3 Gender and violence

The human inclination to aggression is innate and comes from a deep, ancestral pattern of behaviors that are probably hundreds of millions of years old. Animals from across the taxonomic spectrum exhibit aggression, often in predictable patterns. In mammals, especially, we see patterns that directly inform on human aggression. Females are known to be extremely aggressive in defense of their young. Males are known to be aggressive when competing for access to mates. We see both of these behaviors in humans.

One pattern that is generally consistent across mammals is that males tend to be more aggressive than females. This is true of humans, where men commit the bulk of violent crimes and are statistically far more likely to be involved in a variety of violent confrontations. The larger body size of males (roughly 40 percent in terms of muscle mass) reflects the selective force of male-on-male aggression.

The general tendency of men to be more aggressive has been used as a justification for preventing women from serving in the armed forces in many countries. However, it is important to keep in mind that these patterns reflect general tendencies rather than black-and-white differences (see Figure 13.3). Aggression, like most genetically determined traits, falls along a distribution. In terms of men and women, the male distribution and the female distribution overlap, so that some women are more aggressive than some men. In fact,

Box 13.3 (cont)

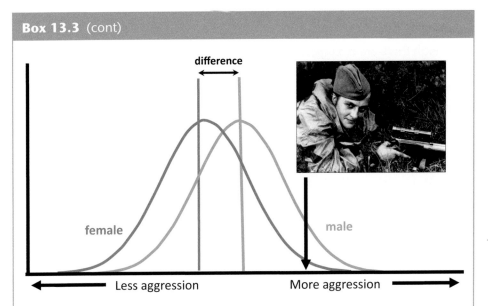

Figure 13.3 The bell curve describes most physiological characteristics of biological populations (height, strength, speed, foot size, hair thickness, body mass, etc.). When we look at male and female human body mass, for example, on average men are larger than women. However, distributions of male and female body mass strongly overlap.

Behavioral traits also fall on bell curves. When looking at males, it is clear that, across most mammals (with a few interesting exceptions, such as hyenas), males are more aggressive. However, there is considerable overlap between males and females for almost all anatomical traits, and this appears to apply to behavioral traits as well.

History has many examples of highly aggressive females, and the photo shows Lyudmila Pavlichenko, a Soviet sniper who killed 308 German soldiers in 1941–1942, during World War II. She is likely to have been more aggressive than the majority of males (she was known for her fearless aggression).

historically, there have been many successful female soldiers when given the opportunity. One of the most successful snipers in World War II was Lyudmila Pavlichenko, who killed 308 German soldiers; many women served as bomber pilots, tankers, and in other combat positions in the Soviet armed forces. Other countries, including the United States, have started to employ women in combat roles, and the women have proven themselves capable, possessing the necessary aggression in combat situations.

Some groups have opposed putting women in combat roles, but these nowadays reflect religious objections or other philosophical perspectives on the roles of women in society, and are not based on the performance of the women in combat. While men will probably continue to serve in the majority of combat-related positions, there are enough women with natural aggression that they are capable of serving effectively in virtually all combat roles in a modern military.

Box 13.4 The role of culture

Although there are biological bases for many human instincts and behaviors, it cannot be questioned that the majority of differences between cultures are simply artifacts of cultural history. For example, the differences between the languages in western versus eastern Europe reflect the extent of the Roman Empire – the empire was strongest in modern-day France, Italy, Spain, Portugal, and Romania, and the modern languages of these countries all descended from Latin. There is no particular way for biology to make a meaningful statement about these languages, and this is the case for most cultural differences in clothing, hairstyles, food, dance, etc. However, sometimes differences in behavior reflect the interplay of biology with culture.

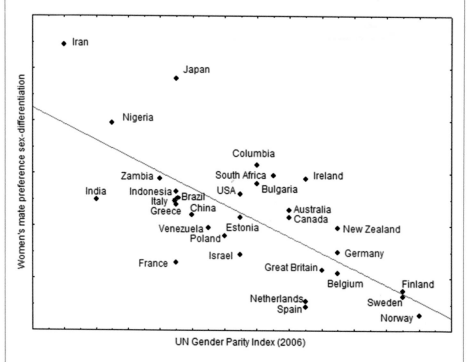

Figure 13.4 Scientists have studied the effects of social conditions on decision-making in an evolutionary context. Here, data from studies in 2006 and 2012 illuminate the effect of economic conditions of mate choice. Women in countries with less gender parity tend to choose men with better economic prospects. This makes sense from an economic perspective, because the women in less egalitarian countries would need to worry more about the availability of resources for their children, whereas the opposite would be true in more egalitarian countries. So the underlying biological impulses are expressed as a function of interaction with environmental circumstances. This result suggests that other impulses (aggression, for example) can also be mitigated by altering social conditions.

Data: Zentner M and Mitura K. 2012. Stepping out of the caveman's shadow: Nations' gender gap predicts degree of sex differentiation in mate preferences. *Psychological Science* 23(10): 1176–1185.

Box 13.4 (cont)

One of the most interesting and important patterns has been recently discovered by social scientists looking across cultures. They have discovered that the tendency of women to select men with good financial prospects is strongly determined by the degree of economic equality in the country (Figure 13.4). In countries like Iran, Japan, and Nigeria, where women make far less than men, women prefer men with strong financial prospects. However, in countries like Finland or the Netherlands, which offer women more economic opportunities, women's preferences shift, and they no longer need to seek men with strong financial resources. This makes sense from an evolutionary perspective, because in more egalitarian countries, and those with good social safety nets, women who can earn good salaries do not need to worry about their mate's economic prospects to ensure the security of their children. So, although the basic instinct of protecting and raising their children is the same, social circumstances can change the way these impulses are expressed.

The way biology and culture interact is important, because it shows how flexible human behavior actually is. Although there are certain underlying biological drivers of behavior, the circumstances of the environments determine how they play out. So biology is emphatically *not* destiny – people react in (generally) rational ways to their environments based on their evolved instincts, but societal circumstances can change the calculation of what is, or is not, rational.

Summary: The Evolutionary Model

Many of the behaviors once thought to be unique to humans, including warfare, sexual aggression, altruism, and culture, are now known to be present in other animals. Needless to say, human behaviors are much more complex than animal behaviors, but, since humans are animals, human behavior *must* be seen as a form of animal behavior. And although human behaviors tend to be more complex, and culture unquestioningly has a strong effect on human behavior, the underlying drivers are no different than those seen in other animals.

Chapter 14: Brain Evolution

You might think that one thing that distinguishes animals from plants and other life-forms is brains, but the simplest animals are sponges, and sponges have neither brains nor nerves. Because they have no muscles, there is nothing they could even do with nerves. Sponges, which had arrived by 600 million years ago, are the most ancient animals, but not long after animals started to evolve nervous systems.

The basic cellular unit of the nervous system is the neuron, and all nervous systems on Earth, from the jellyfish to the spider to the sea turtle to the gorilla, have nervous systems composed of various arrangements and numbers of neurons. Neurons combine electrical and chemical signaling systems to transmit information from one part of the animal's nervous system to another, and to take in sensory information from the environment and send commands to control the activity of muscles and glands.

Nervous systems are a prerequisite for any kind of intentional movement. For example, a jellyfish needs to be able to coordinate the pulsing of its bell to make it swim. A jellyfish has no brain, so there is no central location where decisions are made. Rather, the organism is wired to respond in specific ways to specific stimuli: it moves toward light and away from physical contact. Jellyfish and other primitive animals, such as Cnidarians and sea stars, have a nerve net, in which the nerves are all connected and information can pass to the muscle cells (Figure 14.1). All the muscle cells respond to the stimulus as the information passes through. Individual parts of the organism may respond to touch, but the response is local and in direct response – no "decision" is made.

This is unlike behaviorally complex animals, in which, somewhere in the body, there is a bundle of nerve cells that receive signals in and then pass signals out. This system, in which nerve cells spread outward from a large neuron bundle, is known as a *central nervous system*. Central nervous systems are very old – they are seen in the animals of the Cambrian Explosion, and are visible in trilobites. They are characteristic of all later animals that have bilateral symmetry, including arthropods, nematodes, mollusks, and vertebrates, which all have a central nervous system (this taxonomic group is called

Figure 14.1 Distributed nerve systems are seen in animals that do not have bilateral symmetry and a "head" and a "tail" end. The earliest nerve systems are those seen in jellyfish (left). Jellyfish have no brain – there is no centralized bundle of neurons. Rather, in these animals there is a "nerve net" in which all nerves are connected. Jellyfish have no way of knowing which nerve received the stimulus, and no way to make a specific response to touch on one point, because the nerve net passes the information to all muscle cells.

A much more sophisticated decentralized nerve system is seen in the octopus (right), and some other cephalopods (notably squid). Octopus nerve systems are decentralized, with large bundles of neurons distributed around the entire body, especially down into the tentacles. There is one large bundle in the center of the octopus, but there are more neurons in the nerve bundles distributed down the tentacles.

Bilateria). These bundles of neurons tended to congregate at one end of the organism, which resulted in the cluster of sensory organs (eyes, nose, ears) developing at a "head" end of the organism.

The bundle of neurons that acts as the central switching and processing station is, of course, the brain. The brain takes information in and tells the animal how to respond. This is, in many ways, analogous to the computer in a self-driving car: information on location and speed, and scans of road obstacles, are passed into the computer. A series of decision-making algorithms (if–then decisions) use the information, and then instructions are distributed out to the steering wheel, gas, and brakes. In animals, as in computers, the bigger the brain, the more complex the scenario that can be handled. A simple animal, such as an ant, with only 250,000 neurons, has only a few problem-solving behaviors programmed into its brain. As brains get larger and more complex, more complex problem-solving behaviors emerge. A guppy has 4 million neurons in its brain, and a frog has 16 million, whereas a larger mammal, such as a raccoon, with 2 billion neurons, has many more possible solutions to the problems it encounters. The additional neurons allow for the storage and processing of complex information, so a raccoon can observe problems and decide on the best way to solve them (anyone who has had to keep raccoons out of their garbage cans knows how inventive they are).

Brain Anatomy, Simplified

But the brain is not only for making decisions. Many parts of the brain are devoted toward general body maintenance in which the animal has no conscious decision-making: keeping the heart pumping, constricting blood vessels when cold, dilating them when warm, constricting the muscles of the digestive system to push the food through the intestines, etc. This is why large animals tend to have larger brains than smaller animals. A giraffe simply has more cells in these systems that have to be maintained than a tree shrew, so the parts of the nervous system keeping them going also have to be larger.

There has been academic debate about the patterns and evolutionary origins of the parts of the brain, but, broadly speaking, the vertebrate brain can be broken into three basic anatomical parts:

1.   The brain stem helps maintain respiration, heart rate, and blood pressure, and acts as a main switching station for motor nerves and pain nerves passed from other parts of the body. This is considered to be the most ancient part of the brain, and was certainly present in very ancient animals during the Cambrian Explosion.

2.   The cerebellum helps the body process data received from the spinal cord and other brain structures to ensure fine or complex motor control. This part of the brain has more neurons than any other part in mammals, because of the variety of complex motor behaviors in this group.

3.   The pallium is the part of the brain that deals with higher function (thinking). In mammals this is known as the cerebrum, most of which is composed of the cerebral cortex (the *neocortex*). In humans, this is the largest part of the brain, but this is not the case in all vertebrates, especially reptiles. In general, when we want to know if an animal has greater problem-solving abilities, we examine the number of neurons in the cerebral cortex. For example, an elephant brain has 257 billion neurons overall (including the cerebrum and brainstem), which is far more than the 86 billion found in the average human brain. But the elephant neocortex has only 5.5 billion neurons, whereas the human has some 16–20 billion. This larger neocortex is one reason humans have greater problem-solving abilities. The rest of the elephant neurons are mostly in the brainstem and cerebellum, which reflects the larger overall size of the animal.

Within these three main parts, there are many smaller structures that have specific functions. These parts of the brain often reflect the adaptations of the animals that possess it. There are specialized parts of the brain for processing all kinds of sensory information. For example, the olfactory bulb is where data from the olfactory (smell) cells are analyzed, and animals with a keen sense of smell (such as dogs) have quite large olfactory bulbs. A chance preservation of the shape of the brain of *Tyrannosaurus rex* shows us that the olfactory bulb of this fearsome beast was larger than the rest of the brain. Because of this, some paleontologists have insisted that *T. rex* was a scavenger rather than a hunter, and that it used its olfactory bulb to identify the smell of nearby rotting carcasses (although this is definitely a minority view in paleontology). Modern scent-hunters, like dogs, also have large olfactory bulbs, whereas cats, which are vision-oriented ambush predators, have smaller olfactory bulbs. Humans, and primates generally, also have smaller olfactory bulbs.

Primate and Human Brain Evolution

For reasons we do not yet understand, primates have greater neuron density in the neocortex than other mammalian lineages. For example, the brain of a macaque (a monkey) and a capybara (a very large rodent) have brains roughly the same size – about 80 cc, but the macaque has 6.3 billion neurons whereas the capybara has only 1.6 billion. Over time, the primate lineage trended toward larger and larger size brains (see Box 14.2) as the primates themselves became larger (Figure 14.2). In the ape lineage, brains are quite large – the average chimpanzee skull is close to 350 cc.

Box 14.1 What makes brains become large?

As humans who are highly dependent on our large brains, it seems intuitive that a large brain is good to have. In general, this is true, as they are extremely useful for solving a variety of problems that animals face. However, not all animals will benefit from a large brain, and only some animals evolve large brains.

Certain types of animals tend to have larger brains than others. Hunting animals (e.g., carnivores) have larger brains than their prey animals. An antelope or deer only needs enough of a brain to find leaves or grass to eat. Any more than that is essentially expensive surplus. Remember – brains are metabolically expensive, so having more brain than you will use is actually a disadvantage, especially when food becomes scarce, as when droughts occur. So the trick for an antelope is to have a large enough brain to solve the various problems it faces (finding mates, avoiding predators, finding food, etc.) but no more than that. Of those problems, finding food is pretty easy – grass and leaves are all over the place and they don't move.

Box 14.1 (cont)

But for the predator, the problem of finding food is more complex. A predator has to predict where the prey will be, and has to have a suite of behaviors that will allow it to get close enough to strike. In all animals, there are parts of the brain that process and store olfactory information (smell). In canids (dogs/wolves), these brain structures are especially large so the dog can stalk its prey. In felids (cats), there is an adaptation toward silent movement, as well as especially sensitive visual processing for hunting in low light. These tools, as well as the need to understand, at least to some extent, the behavior patterns of the prey, mean that carnivores needed to evolve a larger brain than the prey animals. The high-quality food (meat) offsets the additional metabolic demand of these larger brains.

Other factors can also drive increased brain size. The ability to manipulate the environment allows animals to gain advantages from the increased problem solving enabled by a larger brain. Humans and other primates, with their strong, opposable thumbs, fall into this category, but so do other animals. The elephant, for example, has a flexible and powerful trunk that allows it to manipulate its environment in many complex ways. Elephants and rhinoceroses are both large vegetation-eating animals. The elephant is roughly twice as heavy as a rhinoceros, but its brain is six times as large. Without some way to manipulate the environment, there is probably no reason for the rhinoceros to evolve a metabolically expensive brain that gives it no real advantage. In other words – what would a smart rhino do that a normal rhino can't?

The other types of animal that tend to have a large brain are the ones that live in large social groups. Mammals that live in groups have complicated familial and political relationships. There are often hierarchies of power and favors are exchanged. All of this political information is critical to the survival of these social animals – studies have shown that animals with strong social ties tend to thrive, whereas more isolated members of the group suffer more stress and are less likely to access shared resources. Social animals are good at remembering and predicting the behaviors of other animals in the group – most of us can remember our favorite teacher from elementary school, or the mean kid from third grade, even decades later, and know the types of behaviors they would have liked or disliked.

This powerful social memory is the product of natural selection helping us compete with other individuals in our groups. People with particularly good social memory often thrive in modern human society, and some politicians are famous for remembering people's faces and names years or decades after meeting them.

Box 14.1 (cont)

The most intelligent mammals are highly sociable: elephants, whales and dolphins, and primates. The larger the group, the more information has to be stored. However, brains are flexible, so even if natural selection acted to increase brains size due to the requirements of group size, the larger brain would have enabled more sophisticated problem solving. In humans, tool sophistication tends to track the increases in brain size – for example, we don't see projectile weapons (spears) until the brain is roughly 1,000 cc.

It is important to remember that brains are anatomical adaptations, just as limbs and organs are, so they are the products of natural selection. Our brains are particularly large, but there are quite a few other animals that acquired large brains for precisely the same reasons we did, and they received the same advantages.

Box 14.2 Nonmammalian intelligence

Bird Intelligence

You may have heard the insult "bird brain" thrown around among your friends (or enemies). The comparison is meant to imply that anything with a brain like a bird is unintelligent, and it is definitely not a compliment. Up until only the last few decades, bird intelligence had been almost universally dismissed, especially in comparison with mammals. However, brain researchers have been reassessing the avian brain, and it turns out that there is quite a bit more to the bird brain than we had thought.

One interesting fact is that, like primates, avians have very neuron-dense brains. In fact, the avian brain is as dense with neurons as primate brains (or denser), so that the brain of a macaw (a type of parrot) has as many neurons as the brain of a baboon. The corvids (crows and magpies) also have large numbers of neurons – ravens (at 1.3 billion neurons) have as many neurons as a squirrel monkey.

The problem-solving abilities of birds is now appreciated far more than in the past. Birds are known to use tools in much the way monkeys do – picking up a stick or rock to dig or break into something. In tests of consciousness, corvids have been seen identifying themselves as individuals in much the way elephants and whales do. And birds are highly trainable, which means that they have a very flexible intelligence.

Box 14.2 (cont)

Some birds are smarter than others, but this is seen in mammals as well. This reappraisal of bird intelligence has led scientists to go back and try to figure out which of the dinosaurs were more intelligent (since birds *are* dinosaurs). The same rules of the evolution of intelligence that apply to mammals would have applied to dinosaurs as well. Predators have larger brains than their prey, social dinosaurs have larger brains, and dinosaurs with prehensile forelimbs (functional hands) have larger brains. The saurischian dinosaur *Troodon*, with a relatively large brain–body ratio, has been proposed as the smartest dinosaur (veloceraptors are also candidates for intelligent dinosaurs). Although we will never know, some scientists have argued that the general increase in animal intelligence means that, if the meteor did not hit the Earth 65 million years ago and wipe out the dinosaurs, sooner or later a dinosaur would have evolved human-like intelligence levels. It is an intriguing possibility, but one that was eliminated by the K–T Extinction.

Cephalopod Intelligence

Another group that has recently received attention for their intelligence is the octopus-squid group. Octopuses, especially, have been observed solving very complex problems in laboratory and zoo settings (including opening food jars by twisting the tops off), and are known to use tools. These animals have relatively large brains for their body size, on par with mammals.

The selective forces that caused these invertebrates to have large brains appear, to an extent, to be similar to those that generated large brains in mammals. Cephalopods exist in a complex environment as predators, subsisting on a wide variety of prey. Importantly (and unlike the fishes), they have appendages that can perform complex tasks. This would be analogous to the hands of primates, the trunk of elephants, and the beaks and claws of birds. With coordinated appendages, selection on a larger brain can benefit the animal, as it can solve problems that would not be accessible to a nonappendaged animal. Sociality is poorly understood in cephalopods, but there may be some degree of group interaction, which is often characteristic of larger-brained animals.

One interesting difference between the brains of terrestrial vertebrates and cephalopods is the organization of the brain. Birds and mammals have a centralized nervous system, with the main neurological bundle being the familiar brain. But in octopuses, the organization of the neurological system is completely different. Although there is a central brain, it does not possess the majority of the neurons in the nervous system. Most octopus neurons are

---

Box 14.2 (cont)

distributed in clumps around the cephalopod body, especially in the tentacles. The tentacles, under the control of these "clumps," can operate independently of the rest of the organism. This means that an octopus can reach out with a single tentacle and receive information that the rest of the octopus may never get – the neuron bundle in the tentacle is actually doing the thinking independently of the octopus brain. However, some information does pass back to the centralized brain, because in an emergency, if a tentacle is damaged by a predator, the entire octopus will try to escape. The decision-making is both local and central. In fact, the octopus body is so decentrally organized, with neurons distributed around the body, that there is not really a clear division between the brain and the body.

Interestingly, this has implications for our ideas of consciousness. Our brain is generally considered to be the location of our consciousness. We are used to thinking of the brain as the central part of our intellectual identity – it receives information from the body and sends information to the body, but the vertebrate brain is, in a sense "separate" from the body; you could at least imagine transplanting one's brain into a different body. But the octopus is not like this – the body *is* the brain and there is no way to separate the two. There is probably no single location in an octopus where the consciousness exists (whatever octopus consciousness is like). Their consciousness is distributed throughout the brain and neuron bundles and strands all over the octopus body.

One prominent researcher has argued that understanding cephalopod intelligence is as close as we can get to understanding alien intelligence. There is no particular reason why we should expect intelligent alien life-forms to have a central nervous system, and the octopus, with its distributed neurological network, is a completely different way to have intelligence. Nature often works in ways we don't anticipate, and the octopus has helped us understand this critical fact.

Early humans had brains roughly the size of a great ape. Humans and apes separated between 6 and 8 million years ago, but even at 4.5 million years ago *Ardipithecus* had a brain that would fit within the normal range of variation for chimpanzees. But slowly the brains of hominins became larger – *Australopithecus*, at 3–3.4 million years ago, had a 450 cc brain, which would not be uncommon in a gorilla, and by 2 million years ago the brain was larger than 600 cc. At that point, the trajectory really started to take off, as humans became more and more dependent on their brains. At 1.8 million years ago the brain is more than 800 cc, and it accelerated steadily over the next 1.5 million years to ~200,000 years ago, when the brain reached its modern size (~1,350 cc).

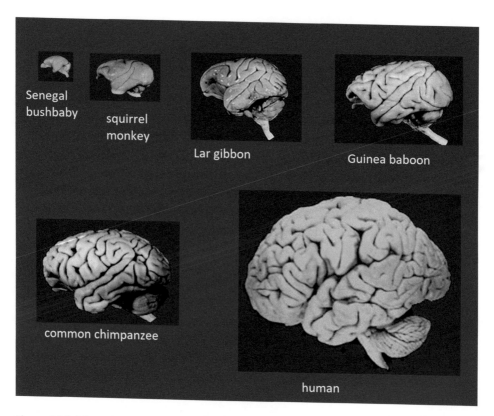

Figure 14.2 Primate brains are remarkably similar to each other. A primary difference seems simply to be size. Even much of the human brain anatomy can be understood as a simply scaled-up primate brains. Under natural selection, size is one of the easier changes to make.

There are two strong hypotheses for how the brain manifests consciousness. One is that it simply "emerges" once a brain become large enough and acquires enough neurons. Since consciousness is not "anywhere," maybe it is "everywhere" (see Box 15.4). Another strong hypothesis is that a social brain selects for the ability to interpret the individual intentions of others so that, at some point, one's own intentionality and individuality becomes recognized. Interestingly, under both hypotheses consciousness is a byproduct of selection for something else (large size in one case, sociality in the other).

Box 14.3 Is there a threshold for human intelligence?

Modern human intelligence has fascinated scientists and philosophers for hundreds of years, and one of the major questions that researchers have long asked is: When and why did humans acquire the modern suite of intellectual abilities? For about 4 million years, the brains of human ancestors have been increasing in size, and we see a steady increase in the sophistication of tools and technologies over this same time period. Stone tools become more complex over time, and other technologies, such as spears, structures, and art

Box 14.3 (cont)

all appear at various points in the fossil record, roughly tracking the overall pattern of increased brain size.

Around 250,000 ago our skulls reached their current shape and size. However, the technology associated with these early modern humans is not different from what we see at 300,000–750,000 years ago, associated with earlier human species. These earlier humans (sometimes called *Homo heidelbergensis*) have brains a bit smaller than ours, but they are still quite large (larger than 1,200 cc), and in fact they have some overlap with modern human brain size. Researchers had expected that there should be some kind of an identifiable change in technology once modern humans evolved. In a way, this is a form of self-fascination or egotism: since we are so intelligent today, we must have been smarter than anything that came before. But this is not what the archaeological record shows. Rather, the earliest modern *Homo sapiens* had technology identical to what had been prevalent for the previous 500,000 years.

Paleolithic archaeologists had long seen a shift to more advanced technologies around 100,000 years ago, including the first appearances of fishhooks, shell-fishing, long-distance exchanges of goods, beads, and art. Since the modern skull had evolved by 250,000 years ago, the best way of explaining this seemed to be that there was some major change in the brain at 100,000 years ago that was not apparent from the fossils. They hypothesized a "rewiring" of the brain that would have enabled all of these innovations. However, over time, as archaeologists found more and more sites with the various technologies, it became apparent that the innovations came on slowly rather than suddenly.

In fact, the steady growth in technologies, rather than a sudden appearance, maps very well onto what we now know about growth in populations. And as populations grew, it would have been easier for innovative technologies to spread. Under this model, the appearance of new technologies reflects their spread rather than their invention. It is extremely unlikely that the first appearance of any innovation would be preserved in the fossil record, so new technologies would only ever be preserved once they were relatively common.

There are two reasons why scientists like this new model so much. The first is that it doesn't require us to invoke some sort of invisible change in brain architecture. There is no reason why we should accept an explanation that the brain suddenly evolved special abilities 100,000 years ago. The other reason scientists like this model is that it reflects what we see in the world today. We have plenty of documented historical examples of technologies spreading, and

Box 14.3 (cont)

they spread faster when the populations are denser, and more slowly or not at all when populations are thinner. So it is consistent with observations of modern human behaviors.

Finally, it reinforces the idea that modern humans do not represent some special case of evolution. Humans have steadily increased in brain size and, while we do have slightly larger brains than earlier species, and are probably slightly better at problem solving, the immediately previous humans were extremely capable problem solvers (they were very successful, surviving in varied habitats across multiple continents for more than a million years). We are not a "special case" of evolution, and we just happen to be the latest model of a trend that has been operating for more than 4 million years.

Box 14.4 Do we only use 15 percent of our brains?

Very broadly speaking, there are two types of tissue in the neocortex. You have probably heard of "gray matter" – this is the part of the neocortex where decisions are made and data are stored. The cells in the gray matter are neurons. The other type of tissue is known as "white matter," which is primarily composed of the parts of neurons that allow for connections between distant cells (axons), plus other cell types (glia) that perform metabolic functions in brains (Figure 14.3). The human brain actually has more white matter than gray matter, and some simple math will help explain it. Imagine you have just two neurons in a brain (although no brain has just two neurons!). Just one axon is necessary to connect them. If you have three neurons you need three axons. But then the number of axons rapidly outstrips the number of neurons: four neurons requires six axons, five neurons requires ten axons, six neurons requires fifteen axons, and so on (the formula is actually: $axons = \frac{neurons \cdot (neurons - 1)}{2}$. If you have 1,000 neurons, you require 499,500 axons to connect them all. And as if that's not bad enough, as axons have to connect among neurons at greater distances, they have to either become larger or better insulated in order to rapidly send messages, taking up even more space. Most neurons are connected to other neurons, so the brain doesn't require all neurons to be connected via axons, but in a brain with many neurons there need to be many, many axons. The human neocortex is actually, overall, about 30 percent more axon than neuron, but this is essential if the neurons are able to communicate across the volume of the brain. There is so much large-scale

Box 14.4 (cont)

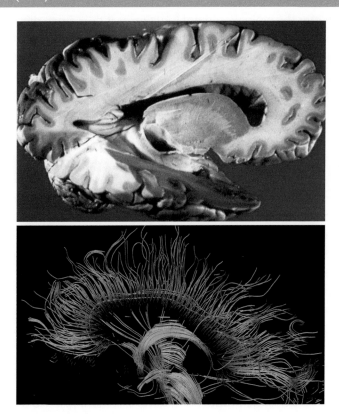

Figure 14.3 The image on the top is a cross-section of the human brain showing the gray matter and white matter. Gray matter is largely composed of neurons, and white matter is axons and other types of brain cells that support the function of neurons. Although neuron count is often used to measure encephalization, as well as provide an estimate of animal intelligence, white matter provides a crucial role. White matter provides the critical networking that allows the neurons to communicate across the brain. The Diffusion Tensor MRI of a human brain on the lower image shows the enormous number of connection paths across the human brain.

communication in the human brain that we have evolved some special circuits for communicating between major sections. The *arcuate fasciculus* is a major circuit connecting the frontal and temporal lobe of humans, which is modified from a similar white matter pathway found in ape and monkey brains, potentially to enhance our ability to use and understand language.

You may have heard the urban legend that we only use about 15 percent of our brains. This probably comes from older brain studies examining white matter and gray matter. Some of these studies calculated that gray matter was only 15 percent of the brain, and this is likely where this idea originates. But of course it is not true – we use both gray matter and white matter, and without

---

Box 14.4 (cont)

the white matter the brain would not function as it does (with different parts of the brain rapidly communicating with other parts). As metabolically complex to maintain as the brain is, and as long as it took for natural selection to generate such a large and complex one, we would never have evolved such an expensive organ unless we used all of it!

Box 14.5 Are humans smart?

We are used to thinking of humans as particularly smart – smarter than any other animal. After all, our brain is the largest relative to the body among all animals, and we have demonstrated our ability to exist anywhere on Earth, and even in space. In the popular media you will sometimes see the human brain characterized as "limitless." Motivational speakers, especially, will sometimes say that "you can do anything with your mind."

Unfortunately, this runs up against reality. There are tests of intellect and memory that chimpanzees can do better than any human. In tests of geographic or locational memory, chimpanzees easily beat humans. Other animals can certainly perceive and analyze things like smell, taste, shape, sight, and sound better than we can.

The human brain is the product of natural selection, just like our livers, eyeballs, gluteus muscles, and finger bones. All of those parts of the anatomy have limitations – we wouldn't say that there is no limit to gluteal muscle strength or liver enzyme production. The same applies to what our brain can do. It is trivially obvious that we cannot analyze as much scent data as a dog – we don't have large enough olfactory bulbs.

Pattern Recognition

But there are also analytic limits. There are certain things that we do poorly. One of these things involves patterns. Natural selection acted to give us extremely strong pattern-recognition abilities because, for much of our evolutionary history, we were quite small and had to worry about large-bodied predators like lions and leopards. Imagine you are a 4-foot-tall hominin 3 million years ago. If you were looking around the forest and saw what looked like a leopard, you would scamper up a tree. If you were wrong, well, you wasted twenty seconds and about as many calories. But imagine the reverse. If you failed to see the pattern and a leopard was actually there, you were likely to be the leopard's next meal. So natural selection "selected" for pattern

Box 14.5 (cont)

recognition and didn't worry too much about seeing patterns where none existed – it erred on the side of seeing patterns.

Today, in our modern world, with information all around us, it is easy to see patterns where none exist. There are innumerable examples, but to give one – in 1976 NASA sent the *Viking* probe to Mars. Photos were transmitted back of the Martian surface for the first time in high detail. One hill (out of thousands) looked vaguely like a human face. Some people immediately assumed that it was evidence of now-extinct life on that planet. This over-recognition of patterns is known as *apophenia*, and it is the reason why people see faces in tortillas and toast (incidentally, subsequent Mars probes showed that the "face" was a coincidence of sunlight at a particular angle).

This is also the reason people believe conspiracy theories. There are small bits of information that seem to coincide, and people create a conspiracy theory to link these unrelated pieces of information together. Because of the instinct to find patterns, people have a need to find an underlying common explanation, even though there may be no actual evidence of the linkage.

The upside to the evolution for pattern recognition is that it may be the reason humans like music and art, especially art that is not directly representational. The universal appreciation of instrumental music, impressionism, expressionism, and abstract forms of graphic art may represent the fact that our brain responds to patterns, even when they are not trying to look or sound exactly like something from the natural world.

The Trouble with Time

Another thing people have trouble with is understanding time. We, as a species, tend to make decisions that provide immediate rewards, even if there is significant long-term cost to those decisions. The incredible burden of credit card debt shows how people will make purchases that provide immediate gratification, even though, with the interest rate, the long-term cost of the purchase might be enormous. People will also eat foods that, while gratifying when eaten, over the longer term may have significant health consequences. In all likelihood, in our evolutionary past, the difficulties of predicting future events meant that there was relatively less selective pressure on long-term planning and delayed gratification. However, today we can make long-term plans very accurately, and our refusal to make decisions in accordance with this shows how maladapted we are to our current environment.

In 2017 the Nobel Prize in Economics was awarded to Richard Thaler, an economics professor at the University of Chicago. Thaler had identified the

Box 14.5 (cont)

problem of short-term decision-making in human economic policies (he calls it *temporal discounting*) and had advocated policies to prevent people from making bad decisions. He convinced large corporations to create retirement plans for employees that were automatic – they didn't have to decide to join the plan, rather they had to decide to quit the plan. This took the short-term impulsive behavior out of the picture. He is credited with creating more than 50 billion dollars of retirement funds for employees of these companies, by preventing them from making bad short-term decisions. In essence, he won the Nobel Prize by preventing people from following their natural, inherited instincts.

The Limits of Human Intelligence

There is no way to predict what we will be able to accomplish, or not accomplish, in the future. But the struggle of modern scientists with generating a comprehensive "grand unifying theory" out of some of the ideas of the subatomic Universe – string theory, Brownian motion, "spooky action at a distance," quantum mechanics, and others – suggests that there may be limitations to human analytic powers that may constrain what we are able to understand. The human brain is the product of natural selection, much as the brain of a dog is. If you try to teach algebra to your family dog, it will never understand. Humans likewise will have constraints, even as we may, *because* of these constraints, have trouble understanding what they will be. After all, a dog is not really able to understand that it does not understand algebra. We may not be able to understand that we do not understand.

This does not mean that we should ever stop trying to understand the Universe, but it does mean that we should have a little humility when framing ourselves in the Universe. We may be nothing more than a "smart dog" compared to the complexities that exist in the Universe. We have about twice as many neurons in our neocortex compared to our closest cousins, chimpanzees. Imagine how much more a creature with twice as many again would be able to understand effortlessly. They might learn string theory and quantum mechanics as children. We would appear as simple-minded as chimpanzees to them, and perhaps they could never hope to teach us their advanced knowledge, even as we could not teach algebra to a dog or chimpanzee. While we may never meet life-forms from other parts of the Universe, they might be like early life on Earth (cyanobacteria), with no neurons, but they might also have something that far exceeds our abilities. Rather than being the most intelligent organism, we are likely to be just be one point on a continuum that ranges far beyond what we see on Earth.

Increase in brain size was probably enabled by our transition to eating meat and animal parts some 2.6 million years ago. The brain is a very expensive organ, metabolically speaking. A chimpanzee uses roughly 8 percent of its metabolism supporting the 350 cc brain, and a human uses ~25 percent of its metabolism keeping the brain fed. Today, we have little trouble finding enough calories, now that we have industrialized agriculture with domesticated grains and animals. But at 3 million years ago, *Australopithecus* was not a meat-eater, and with a diet of natural vegetation could not have supported a large brain. At 2.6 million years ago, humans had started to scavenge meat from kills, and today archaeologists find cut marks on bones from those temporal horizons, along with early stone cutting tools. This is when the human brain gradually started substantial evolutionary growth.

As humans became more and more dependent on meat, their brains grew, enabling more sophisticated techniques for acquiring food. These more sophisticated techniques (such as organized hunting and projectile weapons) would have enabled the acquisition of more food, enabling more brain growth. This cycle became a feedback loop, and humans became more and more *encephalized* (large brained) over the next 1.5–2.0 million years. Humans are juveniles for longer than other mammals and this period has been extended (under the force of natural selection) to allow for the additional brain growth necessary to acquire the additional size. One result of having the brain grow so much after birth is that human brains tend to be learning as they grow, allowing for a great deal of developmental influence.

Generally, the trend was for larger and larger brains, and modern humans (*Homo sapiens*) are not the species with the largest brains. That honor goes to the Neanderthals, but several researchers have pointed out that they appear to have been more heavily built than modern humans, and if we adjust for their greater body mass their brain-to-body mass ratio is like that of modern *Homo sapiens*. The Neanderthal brain is differently shaped than ours – it is longer where ours is tall. The reasons for this are unclear, and without soft tissue, interpretations of Neanderthal brain anatomy is speculative at best.

Today, we are heavily brain dependent, but to be human is not necessarily to be brain dependent. The so-called robust genus of humans (*Paranthropus*) never grew a large brain, and was stable at a roughly 500 cc brain for more than a million years. The "hobbit," *Homo floresinsis*, seems to have actually reduced its brain size because of limited resources. Brain reduction does happen in other lineages when the animals can no longer acquire the resources to support large brains. This is an important reminder that for all the attention the human brain gets, it is an anatomical feature, much like any other, and it is subject to evolutionary forces in the same ways.

Chapter 15: Chaos and Complexity

One of the promises I made at the beginning of this book is that although it is about science, there will be no math. In this chapter I fudge it a little bit – the concepts discussed in this chapter are mathematical, but you don't have to do any actual calculations. However, we do discuss mathematical ideas, so you may have to shift mental gears a bit when reading this chapter.

Virtually all sciences require some degree of math, and we use math to understand patterns that we see in nature; for example, we know that animals with longer legs run faster, and we know this by measuring animal leg lengths and running speed to see if faster animals have longer legs. What this means is that there are mathematical rules that apply to nature, and if we understand those laws we can understand nature better.

It is easier to understand some of the mathematical relationships in nature compared to others. Understanding the Pythagorean theorem is straightforward – you can look at a single right triangle and understand the idea fairly easily. There are only three numbers to keep track of, and they have a fixed and easily calculable algebraic relationship. However, it is not uncommon for people to have trouble with large numbers of things, which is why some students have trouble with statistics. It is also why people have trouble understanding differences between groups – we want there to be a group that has "X" and a different group that has "Y," when in reality the large numbers in those populations means there is no easy way to characterize them. You sometimes hear a child ask questions in a binary fashion: "are they big or small?" or "are we rich or poor?" The answer, of course, relies on knowing how big or wealthy the rest of the population is, and knowing where your "big" or "poor" individual falls on the distribution. The simple question requires a complex answer.

Nature is extremely complex, so our ability to interpret it relies on understanding the interrelatedness of large numbers of things. There have been some impressively complex calculations of complex things made by scientists. In 1977 NASA launched the *Voyager 2* probe. This probe had to be launched so that it would cruise by Jupiter, Uranus, Saturn, and Neptune. *Voyager 2* had to have a very precise trajectory, so that it could use the

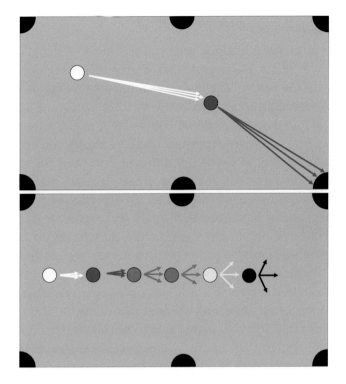

Figure 15.1 Any billiards player is familiar with the problems presented by chaos theory. When the cue ball (white) is struck against a single ball, if it is hit within a certain range of spots, it will predictably go into the pocket. However, in combination shots, the imprecision increases with each ball struck.

Some slight changes alter the angle of the final ball to a minor degree, but with other slight changes, the final ball might be missed altogether. So the result is not necessarily proportional to the variation in the inputs.

This is why accurate long-term weather prediction is impossible. Imagine the imprecision, as well as the prodigious error accumulation, when calculating the temperature, velocity, and humidity of the molecules in the atmosphere (the Earth's atmosphere has roughly 10^{44} molecules).

gravitational field of each planet to steer it to the next destination (while moving at 35,000 mph, no less!). The NASA scientists had to predict where each planet would be in its orbit when the probe arrived, years into the future. But these calculations were successful, well before the use of powerful computers to model planetary orbits.

This logic – that we understand the interrelatedness of events in time and space – is what scientists use to predict other events in the future (the outcome of physics particle experiments) as well as the past (e.g., we understand geology because we can observe the actions of erosion, plate movement, and sedimentation in action today and infer backward). This is even the same logic used to play billiards – by striking a particular ball at a specific angle, you can predictably get the ball into the pocket. These kinds of systems are called *determinist*: the initial conditions determine the result (Figure 15.1).

Chaos Theory

But sometimes, scientists cannot predict the future. If there are too many factors, and each factor has a particular degree of imprecision, there is no way to calculate the future. The classic example of this failure is in weather prediction. Meteorologists can predict the weather with some degree of precision only about four days out. Even with knowledge of temperature, humidity, wind direction, and barometric pressure, there are simply too many molecules of air and water moving in too many different directions for scientists to be able to calculate the results of their interactions. Instead, meteorologists look at the overall weather systems, track their movements, and, based on what similar weather systems have done in the past, try to estimate their trajectories. Despite the power of our modern computers, all we can do is make broad estimates. As you well know, weather predictions are frequently wrong. This is not because the meteorologists are bad at their jobs; rather, predicting the weather with extreme precision is not really possible. A weather system, with innumerable interacting molecules of water and moisture, is simply too complex to be predictable and is therefore said to be *chaotic* (Figure 15.2).

Many natural systems are chaotic – geologists, for example, cannot predict when earthquakes will occur because there are too many interacting variables. Aeronautical engineers, when studying wing shapes on jet planes, found strange patterns of turbulence that did not follow predicted patterns. The biology of large systems of organisms is similarly unpredictable, so that we cannot use ecological information or biological parameters to precisely predict when organisms will speciate or migrate, or when they will go extinct.

Despite the fact that we seem not to be able to predict future conditions in complex systems, it does not mean that these systems are not following a pattern. To use a simple example from weather: when there is high humidity and low barometric pressure, there is an increased *probability* of rain. So there is a tendency that we can observe, even if we cannot make a direct prediction. The system is following the principles of physics, and could *in theory* be predicted from the initial conditions. But we can never know all of those conditions, and could never calculate their interrelated actions if we did. So we try to make probabilistic predictions and be explicit in our imprecision.

One example of the effects of chaos theory on our daily lives occurred in L'Aquila, Italy, in 2009. The city lies in a mountainous, tectonically active region in Italy, and city officials asked local geologists to predict whether an earthquake was likely. The geologists predicted than an earthquake would not occur any time soon, but just a few days later 209 people were killed when heavy tremors hit the city. The geologists were later prosecuted for the deaths of the townspeople (although acquitted). In actuality, their "crime" was

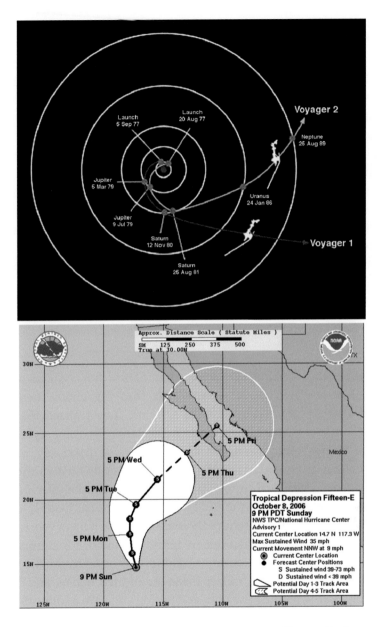

Figure 15.2 The difference between chaotic and nonchaotic systems is illustrated in these two images. When NASA launched *Voyager 2*, it had to calculate a trajectory that would predict the location of Jupiter, Saturn, Uranus, and Neptune over tens of billions of miles of space, roughly twelve years into the future (when *Voyager 2* finally reached Neptune).

Compare that to our inability to predict the movement of a hurricane on Earth over just a few hundred miles during a span of three or four days. In a hurricane, the number of molecules in the air (10 trillion trillion per cubic meter) are beyond the reach of any computer. In fact, scientists never try to model the individual molecules of a hurricane. The variation seen in the past is incorporated in the imprecision of their prediction, which is why, in the image, the potential hurricane path gets "wider" as time passes (this is sometimes called the "cone of uncertainty").

overstating their confidence in their understanding of the risk of an earth-quake. Because it is part of a chaotic system, it is not possible to state with any accuracy the time and place of an event like an earthquake.

One thing to note about chaos theory is that the fact that individual outcomes are not predictable with perfect accuracy does not mean that making predictions with available data is useless. Even chaotic systems have patterns and boundaries. The weather is not directly predictable, based on available models, but meteorologists do know the boundary conditions (the range of temperatures) for winter days when compared to summer days, and they also know the times of year when hurricanes and typhoons tend to occur, even though they can't directly predict the when and where of a specific weather event. This is the case for all chaotic systems. The trick for the scientist making the prediction is to understand the range of outcomes and to make a probabil-istic prediction. This is, of course, what meteorologists are doing when they predict a "40 percent chance" of rain. Unfortunately, for people deciding whether to carry an umbrella a "yes" or "no" answer would be preferred.

Complexity

Chaos theory relies on understanding that nature has systems with large numbers of interrelated parts. As scientists studying chaos theory in the 1970s and 1980s examined these large systems, they realized that in some cases large numbers were the key to the systems. In these systems, large numbers of "things" (they can be gas atoms, or stars, or blood cells, or neurons, or organisms) act in a way that results in something sophisticated and complex "emerging" from a group of simple things following simple rules.

One intuitive example is an ant colony. Ants are social insects, and live in groups that can reach into the tens of thousands. Each ant has a very small brain: an ant might have 250,000 neurons. A house cat brain, for comparison, has 750 million neurons. Yet ants accomplish things cats never could: they construct complex structures, including bridges made out of their bodies, and collaborate on acquiring food, defending the nest, and raising young in a sophisticated labor hierarchy. They also capture other insect species alive (aphids) that they use for a variety of purposes, including harvesting their body fluids (which has been called "insect slavery" or "insect husbandry"). House cats can do none of these things, nor could they ever be taught to do so. But, since ants have brains roughly 1/3,000 the size of a house cat, how do they accomplish all these things?

The answer is that no individual ant could accomplish any of these things on its own – it requires the collaboration of many ants. For example, ants do not, in any sense, "know" that they are building a bridge. They have a set of simple instructions in their brains, and repeated among many, many ants the overall result is the complex structure. To build a bridge, ants follow these

Figure 15.3 Social animals demonstrate how following a few simple rules can produce remarkable results. Ants and termites perform extremely sophisticated behaviors, despite having extremely small brains. In each case, they follow a few simple "if–then" rules. But when those simple rules are repeated many, many times, the result is no longer simple. In the image on the right, the starlings are able to remain a coherent group because each responds in simple ways to the movement of the birds around it, and a few external reference points. There is no lead bird – they are all responding at the same time to the same cues.

instructions: when a small gap is encountered, an ant cantilevers its body over the gap and locks its limbs. The next ant crawls on top of the first, cantilevers a bit more over the gap, then locks its body. More and more ants follow until a bridge is made. So, although the final result looks like a complex structure (and it is!), it was not "designed." It simply emerged out of simple, but repeated, ant behavior (Figure 15.3).

Box 15.1 Chaos theory and human behavior

The unpredictability of systems with large numbers of interacting parts is one of the reasons predicting human behavior is so difficult. The human brain has roughly 90 billion neurons and some 200 million connecting circuits (called axons). In each human, the neurons and axons are wired in unique ways not found in any other human (even an identical twin). This is one of the reasons predicting the response of any one human to specific circumstances is so difficult. In the second half of the twentieth century, several governments experimented with ways to manipulate people so they would respond in specific ways, but the results always showed that individual human behavior is not precisely predictable (in the USA this program was run by the CIA, and called Project MK ULTRA, but the Soviet Union ran a similar mind-control program).

However, even though individuals are not easily controlled, groups of humans show statistical tendencies that can be manipulated. This is exactly what advertisers, propagandists, and politicians do. Advertisers know they cannot get at any one individual, but they are aware that some percentage of the population will respond to, for example, the suggestion that buying a certain brand of car may attract a mate, or that a particular type of shoe will indicate that you are wealthy. Not all, or even most, people will respond by buying the car or shoes, but manufacturers know they are manipulating enough people that it makes sense to pay for the advertising. A politician may label a certain group of people a threat, knowing that a certain segment of the population will feel fear and respond by voting for them.

The unpredictability of individual human behavior, and the statistical predictability of groups of people, mirrors the problems faced by meteorologists. Any particular weather pattern is unpredictable, but, in the aggregate, meteorologists know that storms tend to behave in predictable ways. The chaos of large numbers is a universal law, and it applies across nature at almost any scale.

Box 15.2 Engineering vs. "emergence"

When engineers design things (houses, bridges, cars, televisions, etc.) they start with a specific goal and then create a plan for how to construct it. Typically, engineers want to make sure the object (let's say it's a bridge) meets the basic criteria (is long enough to span the gap and strong enough to hold the traffic), but is also durable enough to last for years and is not too expensive. Then will they make a blueprint for the builder to follow. If the design is good, and the construction is good, the bridge will likely be a success, and people

Box 15.2 (cont)

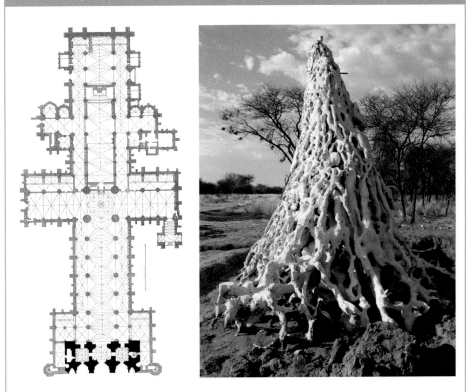

Figure 15.4 As in the cathedral floor plan (left), humans use *designs* to plan out how they construct things (buildings, cars, furniture, computers, etc.), which results in a product that is designed for a specific purpose.

Compare the planned construction of the cathedral to the termite mound on the right (the scientist poured plaster into an abandoned mound, then hosed off the dirt, revealing the passages and chambers inside). When termites build a mound, they do not have a plan. Rather, they simply follow certain rules that are triggered in particular circumstances, until the circumstances change. These can be thought of as "if–then" rules. For example, when termites get too crowded, the crowding triggers a simple construction behavior. When the termites are no longer crowded, they stop. Each termite decides for itself when to build – there is no "master termite" issuing instructions.

will be able to walk and drive over the bridge in safety for a long time. If the design is poor, the bridge may fail (Figure 15.4).

Nature does not work this way. There are no "goals" in nature – organisms simply respond to the environment, and sometimes this results in a solution to a problem. If the response does result in a solution to a problem, the organisms may thrive, while if not, they may die. Consider the ants bringing food back to the colony. If they find food, they will bring it back efficiently. But if the food is

Box 15.2 (cont)

toxic, they will bring it back efficiently and poison the colony efficiently. The inflexibility of the rules is the strength and weakness of ant behavior. They are in no way "thinking" – rather they are iteratively following rules.

The construction of anatomy also follows simple rules. Our vascular (blood) system follows a pattern whereby the vessels bifurcate according to a simple growth rule – the vessels divide according to a simple mathematical formula that increases the bifurcation the narrower the vessels get. Most organisms follow these types of simple rules – trees branch, flowers grow, etc., all according to simple growth rules. No design is necessary. If the result is unsuccessful, the organism dies, and if successful, it survives and passes on its genes.

This last point highlights why engineers cannot really build using the principles of emergence. Nature can afford to repeatedly try a series of unsuccessful solutions. Nature does not "care" if many organisms die while iteratively "finding" the solution. Animals die and species go extinct all the time – that is part of evolution. However, we don't want large numbers of humans to die while we find a good way to build a bridge. Rather than having 1,000 failed bridges, we use engineering principles to design a single bridge that we know will not fail and drop pedestrians into the river below. If we use a little more money or steel to ensure this, that is fine, even if the bridge is a bit overbuilt. Humans have goals in a way nature does not.

Box 15.3 Chaos and complexity in economics

Other than the inherent inaccuracy of long-term weather predictions and the patterns of ant behavior, how else do these seemingly esoteric mathematical models affect us? Actually, both of these ideas describe much of the world around us.

Chaos

Predictions in many systems are extremely inaccurate due to the large number of interacting factors – politics, geology, weather, biology, etc. are all systems that can behave chaotically. It is not really possible to make accurate and precise long- or medium-term predictions in these kinds of systems, although people try.

We discuss this issue in more detail in Chapter 21, but the chaotic nature of economic systems is one of the major reasons why true command-style communism does not work. National economic systems are very complex, with millions of interacting parts. There are millions of workers, and each of those workers, in addition to contributing to the production of the economy,

Box 15.3 (cont)

are consumers of the products of the economy. Each worker needs clothing, food, housing, medicine, transportation, and leisure.

In a communist command economy, a single controlling power attempts to make all of the decisions for the economy. They try to determine what every citizen requires (in terms of food, clothing, transportation, etc.), and then allocates material and labor resources to meet those needs. So, for example, to make cars the central authority determines the demand for cars, and then decides how much steel, rubber, copper, plastic, and aluminum, as well as the various types of skilled labor, to allocate to meet the demand. Decisions about the manufacturing process are also made high in the control system. This is done for literally every single product in the economy – how many shoes, and how many of each size, and for each shoe type and style.

But the problem is that those few individuals are like the pool player hitting the cue ball in a huge combination shot. There is a degree of inaccuracy at each scale of the process, and those errors accumulate down the many layers of the system, so that at the end there is a considerable mismatch between consumer need and the product released. Without individuals at every level of the process correcting the mismatch between the availability of materials, resources, and demand, the errors accumulate enough to make the economic system extremely inefficient. This inefficiency was one of the major reasons why communism in the Soviet Union failed (see discussion of China in Chapter 21).

Complexity

When you see birds moving in a flock, seemingly in coordinated motion together, you are seeing the product of complexity theory – there is no "lead bird." Rather, every bird is following just a few cues. But all together the birds move in a graceful, coordinated pattern.

Classical free market economics follows a pattern that shows emergent complexity. Much like the birds, the actors in a free market economy are making individual decisions that are appropriate for their specific circumstance, and in this way the economy is largely self-correcting. As circumstances change within any one part of the economy, the individuals involved in their businesses (small and large) make individual decisions to maximize their benefit and minimize their risk. Because these people are directly involved with their part of the economy, they tend to be relatively well informed. This pattern, of well-informed self-interest, makes the actions of others relatively predictable. As individuals (and businesses) at all levels of production and consumption follow the general rule of "maximize self-interest," the overall

Box 15.3 (cont)

economy tends to be very efficient. This makes free market economies relatively flexible and able to respond rapidly to changes in demand or resource availability.

Because there are normally multiple companies in competition (in price and quality), there is a constant refining of methods, increasing efficiency. Further, companies are in competition for the best and most innovative workers. From this enormous pool of self-interested, well-informed, and free-acting individuals and companies, we find that overall more people are economically better off, and there is a steady growth in the economy. So from a large pool of independent individuals all acting in their own self-interest, economic growth and health emerges.

But chaos also applies in capitalist economies. A country's economic system has millions to billions of interacting parts. This makes it very difficult to predict any specific outcome. Economists spend hundreds of millions of dollars trying to predict even small parts of the economy. However, the economy has proven essentially impossible to predict with any accuracy. Some investors have made money, but statistical analyses of stock market investment has found that there are no more successful investors than would be predicted by chance. In other words, if monkeys were picking stocks by random, we would see the same number of winners and losers as we have, and a few monkeys would repeatedly, but randomly, chose the right stocks. If a country's economic basics are sound (a productive population with good resources), then the economy will grow, so the stock market does make money for investors, but only because the market grows overall, not because particular investors are able to predict the outcome of a chaotic system.

Box 15.4 The problem of consciousness

One of the longest-standing philosophical questions is how something inanimate (a lump of hydrogen, carbon, oxygen, nitrogen, and a few other elements) can become something that has an inner life. This is the problem of consciousness. Conscious organisms (and so far we know of consciousness in dolphins, great apes, elephants, and some birds) are aware of themselves as individuals, and treat themselves as separate from the rest of the environment. That is, they are not simply responding to the environment but "thinking" about it.

Philosophers have struggled with "where" in the body consciousness lay. The ancient Greeks thought consciousness was separate from the brain and in the heart. Descartes famously argued that consciousness was derived from the divine, and that God communicated with us through the pineal gland, at the

Box 15.4 (cont)

base of the brain (at the time the biological function of the pineal gland was unknown – today we know that it produces melatonin, which helps us regulate sleep). Subsequent researchers have struggled with understanding consciousness, since, although consciousness must reside in the brain, there does not appear to be any specific location in the brain where consciousness lies. Early studies of brain damage found that there is no specific location that, when damaged, eliminates human consciousness. Modern brain-imaging studies have similar results.

Treating human consciousness as an emergent property avoids all of these problems. Instead of looking for a specific location where consciousness resides, or trying to figure out the mechanism by which the brain produces consciousness, complexity theory offers a straightforward solution. Under complexity theory, once a brain reaches a certain size (in terms of numbers of interconnected neurons), consciousness simply emerges. This is a slippery concept to grasp, because we are used to thinking of different organs having specific anatomical functions. Consciousness is not like this – it was never selected for, nor does it have a true biological purpose. Rather, it is an incidental byproduct of selection for increased numbers of neurons in the brain.

This may mean that there is not really such a thing as consciousness – there is just a set of behaviors and senses associated with large numbers of neurons. We may think that we perceive of the Universe in a different way from other animals, and "think" about things differently, but the difference may be one of quantity rather than quality – perhaps other animals have something analogous to consciousness, only in a simpler form. There is no particular reason to think that there is a threshold that divides animals with consciousness from those without. There is likely to be a continuum on which animals fall, depending on the size and organization of their brains.

In experimental settings we can test for consciousness using mirrors – if an animal recognizes itself as unique among other animals we say that it has self-awareness. But this may be too coarse a tool to measure the intellectual life of other animals. Philosopher Tomas Nagel touched on this in his famous essay: "What is it like to be a bat?" Bats use echolocation to find their way in the world, and use auditory information to generate a three-dimensional representation in their brains. We, as humans, do not possess the neuroanatomy to interpret sound waves in that way, so we can literally never understand the world the way a bat does. It is likely that many animals have similar differences in their brains that prevent us from knowing about their consciousness.

This seemingly simple observation – that many complex structures in nature are the result of a few simple rules, followed repeatedly – is extremely powerful. One of the main advantages of this system in organisms is that the results are constantly being refined. If the first ant misjudges the gap, and it is too far to bridge by ant bodies, that ant is likely to die, taking the gene for misjudgment with it. But even when death is not the result, because there are so many ants the results are constantly refined. If an ant finds a food source, it will walk back to the nest, leaving a trail of pheromones for other ants to follow. Other ants will follow the path to the food. But some will return along a slightly different course. If that new path to the food is shorter, the pheromone trail will be stronger because the time since that ant walked along it is shorter. Other ants will follow the shorter rather than longer path because the pheremonal signal is stronger. Over time, the ants will iteratively experiment with all available paths from the food to the nest and converge on the shortest path. By following a few simple rules, the ants find the most efficient way to bring the food home to their nest. Under this model, the number of neurons in the ants is combined, so that the sum of ant neuron processing power, aggregated, exceeds that of the house cat.

This pattern of complexity emerging out of the interrelatedness of many simple things appears to explain many things in nature that had previously resisted explanation. Perhaps the most important is the emergence of consciousness. Consciousness is the awareness an animal has that it has identity: that it exists as an individual, separate from the Universe around it. Humans have this ability, as do dolphins, great apes, elephants, and some birds. The question of why humans have consciousness has historically been one of the great philosophical questions of the Western world, puzzled over by ancient Greek philosophers in the fifth century BCE, and regularly ever since. What is it that makes us individuals? Why would nature want us to have individual identity?

The answer appears to lie in complexity theory. Natural selection does not actually select for consciousness – it serves no specific purpose necessary for our survival that we can detect. Rather, nature selects for large brains (increasing numbers of neurons) which allow animals to solve problems. It appears that, once our brains reach a certain size, and we have a certain number of neurons, consciousness emerges. This has happened in our lineage but also in other animals with requirements for large brains (for solving food-related problems, or dealing with complex social situations). In this view, human consciousness is not unique, and humans brains are simply following a set of rules seen in other animals. When we can apply a model across the animal spectrum, without making a special pleading for humans, it always engenders confidence in the model.

Part II: Science and History

The second half of this book has some important conceptual differences that distinguish it from the first. For the last fifteen chapters you have read about how scientists view the world – the events, and more importantly the causes of these events. The forces of plate tectonics drive the shape of our world; nucleosynthesis is responsible for the distribution of elements throughout the Universe; evolution is responsible for all biological variation; complexity theory explains social insect behavior and human/animal consciousness.

But applying some of these principles to humans has, historically, been controversial. The debate over sociobiology demonstrates the general principle that humans have difficulty applying scientific concepts to many aspects of themselves, particularly behavior. The reasons for this are complex, and there is no real consensus on the topic. My view is that most people, in general, view humans as a special case – not really part of nature. Many religions espouse this view – in the Bible, in Genesis 1:26, it is said that mankind will "rule over the fish in the sea and the birds in the sky, over all the livestock and all the wild animals, and over all the creatures that move along the ground." This verse reflects the general idea that humans are separate from nature and not part of it.

When examining historical events, historians today tend not to apply scientific concepts (although there are exceptions). The common theme in most historical accounts and explanations is the "narrative." A narrative is a recounting of events, but it is typically not framed in terms of scientific hypotheses of underlying causes; rather, it tends to emphasize sequences of events. Many narrative events focus on a single individual experiencing the historical event. If causality is addressed, it is not framed as testable hypotheses.

A historical narrative typically attempts to understand past events from a particular perspective: What was the experience of former slaves after the Civil War? What was it like to be a woman on the western US frontier of the 1870s? How did the generals in World War I decide on trench warfare? How did the Athenians win the Battle of Marathon? Where did Mahatma Gandhi learn to be a pacifist?

These are questions about specific events. In general, science tends not to look at events on such a small scale. Rather, scientific questions about historical events might look at a general *class* of events and attempt to draw broad inferences. Rather than ask about the experiences of former slaves in the USA after the Civil War, for example, a more scientific question might be: under what circumstances do former slave populations integrate well into the society in which they used to be slaves? Rather than ask about the tactics at the Battle of Marathon, a more scientific question might be: Do democratic citizen-soldier armies tend to beat large mercenary forces in warfare, and if so why? Science tends to ask "why" questions; historians tend to ask "what" questions.

These scientific questions attempt to find general patterns to explain phenomena. Some people find this type of research unsatisfying. If you are interested in the behavior of individual generals in the US Civil War, for example, you might not be interested in the scientific hypothesis that agricultural countries never defeat industrial nations in open warfare. The scientific approach reduces the importance of the individual generals and presents the issue of the forces that lead to victory somewhat deterministically. The scientific approach doesn't leave room for heroes or cowards. If an agricultural nation has never beaten an industrial one, one might argue that the actions of individual generals don't matter too much – they represent the background "noise," whereas the industrial might of the North was the determining factor.

This scientific approach is not always popular among historians because it deemphasizes some of the importance of the individual. Since we are human, and we tend to be interested in the actions of other humans, reducing all their behavior to "background noise" can be pretty unsatisfying. However, this doesn't mean that it is incorrect. The fact is, there *are* general scientific principles that can be applied to history, because there are large-scale factors that determine historical events. As scientists, our job is to draw those out by examining patterns in whatever data are relevant. This does not mean, by any stretch, that individual people do not act as heroes in some circumstances, and that they should not be recognized for it. My personal moral hero has long been Hugh Thompson, the US helicopter pilot in the Vietnam War who, virtually single-handedly, stopped the Mỹ Lei massacre. This act had no large-scale significance – it didn't stop the war or have any notable effect on the broad-scale outcome of the conflict. It was unacknowledged and unrewarded for almost thirty years, and that is what made it such a brave and moral act. So, while we might take a more scientific approach when examining the causes and outcomes of the war, we are still able to acknowledge the importance of individuals. Science and history are asking different questions – the heroism of Hugh Thompson is important and interesting, but it is not *explanatory*. Science wants explanations, and in the following chapters I apply explanatory models to historical questions.

Chapter 16: The Neolithic

Hunting and Gathering

Modern humans arrived on the scene some 250,000 years ago and, over the next quarter million years, slowly spread out over the whole world (Antarctica excepted, of course). They outcompeted other human species, and by around 25,000 years were the last humans left after the radiation and subsequent disappearance of some thirty other species over 6 million years.

During this quarter million years they lived a very particular lifestyle known as hunting and gathering (also called "foraging"). All food was from wild plants and animals collected or killed, and all tools, clothing, and shelter were derived from naturally grown or found materials. Overall, this was a highly successful lifestyle, as humans occupied a wide variety of climates, from the desert to the rainforest to the Arctic.

The term "hunting and gathering" (or "foraging") is a general term, but the ways in which people survive in this way are very diverse. It means subsisting off game and wild plants in the rainforest, but also living off shellfish on the coast of South Africa, or harvesting wild honey in Tanzania, or subsisting on seals and small whales in the Arctic. The main characteristic of this lifestyle is that it involves no agriculture – no domesticated animals for meat or milk, and no harvesting of domesticated grains such as wheat, corn, or rice (see Figure 16.1 for today's hunter-gatherers).

This lifestyle was the means by which all humans survived for the vast majority of the existence of our species. We only moved from hunting/gathering some 10,000 years ago. Most of the world transitioned away from this lifestyle thousands of years ago, but it still persists in some parts of the world. The Yanomami of the Amazon Basin, the !Kung San of Botswana, the Hadza of Tanzania, the Penan of Borneo, the Meakambut of New Guinea, and the Inuit of Canada, are all examples of peoples who have retained at least some of their ancestral ways of survival. Because they represent different variants of the way humans survived for most of our existence, they tend to be heavily studied by scientists such as anthropologists, ethnographers, and increasingly, biomedical scientists.

Figure 16.1 For the first quarter million years of our species' existence, we lived as hunter-gatherers. Although this may bring to mind images of San Bushmen in southern African deserts (top), or the Yanomami in the Amazon, in the past hunter-gatherers would have lived across virtually all ecosystems (Inuit in the Arctic, left, and the Hadza of East Africa, right). As the world transitioned to agriculture, foraging peoples have been pushed to the margins, where there is little competition for the land.

There are some interesting characteristics of foraging peoples that our societies have lost, and that tend to be appreciated today. The first is personal time. There is some variation, of course, but foragers can usually meet their daily nutritional needs in just a few hours. This compares favorably to agricultural or industrial societies, where people work between eight and twelve hours a day. Foraging peoples spend much of their time doing things other than working.

And the food acquired during those few hours is usually healthier than food in agricultural societies. Since it is wild-caught, the meat is leaner, and without dependence on cereal grains, there is less low-fiber starch in the diet. Natural plant varieties tend to have more roughage and less sugar than domesticated variants. Obesity is essentially unknown in hunter-gatherers, and dental caries (cavities), caused by refined sugars, are less common.

There is no real material wealth, nor any private property in these societies. There is no "title" to land – all land is essentially common and accessible to others. Since all tools, clothing, etc. are manufactured from natural materials, there is nothing that your neighbor can possess that you cannot. Without domesticated grains, there is no storable food – all food is acquired and consumed daily (or over a few days).

There are, naturally, social implications to the differences in lifestyle. Since there is nothing that a person can accumulate that others cannot have, and therefore no wealth, it is difficult for one individual to accumulate much political power. Typically, all power is in the form of favors done and assistance provided. To become a leader, one must demonstrate political skill and leadership qualities, so essentially anyone with sufficient charisma can become a leader. This type of leadership is called "first among equals," where a leader is respected because of political or military prowess. But a series of bad decisions can turn political fates around, and someone else can become the "first."

Importantly, without material wealth such as land or money, there is no way for any one individual to pass down wealth and power to the next generation. This prevents the consolidation of power into lineages, such as we see in states and nations (nobility and royalty). In every generation, of course, some individuals will accumulate power if they are charismatic or skilled politicians, or good foragers who can share food. But once those individuals die, their offspring must start anew.

The main problem with this lifestyle is demographic. Hunting and gathering can only support a relatively low population density. There are only so many deer to kill in a forest, or berries to harvest. The population is constrained by the resources available in nature. The shift away from complete dependence on natural resources was probably the most important and consequential transition that our species has ever made.

The Neolithic Transition

In various locations around the globe, when local population densities increased enough (see Figure 16.2) so that there was no "new" land onto which growing populations could expand, people made a major transition. Instead of moving into new, unoccupied lands, people started controlling the local

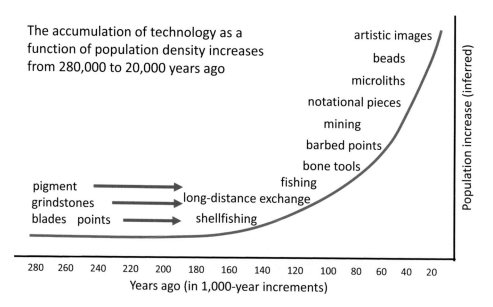

The accumulation of technology as a function of population density increases from 280,000 to 20,000 years ago

artistic images
beads
microliths
notational pieces
mining
barbed points
bone tools
fishing
pigment
grindstones
long-distance exchange
blades points   shellfishing

Population increase (inferred)

280   260   240   220   200   180   160   140   120   100   80   60   40   20
Years ago (in 1,000-year increments)

Figure 16.2 The development of advanced technologies over the last 250,000 years was once presented as being the result of a transformation of brain anatomy somewhere between 150,000 and 100,000 years ago. However, the apparently discrete shift in technology at this time was more likely the result of the poor resolution of the archaeological record. A better way to model the increase in technology is to map it onto the increase in population density over the last 250,000 years. As populations increase, a given individual can access learning from a wider network of individuals, and better ideas can more easily spread and remain in the population.
Figure using data and figure concept from: McBrearty S and Brooks AS. 2000. The revolution that wasn't: A new interpretation of the origin of modern human behavior. J. Hum. Evol. 39: 453–563.

resources. Some plants were encouraged to grow, and others pulled out to make way. Certain animals were found to be easy to control. Over time, these people settled on land where the plants or animals grew well and became, in essence, farmers or herders.

This is a much abbreviated description of the transition, as there was enormous variation across geography and time: the plants and animals were different, and the length of time to complete the transition varied. But one of the most important things to understand is that it happened independently in multiple locations across the globe. There is no one "single origin" where the idea of domestication started and from whence it spread (see Figure 16.3). Whenever conditions were appropriate, the transition occurred. Dates are still subject to change, since archaeologists are always discovering new Neolithic sites, but, for example, by 10,000 years ago there was domestication in Turkey (wheat) and the Andes mountains (potatoes) completely independent of one another. Around 8,000 years ago, rice was domesticated in China, ley in India, and millet in West Africa. Around 7,000 years ago, corn was domesticated in Central America and taro was domesticated in New Guinea.

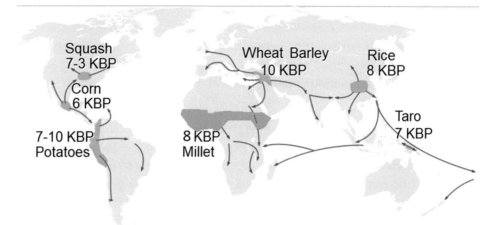

Figure 16.3 In this image, we see the locations and dates (KBP = thousands of years ago) of the earliest agriculture, as identified by archaeologists. Agriculture originated in multiple locations across the globe, entirely independently. As populations became denser, peoples across the globe found ways to domesticate grains and animals, leading to the growth of agricultural societies. Areas without these factors never independently acquired agriculture, as in the Arctic or the desert, for example, there are no domesticable plants or animals.

One important implication of the independent acquisition of agriculture is that it suggests that people across the globe are functionally equivalent, and that environmental circumstances are driving the changes we see in broad population areas. As an aside (and relevant to Chapter 12), it is worth noting that agriculture did not appear in mainland Europe until it was well developed in southwestern Asia (modern Turkey, specifically).

The domestication of plants and animals made food resources much more reliable, which made several important transitions possible.

Positive changes included:

1. Greater population densities. With more reliable food sources, people could have more babies. Hunter-gatherers reproduce relatively slowly, because they can only support relatively few people in their groups. Women in these groups use several methods to reduce their reproductive rate (see Box 16.1). In Neolithic societies, with their predictable and controllable food sources, the reproduction rate can increase dramatically.

2. Food can be stored. This means the society can accumulate resources for various purposes – consuming them over the winter, for example, but also trading them for other things they need. This is, in essence, the origins of "wealth," which is simply a way of accumulating resources.

3. Surplus resources mean that certain people can be released from the need to accumulate wealth, and can do other, more specialized jobs for society. Artisans, for example, might be "employed" by society if it has enough surplus

Box 16.1 Are hunter-gatherers living a "better" life?

Probably the most famous statement about pre-Neolithic societies was made before the term "Neolithic" was coined. Thomas Hobbes, when writing about these societies in his monumentally influential work on government, *Leviathan* (1651), wrote:

> *In such condition, there is no place for industry; because the fruit thereof is uncertain: and consequently no culture of the earth; no navigation, nor use of the commodities that may be imported by sea; no commodious building; no instruments of moving, and removing, such things as require much force; no knowledge of the face of the earth; no account of time; no arts; no letters; no society; and which is worst of all, continual fear, and danger of violent death; and the life of man, solitary, poor, nasty, brutish, and short.*

This was the opinion for much of history, where "civilization" was viewed as an unqualified good, and foraging people were viewed as "savages." However, at the end of the twentieth century, as historians and anthropologists surveyed the destruction caused by "civilization" during the wars of the twentieth century, this perspective was significantly reassessed. Hunter-gatherers were viewed as living in tune with nature and creating no environmental impact. Their societies seemed idyllic and slow-paced compared to hurried modern city life. Without wealth there is no poverty, as anyone can acquire all that is available. And their weapons are simple and handmade, and unable to create the kind of destruction associated with the machine gun, the tank, and the strategic bomber.

Similarly, the egalitarian nature of foraging societies is attractive to some. There is no doubt that, in state societies, the accumulation of wealth by a few lineages has allowed those few people to exploit many others. For example, the accumulation of wealth and land by relatively few large landowners in the southern USA drove the enslavement and forced migration of millions of Africans from the sixteenth to nineteenth centuries. By acquiring land and keeping it in families, those families were in a position to exploit others. This is true throughout written history – the Hundred Years' War, for example, was a struggle between two competing lineages to acquire and maintain power in France and England, which in the process laid waste to much of the French countryside, and resulted in the deaths of 2–3 million people through war, starvation, and disease. Exploitation on such a massive scale could never occur in foraging societies.

Some researchers argue that foraging is the "true" or "natural" human lifestyle, and that our transition away was a major mistake. Viewing the planet

Box 16.1 (cont)

in ecological terms, this perspective is understandable. For those 200,000 or so years we made relatively little impact on the planet. Population densities were so low that our food demands were not great, so the large parts of the planet that are under agriculture today were full of wild game, and we hadn't yet introduced pollutants into the atmosphere or the oceans.

These two views are in conflict, and the debate over the more "ideal" lifestyle continues to this day. But, in order to make an accurate assessment of the merits of the two systems, it is necessary to understand hunter-gatherer societies in a more sophisticated and less romantic way. The truth is, of course, far more nuanced than either view espouses.

The Pros and Cons

Hunter-gatherer groups do have relatively flat power structures. In general, there is no private property, so power is not inherited and nor is wealth; as a consequence, individuals tend to have more power. If one "big man" ruler makes bad decisions, he/she will lose that position. Since there are no stable power structures, the relationships between people are not formalized by society. This makes relationships between people more flexible – for example, there is no formalized "marriage" in which there are specific obligations. If one person is dissatisfied with the relationship, they just move on. No one has a claim on another because there are no laws or regulations on relationships. Without pro-natalist laws espoused by the government, there is no societal control of sexuality.

These seemingly positive aspects of hunter-gatherer societies have led some researchers to argue that foraging societies were, in some sense, "better" because of the greater freedom and egalitarianism. For example, several anthropologists have argued that equality among the sexes is greater in these societies, since there is no central authority regulating the relationships among men and women.

But life in these societies can be quite harsh. Hunter-gatherers cannot support high rates of reproduction, and although their low-fat diet and long lactation schedules tend to suppress ovulation, infanticide is not uncommon. Access to valued resources (e.g., productive hunting grounds) is valued, so warfare between these groups is very common. Death rates from violence for foraging groups, for example, is many times higher than in agricultural groups (see Figure 16.4). There are no outside authorities to mediate between groups – the only recourse is often armed struggle. Also, the rights of one individual are only what that one person (or their close family) can defend. If someone decides they want what you have, and they can take it, there is no outside legal authority to prevent them or punish them.

Box 16.1 (cont)

Figure 16.4 After World War II, some academics (including anthropologists) argued that foraging peoples represented a more peaceful form of humanity, and that violence and aggression were the result of living in an agriculture-based state society. Two discoveries have suggested that this is an idealistic viewpoint. The first is the violent aggression seen in chimpanzees (Chapter 13), and the second is the discovery that foraging societies, past and present, are extremely violent. Studies of the Yanomami, for example, have shown them to engage in a frequency of aggression exceeding that of almost any state society.

At Nataruk, Kenya, a preagricultural archaeological site from 10,000 years ago was discovered in 2012. This site appears to show the massacre of one foraging population by another, as the skeletons show evidence that they had their hands bound and were the subject of traumatic violence to the head, ribs, knees, and hands.

Personal male–female relations are not necessarily egalitarian. Males of all peoples are larger, stronger, and more aggressive. In archaeological studies of the Chumash peoples of the Channel Islands (off California), who were foragers quite late in history, the skeletons of women show cranial trauma

Box 16.1 (cont)

patterns consistent with violent abuse. Ethnographic records of English colonialists in Australia document considerable abuse of foraging native Australian women by the native Australian men. In these foraging groups, female sexuality is still a resource that men will compete over, and the Yanomami of the Amazon Basin will raid nearby villages, killing men and capturing women. These captured women are considered a resource by the men, and their sexual consent is not considered. Much of this is consistent with sociobiological theory, and part of the debate over foragers has been entangled with debates over the biological underpinnings of human behavior.

People have a tendency to idealize the past, and the notion of a "Golden Age," when life was better for all people and the corruption of the modern world had not yet appeared, is strongly tempting. But humans behave in consistent ways, so there is no particular reason to expect that, simply because they earned their living in a different way, foragers would be more fair or kind. We simply need to look at the behavior of our chimpanzee cousins to see the universality of much of our behavior, past and present.

Box 16.2 The health implications of the Neolithic

During the Neolithic, food became more controllable and storable, and there were often enough surplus resources to have a physician class in the society. So we would intuitively expect these people to be healthier than the people who have to find their resources daily, and who occasionally go without when resources are not to be found.

However, that is not what we find when we examine the archaeological record. There are several places in the world where we can examine genetically cohesive populations as they went through the Neolithic transition. When we look at their skeletons, we typically find evidence that the populations were far healthier when they were foraging when compared to the population living off domesticated grains and animals.

One of the main reasons for this is that easily domesticable foods tend to be high in starch but relatively low in nutrients. Foods such as bread, potatoes, rice, and corn are high-starch foods with a lot of energy content, but which provide few proteins, amino acids, and nutrients.

Box 16.2 (cont)

One piece of evidence is simply stature. Hunter-gatherers are typically about five inches taller than genetically related groups of farmers. We also see evidence of anemia in the skeletons of early farmers. Anemia is a disorder caused by lack of protein, and it is especially visible in the human skull, which becomes spongy and develops holes when anemia strikes. The teeth of these early farmers also show hypoplasias, which indicate periods of low protein during youth.

The work associated with growing and harvesting these starchy plants also leaves stress markers on the body. Typically the work is repetitive and heavy – thrashing wheat, grinding corn, or pounding taro, for example – and it is common to see markers of repetitive stress on the legs, arms, and spine. Markers of osteoarthritis are commonly seen in these Neolithic populations but more rarely in hunter-gatherer populations.

Some of these nutrition-related conditions have existed throughout history, quite late into the modern period. For example, the height of Japanese boys increased by 4 inches between 1894 and 1935. This is the period when the Japan became an industrialized, modern country, with significant changes to the country's agricultural methods. Today, the difference between the height of North Koreans and South Koreans is about 4 inches and directly relates to the modern industrialization of the diet in the south, with greater access to protein and nutrients. Of course, this has also led to an increase in obesity – the curse of the modern diet.

Box 16.3 The independent origins of agriculture

One of the interesting things about many phenomena is that events are often repeated when circumstances are right. In biology, we see multiple iterations of the saber-tooth predator, including one marsupial species from Australia. They repeatedly appear in the fossil record and repeatedly go extinct. We don't know exactly what specific behavior the long canines were used for, but the circumstances must have reappeared multiple times over the last 65 million years. The animals with long canines are not particularly closely related – there is no "saber-tooth lineage." Rather, the adaptation reappears when circumstances warrant it.

The appearance of agriculture is a bit like that. It appears multiple times, but these events are unrelated to each other, other than the fact that the same types of circumstances that led to agriculture in one place are the events that led to it in another (as in Figure 16.3). There was not one origin for the idea – it

Box 16.3 (cont)

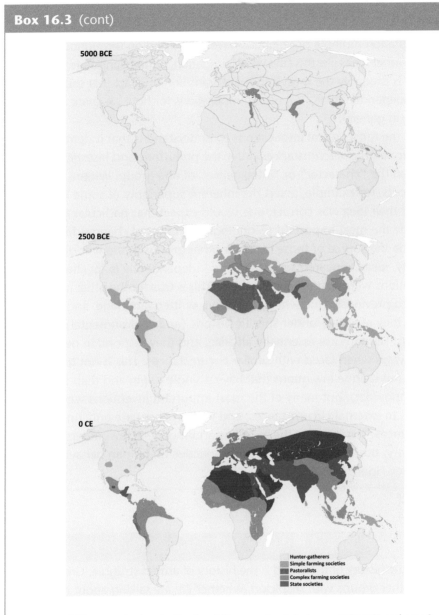

Figure 16.5 The growth of agricultural societies was steady between 5000 BCE and year 0. These populations were easily able to outcompete local foraging populations for desirable land, because they were able to have a large, dedicated warrior class. Asia, Europe, and Africa, with ancient populations and high population densities, were the first to acquire complex state societies. The Americas, which were not populated by humans until roughly 15,000 years ago, had lower population densities, and developed large state societies at a slower pace. Large parts of the Americas were populated by foraging peoples into the nineteenth century, and they persist today in parts of the Amazon Basin, although they are threatened by logging and agricultural development in Brazil.

Box 16.3 (cont)

is an adaptation multiple populations have converged on throughout history. There are other ideas like this (the wheel, irrigation, written language, etc.), but since this is probably the most important transition, the fact that this idea arrived independently has important implications for how we think about humanity in general.

One of the most critical implications to understand is that it demonstrates that environmental circumstances drove the prehistoric and historic changes, rather than the "character" or "intelligence" of one group. White supremacists, for example, assert the inherent superiority of some European groups. If their idea was correct, one would expect that particular group to have made the most important technological and social changes first (and perhaps be the source of those ideas for other groups in the world). But, repeatedly, ideas appeared separately and independently in locations where the conditions were appropriate for them. This is true for the other inventions mentioned previously (the wheel, irrigation, written language, and many others) – they reappear under specific historic (and environmental) conditions. Humans are essentially all alike, and they respond in nearly identical ways when faced with similar circumstances. This is not to say that there are not unique inventions that have a single origin and then spread (e.g., gunpowder), but many of the most important inventions were direct responses to external circumstances, and were appropriate only under those circumstances. This is one of the most important general lessons we can draw from history, but one that a scientific perspective on human behavior would predict (see Figure 16.5).

Box 16.4 What happens when agriculturalists and foragers meet?

Much of the history of the world is the history of armed struggle. Groups tend to view other groups with suspicion, whether we are talking about hunter-gatherer bands or modern nation states. They often have disputes over borders or shared resorces, or one is simply viewed as a group to be conquered and their resources taken.

When both groups are nation states, the conflict is a war, and is settled by battles between armies (see Figure 16.6 for the importance of large populations for nation-state armies). However, when one group is an agricultural nation-state and the other is a group of hunter-gatherers, the result is not typically war. Agricultural nations can afford to pay, feed, and

Box 16.4 (cont)

Figure 16.6 Nazi pro-natalist propaganda. Agriculture, with its dependable food supplies, can support larger populations. Throughout history, having a large population was critical for the success of a society when competing with other societies, because, other things being equal, a large army will defeat a smaller army. To this end, agricultural societies have often encouraged reproduction. This has taken many forms, such as restriction on birth control and encouragement of marriage and reproduction. One notable example was seen in Germany under the Nazis. Under the explicitly biologically expansionistic policies of *Lebensraum*, certain peoples in Germany were encouraged to reproduce, and the Nazis captured large areas in eastern Europe with the intention of expanding their populations into newly vacated agricultural areas. Here we see some Nazi pro-natalist propaganda. German eugenics programs encouraged Germanic ethnic groups to multiply, even while the Nazi regime was sterilizing or murdering other ethnic groups.

Box 16.4 (cont)

train large armies of men whose specialty is warfare. They are typically armed with the most sophisticated weapons available (at the time), and they are skilled with those weapons. They are supported by large logistics trains. Hunter-gatherers are composed of small groups who are skilled at acquiring food, but they are not skilled at organized, armed conflict on a large scale, and their weapons are typically primitive. While individuals may be skilled at bushcraft, hunter-gatherers can offer no real resistance to an organized army.

We can look at multiple examples throughout history of foraging peoples being displaced or destroyed by armies, for example: in the American West, when Native American peoples were killed and displaced; in the fourteenth century in Central Asia, when Genghis Khan's armies captured all of Central Asia; in the second millennium BCE, when the ancient Egyptians conquered southern Sudan; in the eighteenth and nineteenth century, when the Bantu peoples expanded into South Africa, displacing the !Kung San; in the nineteenth century, when the Belgians conquered the Congo basin; and in innumerable other cases throughout history. In all of these cases, the foragers could offer little resistance, and did their best to move away from the armies.

In terms of our modern moral assessments, some cases were particularly egregious. The Mongols killed or made slaves of all people who resisted. The Belgians were known to be particularly cruel in their punishments of newly enslaved Congolese. All native Tasmanians were exterminated by the British settlers. Some cases were somewhat less cruel – sometimes the foragers were forcibly integrated into the agricultural society, as in the "Indian Schools" of the USA and Canada (after all armed resistance was quelled with violence). But there is no question that foragers always come out on the losing side. The best possible outcome is probably to be ignored, as in many Arctic native tribes in the USA and Canada. Even today, in the Amazon Basin, powerful economic interests are cutting down the rainforest for timber and agriculture at a prodigious rate, with little regard for the peoples living there. Assuming you are a student in your late teens or early twenties, there is a good chance that within your lifetimes the last of the hunter-gatherers may no longer be living our original lifestyle, as we did for more than a quarter million years. Whether this is good or bad depends, of course. We may lament the end of a connection to our deep past, but for them it may simply be rational to choose to live in a world with more reliable food, modern healthcare, and potentially a say in how the government treats them.

Box 16.5 Disease and the Neolithic

One of the characteristics of agricultural populations is that people live in close proximity to animals. The ability of diseases to cross from animals to humans has led to some of the most destructive diseases in human history. For example, tuberculosis, a bacterial disease that is today readily handled by modern antibiotics, was a fatal lung disease for humans for millennia. This disease originated in livestock, was subsequently spread among humans, and was a death sentence once caught. This distribution pattern was (and still is) the case for many of the diseases, including leprosy, influenza, Ebola, bird flu, cryptosporidiosis, HIV, encephalitis, smallpox, and COVID-19, among others.

In some cases, these diseases acted as agents of natural selection. European populations, which had been exposed to smallpox for centuries, had acquired partial immunity through selection against individuals who were more susceptible. This selective process killed millions of people in Asia, Europe, and Africa but left the remaining populations relatively resistant. However, since smallpox had not been present in the New World, this immunity was not present in Native American populations, who died in large numbers when coming into contact with European populations carrying the virus. Today, populations of hunter-gatherers are highly susceptible to viruses that urban populations have long acquired partial immunities to. Recently, high percentages of some foraging populations in the Amazon have died when being inadvertently exposed to influenza viruses.

wealth that feeding them imposes no real burden on others in the society. The Neolithic is when we see the appearance of written language, for example, which requires people with specialized skills acquired over several years (scribes). Architecture is another job that appears with the Neolithic. You only see the appearance of interesting monumental architecture with the start of the Neolithic (the Pyramids, Great Zimbabwe, Teotihuacan, etc.)

Negative changes include:

1. Social control. Some people were able to accumulate more resources than others, and, over time, they were able to control land and the food resources the land generated. This is also when wealth became heritable – a parent would pass control of land to their offspring. This type of inherited wealth allowed some families to become extremely powerful over long periods of time. This is the origin of "private property."

2.   Armies/police. The accumulated resources might not only be used for the creation of an artisan class – the same accumulated resources could be used to create a professional fighting force. In hunter-gatherer societies, everyone is a generalist, which means folks learn all skills. In Neolithic societies, there was generally enough surplus to have a special group who spent their time learning how to fight, often making them very effective at imposing the will of those who paid them on the rest of the population. Individuals lost significant power, as these military/police forces were able to force individuals to acquiesce to societal norms.

3.   Reproductive control. In hunter-gatherer societies, there is no general sense that people must reproduce for the good of the group. In these societies, children are a significant burden. People reproduce at the "natural" rate of one child every three to four years, and there is no interference from society. However, in agriculture-based societies there is a social drive to reproduce. This is because, in contests between societies the one with the biggest army typically wins. We often see reproduction encouraged in societies today, and there are often state or religious rules restricting personal reproductive control (laws against birth control or abortion). Although, today, there are no real explicit attempts to increase populations for these reasons, these laws are artifacts of those early societies' efforts at increasing population.

All of these changes set the pattern that we see in the world today. Chapters 17–24 explore the various implications of these changes, but one of the most consequential, from a resource standpoint, is population levels. Until the Neolithic transition, the population was increasing but at a very slow rate. After the Neolithic, the population of the world started its exponential growth pattern. Only 10,000 years ago there were perhaps 20 million people in the whole world. Today, there are cities with that many people, and the overall world population is approaching 8 billion. As we will see in Chapters 17, 23, and 24, demographic issues have been a major driver of world history.

Chapter 17: States and Nations

Probably the most profound effect caused by the development of agriculture was the enormous increase in the human population growth rate. We don't have any way of measuring population numbers worldwide in our more distant past, but estimates have been made based on modern hunter-gatherer populations. Genetic variation also provides some estimates of early demographics. The results of these methods vary, but before 10,000 years ago the worldwide human population was likely to have been between 1 million and 10 million.

Once reliable production of food was possible, the population started to increase steadily worldwide (Figure 17.1). These increasing population densities generated some new problems for these early agriculturalists, and the ways they dealt with them were largely the same around the world. Not surprisingly, these problems had to do with numbers.

In a foraging group, leadership is based on the charisma, judgment, and personal political skill of any one person. The leader personally knows each member of their group, and the relationships among the members of the group, so that they can negotiate various alliances and rivalries. Humans likely evolved their large brains, at least partially, to help store and process political information such as this. A population of 100–200 is the predicted number of an average human's "political database," based on our brain size, when compared with other social primates.

Having the ability to understand the political relationships among fewer than 200 individuals would be adequate for a leader in a foraging "big man" society. However, once the population starts to grow, there is no way to keep track of the relationships. This is similar to the problem with having a larger number of neurons in the brain – the number of connections among them grows far faster than the neurons. The same math problem applies to human relationships. Among 100 people there are about 5,000 relationships, but among 1,000 there are roughly half a million. So, when populations get large, no leader can possibly understand and process the relationships among their group.

The way to handle such a problem is with hierarchies (Figure 17.2). In a hierarchy, each person at their level on the hierarchy only has to know the people directly above and below them on the hierarchy. Information flows up

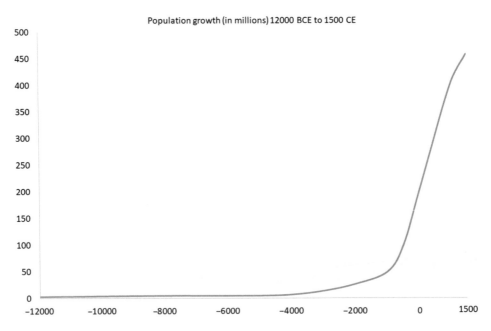

Figure 17.1 Here we see population growth the world over from 12000 BCE to 1500 CE, with the Y-axis representing the worldwide population, in millions. Population growth was slow, over most of the 250,000 year span of our species. But between 6,000 and 5,000 years ago (4000–3000 BCE), agricultural development had spread around much of the globe and populations started to substantially increase. This is when we see the appearance of chiefdoms and states, and the growth of urban centers

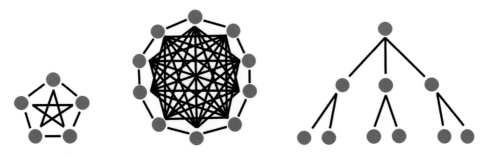

Figure 17.2 In a small population, it is possible for one individual to know all the people in a group, as well as their relationships with others. However, as groups grow, the amount of information becomes too large to store and process. In this image among five people (left), there are ten relationships. Among ten people (center), there are forty-five relationships. And among a hundred people (not shown), there are 5,000. However, in a hierarchy, it is possible to reduce the number of relationships so that one individual only has to know a smaller number of people, but information still passes efficiently through the system. On the right, there are ten people, and only ten relationships, yet information can pass through the system to anyone in the hierarchy. This is why, in large populations, hierarchical systems are necessary to avoid information overload and potential anarchy. Those higher in the hierarchy have more power and freedom than those below, partly because of their ability to access more parts of the hierarchy.

and down the hierarchy, and at each level information is processed and passed to the relevant individuals. Imagine, for example, the leader wants a bridge built. That leader simply has to know the head of the construction group, and that head will pass information down to the bridge-building head, who then passes the information to the foremen, and from there to the workers. The leader never even has to meet the bridge-building workers.

But why do we need decisions to be made this way? The reason is that large populations need certain things to thrive. For example, in small population groups water might be pulled from the river individually by each family. But once the population grows large enough, there may not be enough access to the water. Some people upstream might be watering their crops and taking so much that there is not enough for people downstream. So the group has to either build some form of water body (like a dam or a canal), or restrict the usage of some people. All of these decisions must be made across a large number of people. Without some sort of hierarchy, it is very difficult to make a communal decision. Someone has to sort out the competing interests. Without a hierarchy, you would have individuals and families competing for the water, and there might well be conflict. Even the most egalitarian large-scale societies have hierarchies because anything else would be anarchy.

However, as you may have guessed, there is a problem with hierarchies. In any hierarchy, the people at the top necessarily have greater power than those below. Power, once acquired, is typically retained if at all possible (as also is the case for chimpanzees and other species). Because the people at the top have the power, it can be very difficult for other people lower in the hierarchy to remove the leader. The result of this, of course, is that the power is often passed down to offspring, creating powerful lineages. This is the origin of the most common form of government for more than 5,000 years.

The Evolution of Governance

Across the world, as populations increased, the transitions from one form of government to another followed the same broad pattern. This suggests that there are underlying principles that govern how people relate to each other, both in small numbers and in large. Traditions and customs vary widely across the world, but there are certain patterns that appear to be consistent.

The driving factor is population size. Anthropologists have used the four main types of governmental systems when talking about societies, although there are, of course, cases that fall in between the categories, and it is not necessarily a linear trend in the evolution of governance if conditions radically change. But, overall, patterns of social organization and governance follow fairly consistent patterns (remember: these are broad descriptions rather than prescriptive rules).

Table 17.1 Structural patterns across societies

	Size	Subsistence	Economy	Social structure	Distribution of power	Property
Band	10–50	Foraging	Reciprocity	Egalitarian	Local	Communal
Tribe	100s–1,000s	Pastoralism, horticulture	Reciprocity, redistribution	Minor status divisions	Regional	Communal
Chiefdom	1,000s +	Agriculture	Redistribution	Ranked lineages	Central – lineages by "divine right"	Private property
State	100,000 +	Intense agriculture, industry	Market	Classes, castes	Central – in offices based on law	Private property

Bands. As discussed earlier in this chapter, bands are small-scale foraging populations with a single charismatic leader, and this was the ancestral form of social structure and governance. These bands have simple economies in which goods are exchanged. If one person gets lucky and finds a bee hive with honey, she might be able to exchange that honey for meat from a hunter. Power is not passed down to the next generation and any power that the leader has is strictly limited to the people he or she personally knows.

Tribes. The next step in social organization is the tribe. In a tribe, several bands may identify as a group because of shared traditions, religion, or language, and organization can be temporary or permanent. In a tribe, one of the leaders of a band may have enough charisma and political power to be able to organize the bands, as in against a common enemy. In North America, Native Americans would often join bands into tribes to fight a common enemy. For example, the Lakota Sioux leader Crazy Horse organized independent bands of Lakota for a war against US government troops. More permanent tribal affiliations are seen in the linguistic groups of Africa, who tend to identify themselves by their language. For example, the Afar people of Ethiopia all speak Afar, and while there are individual bands that have local leaders (these groups are called kibelis by the Afar), there is an elected Afar head.

The concept of a tribe is an intermediate step between band and chiefdom, and tribes may be loosely or closely affiliated. In the Western Apache of Arizona and New Mexico, the bands of Apache would offer hospitality to members of other bands, but they never united under a single leader, and they occasionally fought against each other. In countries today, many people have tribal affiliations that represent artifacts of their prestate societies, and often tribes form voting blocs.

Because of the fluidity of tribes and tribal organization, some anthropologists have questioned the term, but it a useful way to think of how smaller groups organize into larger groups without a truly powerful central authority.

Chiefdoms. In some tribal groups one or a few families acquire political power that is retained within the lineage. Power is no longer earned by charisma or political acumen but passed down from parents to offspring. For families to be able to pass down power there must be true inequality in the group. In tribes and bands, people are able to challenge the leader, but once one person or family acquires enough power that is no longer an option for other members of the group.

This is a major transition, because this is the origin of permanent central power structures. In a chiefdom, not only is the power extended through generations, but this is also the origins of central authority. In a chiefdom, the chief has power over all of the inhabitants of the chiefdom.

Anthropologists regard the transition to chiefdoms as the origin of social inequality. Powerful lineages are able to acquire more resources than other members of the group, thereby ensuring that they have the resources needed to retain power. This can take many forms, but typically it means controlling large amounts of land, which gives them the resources to control a warrior class who will fight for that lineage. In fact, the transition to chiefdoms is where we first see the concept of "private property" that is under the control of an individual or lineage.

We see chiefdoms throughout history, as agriculture took hold and populations increased. For example, before the Roman Empire conquered Great Britain and mainland France and Germany, most of those areas were under the control of small chiefdoms that often warred with each other for control of agricultural land. In North America, the Algonquian leader Powhatan (also known as Wahunsenacawh) was the hereditary leader of much of what today is Tidewater Virginia.

State societies. Most modern countries are state societies. In a state society, the leadership can be an inherited position (as in a king) or it can be an elected position (president or prime minister). But the primary difference between a state society and a kingdom is that in a state there are formal institutions, often with written laws and rules. Borders with other states (countries) are generally formal delineations.

Within state societies there are often power hierarchies, so that there may be a lower or working class and many levels of power between them and the ruler. In European countries, these were represented by lesser nobility, such as counts, barons, dukes, counts, viscounts, marquis, and earls. Other countries have or had similar systems – the Chinese aristocracy, up until the twentieth century, had, in descending order of power, had gong, hou, bo, xi, and nan. The Ethiopian aristocracy had negus, abeto, ras, bitwoded, and lij.

The governments of state societies typically have absolute authority, and they exercise considerable control over their citizens (Box 13.2 and Figure 17.3). Because of the degree of control they exercise, they often have large bureaucracies. These large bureaucracies are involved in such activities as accounting, land records, legal disputes, and taxes. Much of what we know about the ancient world comes from early bureaucratic records in languages like Sanskrit and ancient Egyptian.

In addition to control over their citizens, they also have a tendency to attempt to grow. Of course, when a state grows, it can only do so at the expense of neighboring kingdoms and states. This is the origin of imperial conquest, which has been a major factor in world politics since at least 2300 BCE, when the Akkadian Empire expanded across the ancient Middle East.

Figure 17.3 Monumental architecture appears when societies become complex and the leader can accumulate sufficient power to engage in large-scale works projects. Many of these served no societal function beyond the ritual demonstration of power. This happened around the world after the Neolithic transition occurred in each region. In this image we see (from top left): the Pyramid of Kulkucan from Chichen Itza, Mexico (600–800 CE); the Great Stupa at Sanchi, India (125–150 CE); the remaining walls of Persepolis, Iran (fifth century BCE); and the pyramids at Gizah (2500 BCE).

Box 17.1 Power

Power is the basis of any political system. The forms it takes (political, military, economic), and how it is acquired by individuals, distributed among the population, used, accumulated, kept, and lost, are what drives the stories and patterns of history. The patterns tend to repeat themselves whenever similar conditions arise, and because of this there appear to be general universal principles.

In a band or tribe, power is generally bottom up, and if the leaders lose support of the population they tend to quickly lose their position of authority. However, in chiefdoms and states power can be retained because, once permanent control of resources is possible (typically land), they are able to leverage their resources into economic and/or military power. When that occurs, those at the bottom of the power structure have little ability to remove leaders from power. This is the origin of social inequality.

In a chiefdom or state, there is a direct relationship between power and constraint. Individuals closer to the top of the power structure have more power and less constraint, and those below have less power and more constraint

Box 17.1 (cont)

(Figure 17.1). History has many, many examples, but the condition of Russian serfs helps illustrate this. Until 1861, Russian serfs were the peasant class of Russia. They were connected to a piece of land controlled by a local aristocrat and did not have the right to leave the control of their landlord. Serfs could be bought and sold, and a serf sold to another landlord had no right to retain any personal property, nor to their own family (a husband could be sold, and the rest of the family retained, for example). Any serf who fled could be tracked down and returned to the landlord. The landlords had an incentive to keep the serfs powerless (since the serfs worked their land) and the serfs had no ability to acquire power. The patterns of history suggest that, once lost, power is very difficult to reacquire. In fact, it was not until the nineteenth century, when Czar Alexander II of Russia banished serfdom in 1861, that the serfs became free; they did not free themselves.

In any given state, there is a finite amount of power. The more power that the government retains, the less there is available for the members of the society. Leaders throughout history have tended to try to maximize their power. The members of society have naturally tried to reign in the central authority and retain their own power and rights. One famous example of the struggle between central authority and the society is the *Magna Carta*, an English document written and signed in 1215 by King John of England and a group of lesser English nobles. Although it is most often called the *Magna Carta*, the actual name is *Magna Carta Libertatum*, which means the "Greater Charter of Freedoms." The goal of this document was to explicitly limit the powers of the king. The modern notion of *habeas corpus* comes from this document, which prevents arbitrary imprisonment of citizens. Other rights were specifically enumerated in this document – protection of private property, guarantee of legal process, protection from excessive taxation, and others. It was a direct rejection of the notion of absolute and complete authority of the king, which King John had previously accepted as his natural right. John did not willingly sign it, but political circumstances had weakened him to the point where he had no choice.

The struggle over power between citizens and the government has been a central part of history. In most of the ancient world, the power of the ruler was absolute (*authoritarian*), but ancient Athens famously had a participatory democracy in which free male citizens could vote on decisions to be taken by the government (note that women were not allowed to participate, nor were slaves, which Athens had in abundance). In Athens, the free male citizens directly voted on the laws and policies. This is known as *direct democracy*. However, when populations get large, and the administrative bureaucracy gets complex, it is not possible for every citizen to have the time to vote on every law. In ancient Rome, under the Republic, citizens instead elected

Box 17.1 (cont)

senators to represent them in the law-making process (this is known as *republican democracy*, named after the Roman Republic where it was developed). This republican democracy lasted roughly 500 years, until it was lost when Rome acquired its first emperor in 27 BCE. There was no true democratic government worldwide for almost 2,000 years afterwards – until the founding of the United States in 1789.

The struggle over the balance of power between individuals and their government continues to this day, in such debates as government surveillance, taxation, civil liberties (such as same-sex or inter-racial marriage rights, "stop-and-frisk" laws, freedom of the press, rights of citizens to own firearms, freedom of access to government information), and many other issues. Among countries today, there is a wide range with regard to the authority of the government (see Figure 17.4). Authoritarian governments, such as those in North Korea, Saudi Arabia, and China, afford relatively few formal rights to their citizens. More libertarian countries, such as Switzerland, the USA, Sweden, Norway, and Canada make specific legal provisions to maximize government accountability and individual freedoms. In the USA, the US Constitution was specifically written to constrain the government and specified its limitations on power; since that time, many countries have adopted similar written constitutions with the same ends.

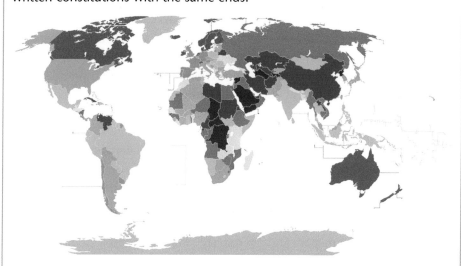

Figure 17.4 The Democracy Index (compiled by the *Economist*, and here mapped) is a visual representation of the balance of power between the top of the hierarchy and the lower tiers. In this map, green represents a more democratic regime, and red a less democratic regime. Rankings are based on several measures, such as access to voting rights by the population, transparency of the government, trust in the government, and the degree of civil liberties in the population. Five hundred years ago not a single country would have been green on this map.

Box 17.2 State coercion

One of the premises of a people coming together and forming a society is that people in that society will not be able to have everything they want or do anything they want. In a system in which everybody is completely unconstrained, you have anarchy, where the strong take from the weak without consequence. So, necessarily, if people are to come together to make a society, there must be laws or regulations that determine what people can and cannot do. Typically, many of the political debates within a country are about what types of laws constrain citizens and what freedoms they will enjoy. In some countries, especially those with written constitutions, there are explicit limits on the power of the government in order to ensure the relative freedom of its citizens.

The government typically coerces citizens through its economic, legal, administrative, and regulatory power in several ways.

Police power. In the United States and England, the right not to be imprisoned without trial is known as *habeas corpus*. The right to live and move about in public without arbitrary arrest or imprisonment is considered one of the core human rights, and has been throughout history. But in many countries, people can be detained by the police without recourse – that is, they have no legal right to challenge their imprisonment. This is common in countries with strong central governments, such as North Korea, China, Iran, and Saudi Arabia, but can also be found in Western countries. When armed factions in Northern Ireland rebelled against the English government during the 1960s–1990s, the English government sometimes responded by imprisoning suspects for extended periods without trial. During World War II, the United States government imprisoned ("interned," as it was then phrased) some 120,000 Japanese Americans without specific evidence of any treasonous activity. However, these were both relatively unusual circumstances (and, in the case of the Japanese Americans, the US government later admitted fault); in most Western democracies, a person must be accused of a specific crime to be imprisoned, and they must be able to defend themselves against the accusation.

In more authoritarian states, the right to possess weapons was historically limited to the police/military and the nobility, often as symbols of their power. The state normally has no incentive to allow a population to be armed, since a disarmed population has less potential to overthrow the government. Because of this, many countries have, and still do, maintain a monopoly on the use of force. The possession of weapons has been a conundrum for many countries – having an armed populace means the leader has an army that can

Box 17.2 (cont)

be assembled more rapidly (in medieval England, owning, maintaining, and training with a longbow was mandatory for military-age males), but the possession of arms in a population can also be destabilizing for the government itself, since armed citizens are less likely to defer to an oppressive central authority. Shifts in technology and modern professional armies have rendered large citizen armies largely unnecessary, so governments since World War II have trended toward a reduction in armed citizens. This issue is oft debated, especially in the USA, which has a tradition of armed insurrection.

Freedom of speech/press. The freedom to express thoughts is closely linked with democratic institutions, and is less prevalent in authoritarian regimes. In many countries (China, North Korea, Russia, Saudi Arabia, Iran, Turkmenistan, and Eretria, to name but a few), the publication of outlawed material can result in fines, imprisonment, or even death (sometimes by assassination, sometimes by execution), and publications that challenge or criticize the government are especially sanctioned. Surprisingly, however, even in the UK there is no specifically enumerated freedom of the press, and debates continue about the freedom of the press in that country.

Although we typically tend to think of freedom of the press as the freedom to criticize the government, it also takes other forms. Up until the second half of the twentieth century, the United States and England, for example, banned publications deemed "obscene." The "Comstock Laws" in the United States outlawed publications relating to contraception and abortion, and were only rescinded in 1957.

Property/taxation rights. The ability of the government to take money or property from individuals is necessary for the government to function. Funds have to be available for militaries, as well as infrastructure such as roads, public buildings, salaries for government workers, and general maintenance of the physical structures of government. However, taxation can impose a considerable burden on individuals, so the amount of taxation imposed on individual citizens, and *which* citizens, is a source of constant debate. The size of the government, and the tax burden it imposes, fluctuates in countries around the world. Dissatisfaction with taxation was one of the sources of the American Revolution, as well as many other revolts and protests from ancient times to the present, including the Jewish Revolt in the first century, the Norman anti-tax riots of the fourteenth century, the Revolt of Ghent (1539), the Japanese peasant uprisings of 1717, 1752, and 1770, and into the twenty-first

Box 17.2 (cont)

century in places like Egypt, Pakistan, Madagascar, Ivory Coast, Russia, and Iran, to name just a few.

Conscription. In many countries throughout history, service in the military has been an obligatory part of citizenship. Citizens are often willing to join the military if the country is under attack, but militaries have frequently been used for wars of adventure, or to pursue political goals not related to defense. In these cases, countries have resorted to conscription to build up the strength of the military. Conscription has been common from ancient times through to today. The Babylonian Empire (seventeenth century BCE) used conscription during wartime, and in the Shang Dynasty of China conscription was used to fill the core of the military after the twelfth century BCE. Many countries, including Sweden, Switzerland, Algeria, Iran, North Korea, and South Korea, still have conscripted armies.

The principles of conscription are still debated. The value to the government's political and military goals are clear enough, but the right of a country to demand an individual face death or injury for those goals is less clear. In the United States, conscription during the two world wars was largely accepted, but during the Vietnam War, when the war was not one of self-defense but about a theoretical struggle against communism, the country largely rejected the draft and there has not been conscription since.

The question of who gets drafted is also something that countries struggle with. Traditionally, most countries only drafted males of a certain age. But recently, with the increased prevalence of women in militaries, the justification for exempting women has been challenged. Already, Norway, Sweden, Israel, and a few other countries draft women, but others are trending in that direction.

Additionally, the burden of the draft has often fallen on the poor. In the Babylonian Empire, the wealthy could pay for a substitute if drafted; this was specifically outlawed by the Code of Hammurabi. Unfortunately, this lesson had to be relearned: during the US Civil War, it was possible to pay to have a substitute in both the Union and Confederate armies. By the time of World War I, the use of substitutes had been outlawed. However, during the Vietnam War, US college students had their draft status deferred. Since college students tended to come from wealthier backgrounds, the draft drew disproportionately from the poor.

Reproductive/sexual rights. Although decisions about sexual behavior and reproduction seem highly personal, and largely outside the scope of government interference, this has not proven to be the case. Many governments have sought to regulate the right to control reproduction

Box 17.2 (cont)

(birth control) as well as determine who is eligible for marriage or legal union. The origins of this may have to do with the pro-natalism of governments looking to increase the population size, although it often presents itself through the lens of religious morality (see Chapter 18).

Although pro-natalist policies in modern times sometimes take the form of tax incentives for parents, there have also been strict laws about contraception and abortion throughout history. In Romania, the dictatorial leader Nicolae Ceauşescu outlawed birth control for most women in 1966, specifically to increase the population. The right to birth control has, relatively speaking, only become legally established recently. In the United States, the Comstock Laws were only overturned in 1957. Today, the official policy of the Roman Catholic Church is in opposition to contraception.

Equally intrusive have been the laws regulating same-sex relationships. In ancient Greece, Rome, Persia, and China, homosexuality was widely accepted, and there was no impulse for the government to outlaw it. However, since the spread of Christianity and Islam, it has been outlawed for much of the last 1,500 years. Only recently has it been legalized in many Western countries, and this may be because most Western countries no longer explicitly desire to increase their populations, along with a general trend toward a lessening of the political power of religion.

Slavery. Perhaps the utmost expression of sovereign power is slavery. Unfortunately, slavery has been present in human societies for as long as there have been societies. Hunter-gatherers have been known to take slaves, as did the earliest agricultural societies. The use of slaves was well known in classical antiquity in the Mediterranean (Greece, Egypt, Rome), western Asia (Mesopotamia, Babylonia, Persia, Assyria), the Asian empires of India and China, and the Native American empires.

Slaves are the lowest rung of any society, and they are generally denied many or all citizens' rights by the government. Slavery has taken various forms throughout history, but slaves typically originate outside the society and are therefore denied the legal protections of the society typically affords its members. For example, the Vikings, Norse raiders of the Middle Ages, would often take slaves when they raided the European and Mediterranean coastline, many of whom they sold or traded. Similarly, when the Natchez of the Mississippi Valley captured prisoners, they would typically kill the men and keep the women and children as slaves. In the wars between the Byzantine Empire and the Ottoman Empire, captured prisoners were condemned to be

> ## Box 17.2 (cont)
>
> galley slaves, where they would spend the rest of their days rowing heavy warships.
>
> ### Slavery in the USA
>
> Social scientists who have examined the overall history of slavery have concluded that, until the early nineteenth century, it was nearly universal. In fact, it is difficult to find a single society prior to that time that did not have slaves at some point in their history. The United States has a very particular history with slavery, and this history resonates strongly in the country today. Slavery was ubiquitous around the world when the USA was founded, so it is not particularly surprising that slavery was used in the plantations of the South. But the United States was also founded during the Enlightenment (Chapter 19), when much of Europe was challenging the morality of slavery, and the idea of civil and human rights was becoming prominent. Following the founding of the USA, slavery was outlawed in Denmark (1803), Sweden (1813), Prussia (1815), Canada (1819), Chile (1823), Mexico (1829), France (1826), Greece (1832), England (1833), the Catholic Church (1839), Tunisia (1846), Brazil (1851), and many other countries and regions. All of these countries outlawed slavery peacefully (although Haiti revolted against France and banished slavery in 1804 – the only successful slave revolt in modern history).
>
> But in the United States, southern plantation owners strongly resisted, and they did so out of economic self-interest. At the outset of the US Civil War, roughly half of the equity (economic assets) of the Confederacy was in the market value of some 3 million persons held in bondage. The southern plantation owners resisted emancipation because they did not want to lose the money those people represented. During previous eras, when slavery was more commonplace, economic justifications were presented starkly – in ancient Rome and Greece, people who owed money could find themselves enslaved to pay off the debt.
>
> In the United States, the principles of the Enlightenment clashed with the brutal economic realities of the plantation system, and various other reasons were offered for maintaining slavery. The two main ideas were: (1) biological (the supposed "natural" inferiority of Africans to Europeans – see Chapter 12); and (2) religious (some southern clergy tried to justify slavery by appealing to the "curse of Ham," in which one of Noah's sons, Ham, is cursed, and is held, by some traditions, to have founded the population of Africa). Both claims were absurd, on their merits and as justifications, but the issue was not settled until the Confederate defeat in 1865.

Box 17.2 (cont)

The US Civil War represented the end of slavery in much of Europe and the USA, but it remained a presence in some places, such as Bhutan and Ottoman Turkey, into the twentieth century (Oman did not outlaw slavery until 1970). Today, all countries explicitly outlaw slavery, although it is present in many countries where enforcement is lax.

Box 17.3 The Age of Empires

Human groups have always found themselves in conflict with other groups. This is not surprising – chimpanzees do the same thing. Human groups attack other groups for the same reasons chimpanzee groups do – to acquire resources. But it is not until we get large agriculture-based state societies that we get large-scale warfare. When states war against each other, the victors have traditionally taken control, and sometimes outright possession, of the land and resources of the losers. When states (countries) expand and take over other countries, we call them empires (see maps in Figure 17.5).

Today, the concept of an empire is closely associated with what we call "imperialism" or "colonialism." This is because, today, many of the borders of the world are where they are as a result of the fairly recent (sixteenth to nineteenth century) imperialism of European powers. Today's language distribution also reflects this recent imperialism – Spanish is spoken in Peru, French in Morocco, and English in Kenya, because those countries were conquered by Spain, France, and England (see Figure 17.6).

However, imperialism has been a constant for at least 4,500 years. The Akkadian Empire, in modern-day Iraq, under King Sargon conquered much of the modern Middle East. For the next four millennia, empires rose and fell. To give a sense of the fact that there was never a period when empires were not rising and falling across the globe, here is a partial list of powerful empires: the Assyrians (2330–2150 BCE), Babylonians (1900–1600 BCE), Hittites (1600–1270 BCE), Egyptian Empire (New Kingdom, 1550–1070 BCE), Zhou Dynasty (1046–256 BCE), First Persian Empire (also known as the Acheamenid Empire, 522–330 BCE), Nanda Empire (450–350 BCE), Alexander the Great (334–323 BCE), Roman Empire (including the Byzantine period,

Box 17.3 (cont)

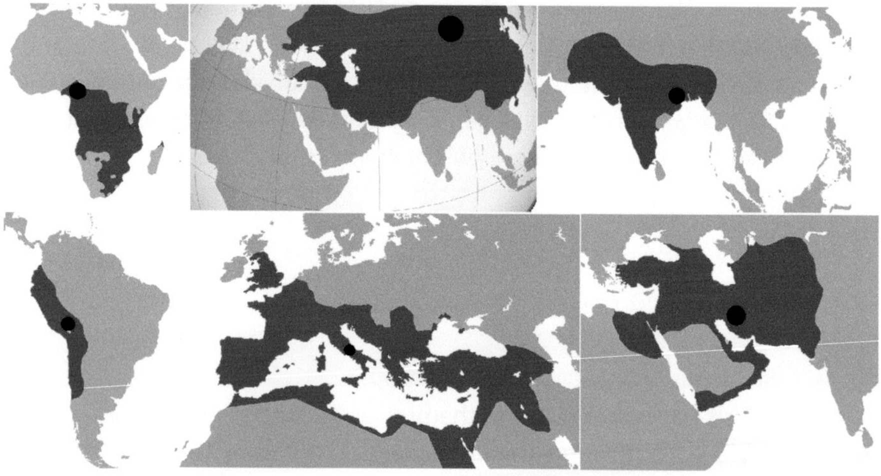

Figure 17.5 Throughout history, peoples have expanded and displaced their neighbors. In this image we see six premodern populations across the globe, from over the last 4,000 years, that expanded to control areas far beyond their origins. Starting at the top left and moving horizontally, then down, we see the furthest extent of the following groups: Bantu, Mongol, Maurya, Inca, Roman, and Persian. The black circles represent the geographic origins of the populations.

509 BCE–628 CE), Xiongnu Empire (300 BCE–100 CE), Han Dynasty (200 BCE–220 CE), Axumite Empire (150–940), Jin Dynasty (265–420), Ghana Empire (500–1200), Gupta Empire (320–550), Islamic Empire (the four major caliphates up to the Ottoman Empire, 632–1260), Ethiopian Empire (1270–1935), Mali Empire (1230–1600), Mongol Empire (1206–1368), Ottoman Empire (1300–1918), Spanish Empire (1500–1830), French Empire (second colonial empire, 1830–1960), British Empire (1650–1997), Zulu Empire (1818–1879), Empire of Japan (1894–1945).[1]

Today we tend to look at the modern countries that formed part of the colonial empires of England, France, Spain, and the Netherlands as uniquely victimized by their colonial powers. However, a perusal of ancient and

---

[1] Note: these dates are often the subject of debate among historians, but I use dates that reflect the rise of the empire, its time as a regional or international power, and its decline, rather than the founding dates of a country or birth of a ruler. Some dates are estimates, or reflect imprecision, and are rounded to the nearest decade.

Box 17.3 (cont)

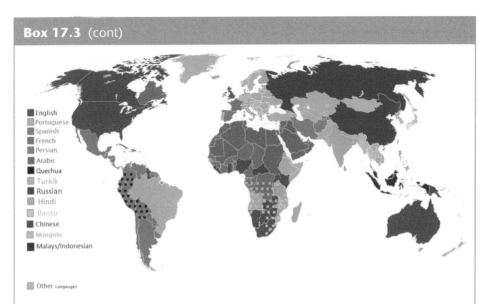

Figure 17.6 The languages that span the globe reveal the empires of the past, some quite distant. In this image we see the recent (twentieth-century) imperialism of the Soviet Union, with Russian control over much of northern Eurasia, then back into the nineteenth century the English and French, then going back a few hundred years more the Spanish in Latin America. From further back, we see the Arabic over North Africa, from the Arab conquests taking place from the sixth century through the thirteenth century. The Mongols spread their language in the thirteenth century. In South America, Quechua, the language of the Inca Empire, is still spoken across western South America, and in Africa, Bantu dialects spoken across sub-Saharan Africa are the remnant of the Bantu expansion from roughly 1000 BCE to year 0. Chinese, Turkish, and Indian languages are similar evidence of past invasions and empires.

medieval history books shows that conquering other peoples and controlling their lands has been a part of the expression of power ever since people have had governments.

Probably the most powerful empire, in terms of military and geographic power, and the most destructive, was the Mongol Empire. The Mongols were horse people from the steppes of Mongolia, and under the leadership of Genghis Khan (1162–1227), captured most of mainland Asia, as well as parts of the Middle East and eastern Europe. Expansion only stopped with the death of Genghis Khan's son, Odegai, in 1241, although his descendants retained parts of the empire into the fifteenth century. During the expansion of the Mongol Empire, some 16 percent of the Earth's land had been conquered and 30–50 million people had been killed, which amounted to as much as 12 percent of the world's population at the time. Half of the population of Hungary was killed by the Mongols, and some 75 percent of the people of Iran

Box 17.3 (cont)

died by violence or starvation. A single punitive expedition to western China directly resulted in 1.5 million dead. Among their most infamous acts was the burning of the famous House of Wisdom in Bagdad, which was the greatest library of the time. It was destroyed when the city was sacked in 1258. Entire regions were effectively wiped clean of their populations, with dams and irrigation canals destroyed and fields burned. In the depopulated areas, regrowth of forests, and the consequent cleansing of the air-borne carbon, has been identified by environmental scientists as being the product of the Mongolian conquest.

Until modern times, most of the world accepted the principle of "woe to the conquered," which is, essentially "might makes right." There was no greater underlying principle of justice in international relations, and there were no international bodies to mediate among international disputes. Disputes among countries were commonly settled by military conflict, not least because any display of weakness could invite invasion by a stronger neighbor. It was only in the eighteenth and nineteenth centuries that Enlightenment philosophers started to question the philosophy of strength, and it was not until the twentieth century that any international bodies evolved to prevent such conflicts. The outrage expressed around the world at the German and Japanese invasions of the 1930s and 1940s reflected this shift in attitudes from premodern times.

The Age of Empires ended with World War II. Although parts of Africa and southern Asia had been colonized for more than a hundred years, the colonial powers of France and England had been forced to turn to the populations of these countries in their fight against the Axis. Soldiers from India, Sudan, Morocco, and Algeria all fought alongside English and French soldiers in battles against Germany and Japan. One of the clear lessons these peoples learned was the value of freedom, not least because they were fighting for the freedom of their colonizers. One by one, after the war, almost all of the colonies in these regions demanded, and ultimately earned, their independence (although it was not always readily granted). In France and England, it became impossible to justify ruling over another people, when France and England themselves had just fought a world war to avoid being ruled by Germany and Japan. Today, the world is relatively (but not absolutely) free of colonialism, and the world no longer accepts the philosophy that the military strength of one country is enough to justify ruling another.

Box 17.4 The origins of "rule by divine right"

One important development that occurs with the appearance of chiefdoms and states is that, frequently, the leadership in a lineage is closely tied to the religion of that group. This is where we first see the concept of "rule by divine right." One way to justify the legitimacy of a particular lineage is by arguing that the leader is literally "chosen by God." For example, in traditional Polynesian societies, the ruling families were considered to have had greater "mana" (spiritual energy) than other members of society. The chiefs were believed to be able to trace their ancestry back to the times of the ancient gods, and were themselves a present manifestation of the ancient Polynesian gods. In another tradition, well known from the Old Testament, David is "anointed" (oil is poured on him) as king of the Jews, which establishes a degree of divinity. This tradition of anointing the king was followed by many European monarchs, for similar reasons.

The advantages, for the ruler, of claiming leadership by divine right is pretty clear: for anyone to overthrow the leader is also to reject and overthrow the religious system itself. If a god has chosen the leader, then it can be very difficult for people to justify rejecting that leader. To go against the leader is to go against their god. This couples political revolution with sacrilege.

The close relationship between the power of a chief or monarch and the religious order has been very powerful throughout history, in both chiefdoms and states. In ancient Egypt, the pharaoh was both the head of the state and the head of the Egyptian religion. In ancient Rome, after the end of the Republic and the establishment of imperial rule in 27 BCE, the emperor was considered to be divinely chosen (first by the polytheistic deities of the Roman pantheon, and later, after 380 CE and the establishment of Christianity by Theodosius, the Judeo-Christian god).

In Imperial Japan, well into the twentieth century, the emperor was the head of the Shinto religion, and was still considered divine; his authority could not be questioned. During World War II, the government took full advantage of this by demanding the complete loyalty of each citizen to the government (after the war, Japan's new constitution separated the church from the state, and Hirohito declared himself no longer divine). Up until the beginning of the twentieth century, the sultan of the Ottoman Empire was the head of both the government and the religion.

The first country to explicitly reject rule by divine right was the United States. In addition to dismissing the divinity of King George III of England, the First Amendment of the US Constitution instituted independence, by law, of the

Box 17.4 (cont)

government from any particular religion. Over the next two centuries, countries across much of the world slowly followed suit, to varying degrees, but even today the effect of religious authority is strong in many countries.

Today, the head of the British monarchy (Queen Elizabeth II) is also the titular head of the Church of England, and this is the case in Sweden and Denmark (in these cases, religious power is not strongly exercised by the monarch). However, in some countries, such as Iran, the ruler (called the Supreme Leader in that country) has powerful political and religious authority. Currently, countries across North Africa and the Middle East have Islam as their formally established state religion, but a few other countries across the world have state religions as well (although most are more in name than enforcement): Christianity is the state religion of Zambia, Buddhism is the state religion of Cambodia, Bhutan, Myanmar, and Sri Lanka, and Catholicism is the state religion of Costa Rica, Lichtenstein, Malta, and Monaco.

Chapter 18: Religion and Philosophy

If an alien were to arrive from a distant solar system and study humans, one thing that would undoubtedly stand out is the fact that there has never been a human society, as far as we know, that does not have some form of religion. Throughout history, religion has been a near universal presence, and has played a major role in transitions, social movements, and wars. Today, there are more than 4,000 identifiable religions on Earth, and many hold incompatible views on a variety of subjects, from morality to the origins of humanity. How does science understand these seemingly disparate worldviews?

In this chapter we study religion from above, as a phenomenon, as if we were those aliens. As I said in Chapter 1, science as a discipline cannot make statements about the presence or absence of the divine, but it can make statements about how religion manifests itself in human societies.

Earliest Religions

We only have a written record back to about 5,000 years ago, in modern Iraq. By that time, religion in the region was already well organized around a pantheon of gods. But the archaeological record of behaviors that we think show evidence of religion or spirituality go back as far as 90,000 years ago. In modern Israel, a modern human burial at Quafzeh shows a mother and child buried together with ochre. Ochre is a pigment that is associated with ritualistic and artistic behaviors, and is commonly seen in ancient burials, such as those at Lake Mungo in Australia and in Europe between 40,000 and 10,000 years ago. Burials often show evidence of religious or spiritualistic behavior. A burial in Israel at about 12,000 years ago is full of grave goods that would support the interpretation of the deceased woman as a shaman (she was buried with tortoise shells, an eagle wing, a boar leg, a cow tail, a marten skull, and a human foot).

Some of the Paleolithic art from around the world is consistent with the type of shamanism we see in hunter-gatherers today. Rock art in Australia and Europe shows hybrid creatures that are easily interpreted as some sort of shaman or spiritual creature, as well as figurines like the Venus of Willendorf (Figure 18.1).

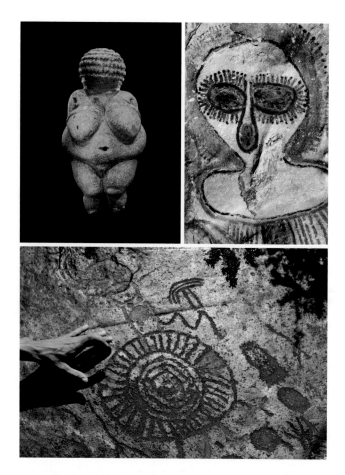

Figure 18.1 Although we have no documentary evidence for religion from preliterate societies, we can look at their art for similarities to modern hunter-gatherer religions. Here we see rock art from Europe (left), Australia (right), and Africa (bottom) in which the figures are presented in ways consistent with shamanism. The earliest rock art with shamanistic qualities is found in Australia, and is roughly 30,000 years old.

We will, of course, never know the significance of these artifacts. They appear to the modern eye to be very similar to what we see in hunter-gatherers, but we will never be able to interpret the actual religious significance of these artistic expressions (although that hasn't stopped people from trying). Since we can't make direct interpretations, probably the most important implication of all of this material is that religion is an ancient phenomenon.

As population densities increased over the last 10,000 years, and societies changed, religions changed as well. In many ways, the evolution of religious practice mirrors the evolution of societies. Anthropologists debate the definitions and categories of religions, but it is possible to look at societies and identify three general types of societal religious practice, and they mirror the scale of the societies that have them:

1. *Shamanistic.* Very small-scale societies typically have a shamanistic religion. In this system, the religion is not organized around any kind of specific god or gods. Rather, some member or members of the society claim to be able to communicate with various spiritual elements of nature, for example, to predict future events or heal the sick. This is the type of religion seen in hunter-gatherers today, and we infer that it would have been present deep into the Paleolithic. In shamanism, spiritual or animistic forces are often associated with various natural phenomena, including the weather, mountains, plants, and animals. However, there is generally not some underlying set of rules or principles that are shared across the various societies, and practices will be idiosyncratic. Since this type of religion is associated with preliterate societies, there are no texts associated with them, and there is no central authority whatsoever.

2. *Communal religion.* Once societies have acquired plant-based agriculture or pastoralism (herding of animals), they increase in size and density. Because these groups acquire long-term identity as a group, they will acquire religious practices that are shared among the broader group. Within these societies, particular lineages may have religious power or authority. Religious practice in these groups tends to be based on local custom or tradition, and will speak to the specific goals for that society. Examples include local fertility gods or agricultural deities that are worshiped to help the society grow and feed itself, or the worship of a past warrior or traveler who was particularly important in the memory of that society.

3. *Ecclesiastical religion.* In complex agricultural and state societies, particularly those with a powerful central government and written language, the most common form of religion is ecclesiastical. Ecclesiastical religions are more formal, with institutional structures and internal hierarchies of power. They often rely on written texts that institutionalize specific rules and rituals. Often they attempt to maintain or expand power by eliminating competing religious practices. Perhaps most importantly for world history, they establish themselves within the general power structure of the society, so that the power of the religion parallels the power of the society. Examples include the Roman pantheon, the monotheistic Judeo/Christian/Islamic tradition, Hinduism, Egyptian polytheism, Zoroastrianism, and many others.

Once a religion acquires its own power structure and hierarchy, it parallels governmental systems, in that there are ranks, and higher-ranking members obtain more divinity and power as well as sacred knowledge (Figure 18.2). Further down the hierarchy, members have less. The possession of sacred

Figure 18.2 Sacred knowledge is characteristic of ecclesiastical religions, with access granted only to a privileged few. The Nilometer, represented in these three images, was a device (or, properly, an architectural structure) used to measure the height of the Nile, in secret. The priests of the ancient Egyptian religion collected data on the depth of the Nile over centuries, and determined that there were long-term cycles to the river flow. This allowed them to predict, with some degree of accuracy, how much the Nile would flood. This ability lent the priesthood legitimacy to the Egyptian public, and suggested that they might have insights into the spiritual plane.

knowledge has long been used as a form of leverage over members lower in the religion, and often the knowledge has been very carefully protected (Box 18.2). One famous example was the translation of the Bible into English by William Tyndale in the sixteenth century. Previously, the Bible had been printed in Latin, and the text was incomprehensible to the bulk of the population of English speakers (only some of whom could read at all). This gave the Catholic Church a monopoly on the interpretation of scripture, and the ability to invoke the Bible as a justification for any number of actions. Once translated into English, any English speaker who could read could make their own interpretations. This threat was unacceptable to the Catholic Church, and Tyndale was condemned as a heretic and burned at the stake in 1536.

Polytheism and Monotheism

Religions can be categorized as *polytheistic* (worshiping many gods) or *monotheistic* (worshiping one god). Although today much of the world is monotheistic, this has not always been the case. As far as historians can tell, ancestrally, every religion that is monotheistic today comes from a polytheistic religion in its past. The polytheistic gods of a particular religion (called a *pantheon*) probably evolved from regional fertility gods and agricultural deities, as larger regions were consolidated into large agricultural societies and states.

Within a pantheon, different gods often have influence over different aspects of nature and society – there is often a god and or goddess of the Sun, the ocean, the harvest, war, love, commerce, fertility, etc. Often there is no fixed number of deities, as local or lesser gods might be recognized. In Mesopotamia, there are at least 2,000 known deities; in the ancient Egyptian religion, there are more than 1,000 gods known from writing of the time. The ancient Greek pantheon of twelve is well known, but there were many lesser deities.

One aspect of these ancient religions is their acceptance of other religions. Although there might be an official religion for a given state, other regions were expected to have their own gods. The idea of a region attacking another specifically to advocate the beliefs of their religion is not known during this period. When one state conquered another, it was common to permit the continued worship of the local gods, although new monuments to the gods of the conquerors might be built. This does not mean that there was no enforcement of religious rules. In many places, rejecting the belief in local gods could lead to execution. Famously, Socrates was executed in Athens in 399 BCE for "impiety," and influencing the youth to reject the Athenian gods.

The early Christians in Rome were executed, not for being Christian, but for calling on others to reject the polytheistic Roman religion (which was closely affiliated with state power in Rome). But the later pattern of wars premised on religious differences has not been identified by historians.

Today, much of the world follows a monotheistic faith, but that may not be a good representation of the world pattern. For most of human history, humans were polytheistic (Box 18.1). Today, the only large-sale polytheistic religion is Hinduism, but it has more than a billion adherents (about 15 percent of the world) and, as the world's third largest religion, shows no signs of disappearing.

Box 18.1 The evolution of monotheism

Today, in much of the world (outside India especially) the dominant religions are monotheistic (see distribution in Figure 18.3). These include the religions that descended from Judaism – Christianity and Islam – but also Buddhism, as well as Zoroastrians, Sikhs, Yazidi, Rastafari, and worshipers of Aten, among others. However, looking back through history, monotheism may be the exception rather than the rule.

The Religions of the World

Christianity
Islam
Hinduism
Buddhism
Judaism
Chinese religions
Korean religions
Shinto
Folk religions
No religion

Figure 18.3 The two largest religions on Earth are Christianity and Islam (in all their various forms and sects). These two religions have a strong proselytizing tradition, and their geographic distribution reflects the fact that, as their cultures spread around the globe and colonized, they converted the local populations, either by force or by persuasion. Religions that readily accept converts, such as Hinduism and Buddhism, have large populations, but religions that reluctantly accept converts, or not at all, such as Zoroastrianism, are in danger of disappearing entirely over the next few centuries.

Box 18.1 (cont)

Anthropologists, historians, and scholars of religion have identified certain patterns that seem to appear consistently in the evolution of religions. Perhaps the most relevant for today is that modern monotheistic religions, as far as we know, have all descended from polytheistic religions. This happened to ancient monotheistic religions (Zoroastrianism, Atenism) but applies to Judaic/Christian/Islamic monotheism as well. Here I describe some notable examples.

Atenism

The traditional religion of ancient Egypt originated some time prior to 3000 BCE. It was a polytheistic religion, with a pantheon of more than a thousand gods, some of whom are familiar from the popular media. For example, Ra was the god of the Sun, Horus was the god of the sky, Set was the god of the desert, and Osiris was the god of fertility. This religion persisted as a polytheistic religion for thousands of years until Christianity (after the fourth century), and later Islam (after the seventh century), became the dominant religions of the region.

However, in the eighteenth dynasty (roughly 1300 BCE), the ruler Amenhotep IV, more popularly known as Akhenaten, changed the state religion of Egypt from the ancient and traditional polytheistic religion to a monotheistic religion, with all worship to be of the Sun god Aten (traditionally the actual Sun, and under the god Ra). Akhenaten moved the capitol of Egypt from Memphis to a new city, Akhetaten, and separated himself from the traditional Egyptian priesthood. He suppressed worship of the old gods and converted the state religion of Egypt to Atenism, with himself as the direct and only intermediary between Aten and the temporal world.

Akhenaten, with his wife Nefertiti, only ruled for seventeen years before he died. After his death, his son Tutankhamen moved the Egyptian capital back to Memphis, and the state religion returned to the traditional polytheism. Some authors have suggested a link between the monotheism of Atenism and the later appearance of monotheism in the Jewish peoples, but such a connection has been difficult to demonstrate conclusively.

Zoroastrianism

Zoroastrianism is one of the most ancient religions in the world, possibly dating back to 1000 BCE in ancient Persia. It was the dominant religion of the Persian peoples during the height of the Persian Empire, and served as the

Box 18.1 (cont)

official state religion of Persia until the Muslim conquest in the seventh century. Zoroastrianism is still practiced today in Iran and nearby countries, although they are the subject of a modest degree of persecution, as Islam is the state religion of modern Iran.

The origins of Zoroastrianism are poorly known, because the written records in the great library in Persepolis were burned by Alexander the Great's troops when they sacked that city in 330 BCE. However, it appears that a historical figure named Zoroaster, who lived sometime between 1000 and 2000 BCE, came to represent the powers of the ancient Persian gods, and remade the religion into a monotheistic variant of the ancient beliefs.

Zoroastrianism is significant in modern religious history. When the ancient Persian Empire conquered what is today the Middle East, the Jews were released from the Babylonian Empire. The Zoroastrianism of the Persians under Cyrus proved to be a powerful influence on the establishment of monotheistic Judaism.

The Origins of Judaic/Christian/Islamic Monotheism

Much of the world is monotheistic (largely Christianity and Islam), and this monotheism traces back to the origins of Judaism. The people we now call the Jews were part of a larger group of tribes of Bronze Age Semitic peoples, all of whom shared a language in what linguists and anthropologists call the Canaanite region (modern-day Israel, Jordan, Lebanon, coastal Syria, and a small part of southern Turkey). Over time, kingdoms waxed and waned, but they included Israel, Judah, Edom, Ammon, and the Philistines and Phoenicians.

Across this region, people worshiped the dozens of Canaanite gods (Figure 18.4). During the end of the Bronze Age and the beginning of the Iron Age in the region (roughly 550 BCE), this region was invaded multiple times by various powerful empires originating in modern Egypt, Iraq, Iran, and Turkey. One of the invasions, around 540 BCE, was by the Persians under Cyrus the Great. The Persians were monotheistic Zoroastrians.

Historically, the tribes of Israel and Judah worshiped the Canaanite pantheon, but at some point during their exile (after capture by the Babylonian Empire some time near 580 BCE) they began to worship one of the gods, Yahweh, over the other gods. This transition was not immediate, and there is considerable scholarly debate on the origins of Judaic monotheism, but it is clear that after the Persian invasion of the region, and their subsequent liberation by the Persians, they became monotheistic. Although they initially

Box 18.1 (cont)

Figure 18.4 Monotheism evolves from polytheism. The Hebrew god Elohim (Yahweh) originated as part of the Canaanite pantheon. Here are several of the Canaanite gods (from top left,): Ba'al, the god of thunder; Ashera, the mother-god; Melqart, the god of the underworld;

Box 18.1 (cont)

acknowledged other gods, and even permitted worship of them, over time this was suppressed, finally resulting in the acknowledgment of only one god.

One artifact of this shift, and the enforcement of it, can be seen in the first two commandments: (1) "I am thy God" and (2) "Thou shalt have no other god." These commandments acknowledge the potential for worship of other gods and, although never stated directly, sees them as potential competition. Biblical scholarship places the formalization of the Ten Commandments relatively recently, between 700 and 550 BCE, and this likely represents the time period when the early Jewish peoples were choosing to focus on Yahweh over other deities.

The subsequent outgrowth of Christianity from Judaism, and the later appearance of Islam, all stem from this initial shift to monotheism in these Semitic tribes during the sixth century BCE. But the evolution of monotheism from polytheism is probably a consistent pattern in the evolution of religions.

Buddhism

For Westerners who are accustomed to religions with firmly established foundational narratives (the stories of Moses, Jesus, and Mohammed, for example) that are defended by religious authorities (sometimes vigorously), Buddhism can be difficult to understand. In some sects, Buddhism is more a philosophy than a religion, but other sects treat Buddha as a true single, divine being. These many Buddhist sects use different narratives to describe their origins, but scholars generally agree that the historical origins of Buddhism lie in early Hinduism (then called Brahmanism) around 500 BCE. Because there is no central authority, nor a single foundational narrative for this religion, understanding its true historical origins is difficult. However, in some traditions,

---

Figure 18.4 (*cont.*) and El, the chief god, analogous to Zeus (in the Greek pantheon) or Jupiter (in the Roman pantheon). Gods were regularly shared or borrowed among ancient civilizations – for example, Ba'al appears in ancient Mesopotamia and in one of the earliest epic stories, the Ba'al Cycle, which dates to roughly 1300 BCE. Ashera appears in the same cultures, as well as Bronze Age Turkey (the Hittites).

Over time, two of the Canaanite civilizations, Judah and Israel, devoted their worship to El, which ultimately became Elohim. Naturally, there is considerable and long-standing debate over the convergence of El and Yahweh, and the Old Testament is inconsistent with the usage of El (see Box 18.1), but Yahweh and El seem to have been synonymous by the time of the Persian occupation of the Middle East (fourth to sixth century BCE). Ashera (second from left), previously the mother-god, and wife of El, was notably removed from Hebrew tradition during this period, and this is the time when Judaism is considered to have become more patriarchal. The statues present in this figure are precisely the sort banned by the Third Commandment ("Thou shalt have no graven images").

Box 18.1 (cont)

the historical figure Siddhattha Gautama Buddha is considered an avatar of one of the three most powerful Hindu gods, Vishnu.

While there is considerable debate over the origins of Buddhism, and the position of Buddhism relative to Hinduism, it is clear that the two traditions are closely linked, and that Buddhism was in some ways a reaction to aspects of inequality in Hinduism at the time, such as the caste system. Much of the debate today (and throughout history) is presented by advocates of various faiths or traditions, and the historical record is poor so there is no consensus other than that the two religions have long been intertwined by tradition and history. But one clear conclusion is that Buddhism, whatever its relationship to Hinduism today, comes out of the polytheistic tradition of early Hinduism.

Box 18.2 Religious conflict

Conflicts between Religions

Conquest has always played a role in the spread of religions, as the conquerors' religions would be the state religion of the empire. However, in the ancient world, empires were relatively tolerant of the various local religions. When Cyrus the Great conquered the ancient Middle East, he made no attempt to interfere with local practices. This was the case with Alexander the Great, the Maurya Empire, and the Roman Empire. This practice was not universal, however. Sometimes, specific religions would be persecuted, as by the Hun conqueror of northern India, Mihirakula (515–540 CE), who persecuted Buddhists and was known to destroy their temples. Christians in Rome who refused to acknowledge the Roman state gods were also persecuted.

But conflict centered on religion is most characteristic of proselytizing religions (see Box 18.3). Some religions have, as part of their belief system, a mandate to expand. This expansion comes at the cost converting people from competing faith systems. Naturally, the authorities of the other faiths may push back. The resulting conflicts have framed many wars over the last 2,000 years.

Perhaps the longest-lasting large-scale religious conflict was the struggle between Christianity and Islam. Christianity was established as the state religion of the Roman Empire by the Edict of Thessalonica in 380 CE. Western Rome had declined in power when Islam was founded, around 600 CE, but the Eastern Empire was still powerful and Christianity was the state religion. Islam spread around the Middle East and North Africa, largely because of the military

Box 18.2 (cont)

conquest of these regions by Arab armies. By the middle of the seventh century (roughly 650 CE), the Rashidun Caliphate covered all of the Arabian Peninsula, Egypt, modern Tunisia, Iran, Iraq, Syria, Israel, and Jordan. But they did not conquer the central part of the Eastern Roman Empire (modern Turkey).

The military expansion of the Arab Rashidun Empire, at the expense of much of Christian Eastern Roman territory, set up a conflict that could readily be framed in religious terms. Over the next 800 years, the Muslim empires of the Middle East battled the Eastern Roman Empire, and both sides would call for members of their faith to join the struggle against the other (called the "Crusades" in Christian Europe), ending with the fall of Constantinople in 1453. There are remnants of this conflict that are very visible in international politics today.

But there have also been wars within these two religious faiths based on disagreements among sects. The Thirty Years' War (1618–1648) was fought in central Europe between Protestant Christians and Catholic Christians, and took some 15 million lives. It was partly a conflict over political power and economic strength, but by framing the struggle as a religious war it became easier for the leaders of the countries involved to justify the war to their populations (who bore the brunt of the war), raise armies, and sustain political support for the war.

Within Islam, the two major sects are the Shia and Sunni, and these two divisions come from a disagreement over the rules of religious succession following the death of Mohammed in 632 CE. These two religious sects also represent different political lineages, so the divergence was simultaneously a conflict over religious and political power. Even today, conflicts between countries are framed in the context of the schism between these two religious sects, as in the Iran–Iraq War (1980–1988).

Perhaps the greatest sways of religion resulting from military conquest can be found in the history of the Indian subcontinent. Buddhism was the dominant religion in India until the seventh century, when the Gupta Empire installed Hinduism as the state religion. Buddhism was further marginalized when it was vigorously persecuted by invading Muslim armies and is now a tiny minority. Starting in the seventh century, Hindu and Muslim armies fought for territory, cities, resources, and followers in the Indian subcontinent, and the conflict persisted, on and off, over the next 1,000 years. When the British placed the Indian subcontinent under colonial rule in the eighteenth century, it was a mosaic of Muslim and Hindu regions reflecting the fluctuations of past empires. The conflict between Hindus and Muslims finally resulted in the

Box 18.2 (cont)

partitioning of the region into India (Hindu) and Pakistan (Muslim), following the end of British control in 1947. Bangladesh, already geographically separated from the rest of Pakistan, later separated to form its own country, partially to form a secular country rather than be in an Islamist state.

Box 18.3 Proselytizing religions

If you look at Figure 18.3 you will notice that the majority of the world is made up of Christians and Muslims. These are both relatively young religions – Judaism is at least 700 years older, and Zoroastrianism and Hinduism are older yet. The reason so much of the world is composed of adherents to these two religions is that, within Christianity and Islam, there is a mandate to convert (proselytize) people of other faiths. Inherent in the faiths themselves are the beliefs that their religion alone allows a path to the divine. Conversion by these two religions could be peaceable persuasion, in the case of missionaries today, or by force, as was used by the Catholic Church during the colonization of Latin America and by the Muslims during the conquest of North Africa.

Other religions, especially those without much central governance, have not generated missionaries seeking to convert others. Hinduism has traditionally not sought to convert, but there are branches that do (e.g., the Krishnas). Buddhism does not have a tradition of proselytizing, and the Dali Lama discourages missionary work on behalf of Buddhism. But both religions readily accept converts, and in some cases large populations have converted to both religions as political leaders were converted. Some religions do not normally accept converts. Judaism, Zoroastrianism, and the Druze typically will not accept converts. This is one of the reasons why there are so few adherents to these religions (persecution and forced conversion away from these religions is another).

But new religions crop up often, and it is difficult to predict their growth. Two religions that have grown significantly since their relatively recent founding are the Mormon Church (also known as the Church of Latter Day Saints or LDS) and Scientology. The LDS was founded in the first half of the nineteenth century and now has some 15 million members. LDS missionaries work all over the world, actively attempting to convert members of other faiths. The Church of Scientology started in 1953 and, although controversial

Box 18.3 (cont)

and outlawed in several countries, has expanded from its origins in southern California to roughly a dozen other countries.

Both the LDS and Scientology have the traditional missionary mandate – they claim to offer the true path to knowledge, enlightenment, the divine, etc., and they both have strong central governing bodies. These two characteristics are usually what we see in religions that grow, but the expansion of Christianity and Islam took hundreds of years, so we may not see significant changes in our lifetimes. In some places, such as the USA, Australia, New Zealand, Europe, and Latin America, rejection of religion altogether has grown, and new religions must now contend with this as they attempt to expand.

Box 18.4 Roles of religions in societies

One of the reasons religion is so universal is that it serves several societal roles. These roles have shifted over time, as alternative ideas have been offered (such as secularism), and religion has had to contend with competition in the intellectual marketplace.

Explain Natural Phenomena

As far as we know, religion has been explaining patterns in the natural world for as long as there has been religion. Hunter-gatherers invoke religious explanations for natural occurrences like thunder and lightning, rainstorms, droughts, the rising of the Sun and the Moon, shooting stars, and many, many others.

However, this role has been significantly eroded in the last few hundred years by the explanatory power of science. Sometimes science and religion come into conflict because scientific explanations contradict traditional religious explanations (see Figure 18.5). The most famous example is probably the Roman Inquisition's persecution of Galileo because of his declaration that the Earth orbits the Sun, which contradicted church doctrine at the time (the Catholic Church has since apologized). But it continues today – some religions (fundamentalist Christianity and Islam, notably) vigorously oppose the evolutionary explanation for the appearance of humans, and, famously, John Scopes was prosecuted by the State of Tennessee in 1925 for teaching evolution in his high school biology class. The US Supreme Court has since struck down any law or ordinance outlawing the teaching of evolution, on the

Box 18.4 (cont)

grounds that it violates the US Constitution's separation of church and state. But not all countries have such a separation, and in some countries the scientific explanation is outlawed.

Provide a Framework for Shared Morality

Most organized ecclesiastical religions have specific rules to which followers must adhere. In the Abrahamic religions (Judaism, Christianity, and Islam), the most important rules are the Ten Commandments. Other religions have similar sets of rules: the Five Precepts of Buddhism, the Four Aims of Hinduism, the Forty-Two Negative Confessions of Maat (in the ancient Egyptian religion), etc. These sets of rules tend to be similar, and have rules against murder, thievery, and dishonesty, and an obligations to care for others.

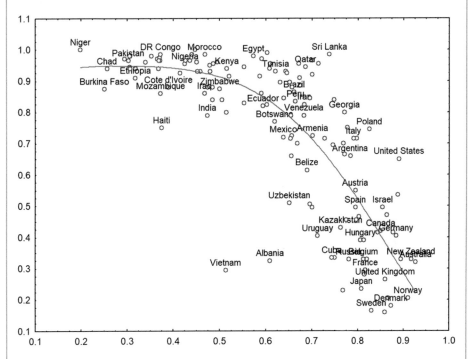

Figure 18.5 As societies become more educated, they tend to acquire more scientific and philosophical literacy. Since science and philosophy can, in some respects, replace some of the purposes of religion, it is natural that more formally educated societies tend to be less religious. In this graph, the UN Education Index (as of 2015) is on the x-axis, and the importance of religion (as measured by the Gallop Poll in 2009), is on the y-axis. This figure shows that religiosity is inversely correlated with education among world societies. Societies above the red line are more religious than predicted for their education levels (e.g., the USA, Poland, Israel, Qatar, Egypt, Morocco, and Sri Lanka), whereas countries below the red line are less religious than predicted (e.g., Vietnam, Albania, Cuba, India, and Haiti).

Box 18.4 (cont)

These rules have helped unify these societies, since it gives all followers a shared set of laws and beliefs. However, today, societies function under a set of civil laws, and although those laws may be based on the religious rules of the society, in many countries they may be based on other philosophical ideas of morality (for example, in Western countries the philosophical ideas of the Enlightenment, or in communist countries the ideas of Marxism). In some societies, breaking religious rules can lead to punishment (including execution), but in most modern countries only violation of the civil code actually leads to sanctions from the state, and over the last few hundred years religious rules have held less and less sway.

Promote Social Solidarity

Religions can provide an identity for groups of people, so that they can feel a sense of identity and affiliation. Much like other social primates, human tend to identify quite closely with their "in-group." As social animals, humans probably have, quite literally, a biological need to affiliate with other humans (this is why solitary confinement is considered such a draconian punishment – and why it can lead to mental illness). Much as people identify with their political group, and their local sports teams, they also identify themselves by their religion.

However, there is a downside to this "in-group/out-group" aspect of religion. It is because of this that leaders have been able to exploit this impulse and rally members of a particular faith to pursue political goals, some of which were quite negative. The "Crusades" is an example – leaders rallied Europeans to wars in the Middle East for their own political goals. By employing in-group/out-group identity it was easier to justify the war, despite the fact that, for an average English or French peasant, the war between the Byzantine Empire and the Seljuk Empire could not have been more irrelevant. Modern examples include the Israeli–Palestinian conflict, the Pakistan–India conflict, and the Lebanese Civil War.

Promote Notions of Charity

Most religions have specific rules or traditions that encourage members of a faith to help others in need. Many modern charities were established by specific faiths as part of this mandate. For centuries, these religious organizations were the only formal institutions of charity for the poor or sick. Today, many of these functions have been taken over by state governments, which provide social services and emergency care for its citizens, although the

Box 18.4 (cont)

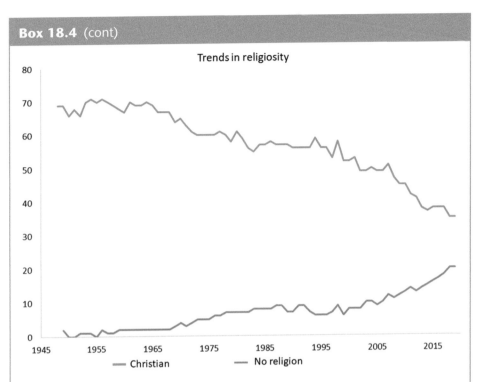

Figure 18.6 The increase in education and the growth of science has paralleled a general decrease in religiosity. Here, Americans were surveyed on their religion, which shows a general decrease in religiosity (in this survey, Christianity is decreasing, while those who responded "No religion" have increased).
Data: Gallup.

charities themselves are still widespread and provide considerable utility to the world.

How Do These Roles Apply in the Modern World?

One of the challenges faced by religions in the modern world is that, to a large extent, these roles can be provided by other institutions. The first role, the explanation of natural phenomena, has been almost entirely supplanted by science. Very few people still turn to religious texts to understand, for example, the position of the Sun in the sky or the cause of rain. The origin myths of some religions have proven the most resilient, as there are significant minorities of people who reject the scientific explanation for the origin of the Universe (the Big Bang) and evolution by natural selection.

Modern societies provide notions of solidarity (citizenship), sets of laws based on morals, and social services. Following the Enlightenment (Chapter 19), societies have had a difficult time justifying preference for one religion over another, and choosing to favor that particular set of values.

> ### Box 18.4 (cont)
>
> In most Western societies, there has been a general trend toward a decrease in religiosity in these populations, as people do not need to turn to religion for the benefits it has historically provided. The long-term result of this trend will likely have many consequences, and it is one of the most significant major shifts we have seen in the world over the last 300 years (Figure 18.6).

The Rise of Philosophy

The rise of religion as a power, particularly as linked to state authority, has given some people pause. It is clear from antiquity that leaders exploited the faith of religious followers for political ends, and sought other ways to understand the world. Almost all religions have creation mythologies and make statements about the physical world, allowing early thinkers to test the predictions of the faith. It is in these intellectual conflicts that we see the rise of science, empiricism, and rational philosophy.

The ancient Greeks gave us the term "skeptic" to describe someone who insists on testing ideas, and who rejects the type of dogmatic assertions that characterize faith. The Greek philosopher Pyrrho (roughly 360–270 BCE) is often considered to be the founder of philosophical skepticism, in which ideas must be tested and any supernatural explanation is rejected in favor of a naturalistic one. Skepticism was a major philosophical school in Greece and Rome, but largely disappeared when Christianity became the state religion of Rome.

But the ideas of skepticism are a natural outcome of a literate and intellectually open society, so it has appeared independently across the world. The Chinese philosopher Wang Chong (first century CE) rejected Chinese Confucian spiritualism and the mysticism of Daoism, instead offering rationalistic explanations for natural phenomena. Slightly more esoteric was the ancient Indian school of Ajñana, which was a generally skeptical school with regard to most forms of knowledge.

These schools of skeptical philosophy played no significant role in history because they posed no real threat to the religious or political power structures of their time. From the collapse of the Western Roman Empire in the fourth century, and for roughly 1,000 years afterwards, western Europe was largely illiterate. The great libraries of Greek and Roman antiquity (the library at Alexandria, Egypt, was the greatest in the Mediterranean region) had been lost, and the only literature was both produced and read by the Catholic Church.

However, the works of the Greek and Roman philosophers had been preserved in the great Muslim libraries in Cordoba, Cairo, and Baghdad, and as Europe left the Dark Ages and entered the Renaissance (fourteenth century), scholars began bringing back copies of the works of the ancient philosophers (Box 18.5).

Among these works was ancient Greek skepticism, and as Europe grew wealthy and flourished during the Renaissance, and more citizens became literate, scholars read and disseminated these ideas. Both philosophers and early scientists began challenging the Catholic Church, and the church responded harshly. The Spanish and Roman Inquisitions persecuted any ideas that contradicted church doctrine. Famously, the astronomer Galileo Galilei was threatened with torture for arguing that the Earth orbited the Sun.

But suppression of ideas is difficult, and during the Renaissance and in the Enlightenment period that followed (seventeenth to nineteenth centuries), philosophers, scientists, and politicians began to challenge the specific ideas of religion (Box 18.6), and perhaps more importantly the linking of religious authority to political authority (Chapter 19). The ultimate expression of skepticism and enlightenment was the founding of the United States, in which a mandated separation of any religious authority from the secular government was written directly into the country's founding document, the US Constitution.

Box 18.5 Philosophical alternatives to religion

When we look at the roles of religion in society (Box 18.3), we see that religion has been displaced almost entirely from explaining the natural world. Even the most religious among us would be hard-pressed to challenge science's authority in fields such as astronomy, geology, biology, and meteorology, all fields in which religious explanations once held sway. Similarly, people can self-affiliate using any number of potential identities (citizenship, ethnic origin, language, etc.). And charitable assistance is often provided by governmental institutions in the form of social services.

But morality is different. Traditionally, one of the key roles of religion in societies was the expression of a moral code. In fact, even today, many theologians argue that morality is derived from religion, and that without a foundation in religious morality (e.g., the Ten Commandments) there can be no way to generate a coherent morality for society. Philosophers have long been aware of this argument, and over the last several thousand years there have been many arguments presented for a logically coherent moral code.

Box 18.5 (cont)

Remarkably, the works of these many philosophers across Asia, Africa, and Europe can be broken down into two basic arguments. The simplest might be called "do the greatest good for the greatest number." This type of argument is known as *utilitarianism*. The basic premise is that in order to decide whether an act is moral, try to determine the implications. Of any two potential actions, choose the one that does the least total harm and the greatest possible good. This philosophy has deep origins, and was expounded by several ancient Greek, Buddhist, and Hindu philosophers. Probably the most famous utilitarian in recent history was John Stuart Mill (1806–1873), who was a strong advocate of this philosophy and used it to justify his vigorous opposition to slavery.

The other argument can be reduced to the "Golden Rule": act toward others as you would have others act toward you. This position is closely associated with Immanuel Kant (1724–1804) who called it the "Moral Imperative," and phrased it (somewhat awkwardly) as: "Always act according to that maxim whose universality as a law you can at the same time will." This rule has even deeper ancient roots, and is visible in the writings of ancient Egypt, Persia, Greece, India, and Rome, and in virtually every religious tradition. Kant argued that the Moral Imperative was, in some profound way, inherent to his being: "two things fill my mind with ever-increasing wonder and awe: the starry heavens above me and the moral law within me." Given what we know about primate social systems, and the likely evolutionary origin of pro-social behaviors, this may be true.

At one time, philosophy was the primary alternative to religion in such areas as the description of the natural world, explanations for historical phenomena, arguments for issues of human behavior like free will, analysis of religion, and positions of morality. Over the last three or four centuries, science has whittled down the position of authority for philosophers in most of these areas. However, morality is generally excluded from the purview of science. Science may be able to explain the evolutionary origins of primate behavior, and why particular actions are regarded as moral by humans, but science never offers moral arguments. To this day, religion and philosophy are the two choices for moral arguments, and both types of argument are visible in the public debates you see on the daily news. Current political arguments over such issues as the legality of abortion, gun control, environmentalism, and even taxes are framed in terms of the moral arguments of religion and philosophy, and Kant's Moral Imperative vs. Mill's utilitarianism.

Box 18.6 Trends in religiosity

For much of the history of humanity, nearly everyone was, to some degree, religious. Before the arrival of science, religion offered explanations for the path of the Sun through the sky, the origins of the Universe and humanity, as well as the cycle of the seasons. People were invariably skeptical of some explanations, but in general the world was a religious place, and belief in a god or gods was to be expected in virtually every person you might come across, from 100,000 BCE to 1500 CE. Certainly there was a lack of alternative explanatory models.

But there have been skeptics in every corner of the world where intellectual thought flourished – ancient China, India, and Greece, and medieval Baghdad, among others. But they never presented a strong alternative to religious explanations because of a lack of explanatory power for natural phenomena.

This changed with the arrival of the scientific method in Europe in the sixteenth century (actually, it had arrived much earlier, during the Islamic Golden Age, but was suppressed by the religious authorities – see Chapter 19). Because Europe was composed of so many different countries (more than fifty separate polities in Europe in 1500), ideas could not be easily suppressed by a central authority. As Europe pulled itself out of the Dark Ages, and rediscovered the ancient Greek and Roman works of skepticism (stored in the great Islamic libraries), they began to question many of the tenets of religion.

In some cases, this led to religious schism. The creation of Protestantism was driven by Martin Luther's desire to reform the Catholic Church. But others turned away, to varying degrees, from intellectual deference to religion altogether. The attempted suppression of science, such as the Roman Inquisition's persecution of Galileo, brought the conflict into relief and forced people to think about the two contrasting ideas. Over time, science presented more and more evidence that was consistent with nature and did not rely on a supernatural explanation (Lyell's geology, Darwin's evolution by natural selection, Newton's astronomy and physics). Parallel to science was a group of philosophers generating morality from logic (see Box 18.4). Together, these trends made it clear that science and philosophy might make religion unnecessary altogether, although many thinkers struggled with discarding their religious traditions (Figure 18.5).

Over time, there was a slow disentangling of religion from various aspects of civil society. In the eighteenth century, several countries declared themselves formally divorced from religious authority (the USA and France were the first), and writers in Europe and the USA offered strong critiques of the linkage between government and religious authority. This trend has continued over

> **Box 18.6** (cont)
>
> the past 200 or so years, and nowadays there are substantial proportions of people in countries across the world who have abandoned religion entirely (Figure 18.6), even if they recognize some cultural practices (e.g., holidays).
>
> In the United States, there have been strong social reactions to the decrease in religiosity among Americans. There have been at least three "Great Awakenings" over the last 250 years, as religious authorities have tried to reestablish authority over an increasingly secular society. In Europe, the continued establishment of official state religions has led to less religious backlash against secularity, but the trends toward a decrease in religiosity is even stronger there.
>
> However, in some countries there have been more violent struggles over the trend toward secularity. By the 1970s, Iran was a modern, secular country. The 1979 Iranian Revolution overthrew the secular (but decidedly corrupt) government, and although the revolution was initially political rather than religious, powerful Islamist clerics took control of the revolution. These clerics established Islam as the state religion of Iran, and violently repressed any anti-religious or anti-Islamist speech. The current government of Iran is perhaps the clearest example of a strong reaction to the secular trend, but it is visible in countries across the world wherever religious authority is challenged and religious leaders feel threatened.

> **Box 18.7** Critique of the roles of religion
>
> Ever since the Greek skeptics, and extending through David Hume and modern writers, the existence of a god and arguments for the construction of the Universe by a deity have been challenged on various logical grounds. But several writers have critically assessed the *role* of religion in society. Two positions are especially notable, historically.
>
> Karl Marx, the writer of the *Communist Manifesto* and *Das Kapital*, two hugely influential works in the area of economics, and considered the father of communism, thought of religiosity in working people as an expression of their despair. He famously wrote:
>
> > *Religious suffering is, at one and the same time, the expression of real suffering and a protest against real suffering. Religion is the sigh of the oppressed creature, the heart of a heartless world, and the soul of soulless conditions. It is the opium of the people.*

Box 18.7 (cont)

For him, religion was, like an opioid drug for an injured person, obscuring pain without curing anything. He made no statements about whether there was, or was not, a god. Rather, he tackled the way religion offered the hope of relief from suffering (in a heaven or something similar). If people are focused on a theoretical heaven, they are less likely to try to change their material lives. Since he wanted workers to rise up and change their working conditions (and change the economy to communism – see Chapter 21), religion stood in the way.

A similar critique of the role religion played in society was offered almost a hundred years later by Sigmund Freud, the creator of modern psychoanalysis. In his book *The Future of an illusion* (1927), he understood religion to have served the explanatory roles seen in Box 18.5, and argued that people clung to religious explanations in the modern world because they were unable to accept the implications of the scientific explanation of the world (e.g., that there is no afterlife). He noted the tension between the religious and the scientific explanations for the Universe. For him, people retained their religious beliefs, even when contradicted by the scientific evidence, because the implications of the scientific explanation were too psychologically difficult to bear. In a scientific context, humans, as individuals, live brief lives, and are composed of the matter of the Universe, organized for a short time in their bodies. This reduces, dramatically, the significance of any one human life relative to the "crushingly superior force of nature". Freud felt that this implication was too hard for people to accept; therefore, they turn to religion, which promises to place a value on their existence.

For Freud, religion therefore stood in the way of the development of a rational world, which he sought. He was not particularly sympathetic to the religious beliefs of the lower classes, but he feared what would happen if large numbers of people, who had previously been kept in check by the morality of religion, were to suddenly abandon their beliefs:

> If the sole reason why you must not kill your neighbor is because God has forbidden it and will severely punish you for it in this or the next life then, when you learn that there is no God and that you need not fear His punishment, you will certainly kill your neighbor without hesitation ... either these dangerous masses must be held down most severely and kept most carefully away from any chance of intellectual awakening, or else the relationship between civilization and religion must undergo a fundamental revision.

Fortunately for the world, even as religiosity has waned across the world, Freud's dire predictions have not come to pass.

Box 18.8 The biological origins of religion

As you can probably tell from this chapter, scientific researchers tend to take a top-down view of religion. They are interested in the universal explanatory reasons that explain religious behavior throughout history. This is done by examining the broad-scale patterns over thousands of years and looking for similarities and differences that might provide explanations for the appearance, say, of monotheism over polytheism, or proselytizing versus nonproselytizing religions. The role of culture cannot be overstated here – people tend to follow the customs of their families and believe what their parents tell them, which likely accounts for most of the general trends in religious belief.

However, from a purely biological perspective, the phenomenon of religion itself requires an explanation. Assuming for the moment that scientists will never accept a supernatural explanation, we need to explain why humans have religion in the first place. There are interesting patterns: the moralities of the religions tend to be very similar, but the explanations of things like natural phenomena are completely incompatible. So, although the underlying rules may be shared, the actual beliefs (in the creation myths, the number of gods, the roles of the gods, past events, etc.) are in complete disagreement. What do we make of this?

Since we are animals, one thing we can do is look at other closely related organisms. It turns out that the rules we find in religions probably predate our species. Behaviors like murder and dishonesty are discouraged in chimpanzee societies, and pro-social behaviors encouraged, as other chimpanzees learn very quickly whom they can trust and ostracize chimps that do not reciprocate or who are too violent. They also engage in altruistic behaviors that, when reciprocated, engender trust. If you look at the rules in the various commandments, those are the kinds of rules that make for a functioning society, in humans and in chimps. So there may be an evolutionary explanation for the "commandments" of various religions.

However, what do we make of the belief in the supernatural? Supernatural explanations are generally discounted in science, for the reason that scientists need to be able to test any idea and are limited to the material (temporal) plane. But many people quite fervently believe in a supernatural existence, and people have regularly reported religious or mystical emotional feelings.

One approach has been to determine if there is anything inherent in the human brain that lends itself to religious experience. After all, the human brain is many times more complex than any other seen in nature, and it is possible that the complexity of the human brain may lead to a predisposition toward religious feeling.

Box 18.8 (cont)

Neurologists have studied the brain and found that certain parts of the cerebrum are associated with religious sensation. Examining brain scans of epileptic patients, for example, has identified patterns of excessive brain activity in the temporal lobe in association with sensations of religious experience. Magnetic stimulation of the human temporal lobe has been reported to produce mystical sensations in the subjects (the so-called "God Helmet" experiments). Damage to specific regions of the temporal lobe can produce hyper-religiosity and hallucinations. Many religious orders stimulate religious experiences through physical means – either psychedelic drugs or through fasting for long periods of time. Both can have a significant effect on brain chemistry, potentially making the brain more susceptible to religious sensations.

Researchers studying this field have generally argued that there are conditions in the brain that predispose them to looking for religious explanations. For example, human consciousness itself may tend to cause people to look for consciousness in other things, including inanimate objects or natural forces. The problem-solving power of the brain may cause us to look for explanations; in the absence of good explanations for seemingly random events, we may try to find an explanation where none exists beyond sheer randomness (this is also why humans tend to generate conspiracy theories).

If religiosity is simply the product of a large, complex brain looking for explanations, why, then, do we not see it manifested among any of the other animals that are known to have consciousness? Interestingly, a fairly recent study (2015) found chimpanzee activities that might be best explained as "superstitious behavior." Chimps walking past a particular tree have stacked stones at the base of the tree. They toss the stones at the tree, then run away. There has been no other good explanation for the behavior, as it does not match any known display behaviors, and was done by males and females. Other animals have shown something akin to ritual when members of their groups have died, or under particular climatic circumstances. Naturally, the conclusion that chimpanzees or other animals are showing evidence or primitive religion is controversial (we certainly can't ask them what they are thinking), but it hints at something beyond the basic drives of food, reproduction, and shelter. Moreover, it suggests that religious experience may be amenable to scientific, rather than supernatural, explanations.

Chapter 19: The Enlightenment

Throughout recorded history, over some 5,000 years and across the globe, kingdoms, states, and empires have risen and fallen. The changes in power were often accompanied by changes to cultures, including language, religion, and art. Some cities, like Damascus, have been conquered and reconquered multiple times throughout history (by the Assyrians, Hittites, Egyptians, neo-Babylonians, Alexander the Great, Romans, Arabs, and Ottomans). As lands were conquered, the people would be subject to new rulers, and over time would reacquire economic equilibrium, depending on the privations imposed on them by the conquerors. But from the broad perspective of history, the patterns of the world were relatively constant. Even as the rulers, languages, and religions might shift, the traditional modes of life remained constant: agriculture was the dominant activity for most people and there were urban areas where learning would accumulate. Knowledge was acquired and lost, as populations grew and declined, and libraries were built and then neglected or destroyed. Information was transmitted slowly in most cases, as travel was dangerous. Events in distant lands had little effect on the lives of most people, who were focused on the day-to-day job of ensuring their families had enough to eat.

But starting in the fifteenth century, in Europe, a fairly unique transition occurred that still has an enormous effect on most people's lives today. In this chapter, we explore why this was such a unique event in world history, and why it occurred where it did.

Europe

For the vast majority of world prehistory and history, Europe was a backwater. The great empires of antiquity and the medieval era tended to form at a geographic nexus. It is no accident that the Akkadian, neo-Assyrian, Egyptian, Babylonian, Persian, Greek (under Alexander), Roman, and Muslim empires all occurred in roughly the same geographic region. The Mediterranean region lies at the intersection of three continents (Africa, Asia, and Europe) and the Mediterranean Sea allows access to both the Black Sea and the Atlantic Ocean, as well as a short overland route to the Red Sea and the Indian Ocean (before the construction of the Suez Canal).

Other great civilizations had similar geographic advantages: India has land routes to China and the Mediterranean (the famous Silk Road), and is a short ocean-voyage from North Africa. In ancient China, the population density and advantageous natural resources (especially the Yellow and Yangtze Rivers), along with the large coastline, made China a cultural hub, with many large urbanized areas even in antiquity (by 2000–1500 BCE).

In comparison, Europe was relatively sparsely populated by small tribes and chiefdoms up until its conquest by the Roman Empire. Europe resisted Roman imperialism, and even after submission was never part of the core of the empire. England and the Germanic areas, for example, never adopted a Latin language. Europe was sparsely populated, without *any* of the great cities of antiquity. For comparison, Rome had a population of 1 million in roughly 100 CE, a number not reached by London or Paris until the beginning of the nineteenth century.

It is worth noting that the Greeks and Romans would not have considered themselves "European," as they were part of the sophisticated Mediterranean culture that extended from Carthage (modern Tunisia) eastward to Anatolia (modern Turkey), and did not share any cultural or linguistic affiliations with peoples living in modern-day France, Germany, Austria, Switzerland, Sweden, England, Spain, Belgium, etc. (many of whom they considered uncivilized savages). Romans lived in Egypt, Anatolia, and across the Middle East, which at the time was all part of the Roman Empire. As Rome fought the various empires of the era (Persia, Carthage, etc.), they considered all of these areas to be part of "civilization," with which they shared values, even as they fought as enemies. Many Romans were literate, and the society produced sophisticated artists, writers, engineers, scientists, historians, and philosophers. Europe was producing very little comparable culture, whereas other Mediterranean cultures were. The affiliation of later Renaissance Europeans with ancient Rome and Greece is the case of adoption, rather than actual intellectual or cultural descent.

Upon the fall of the Western Roman Empire in the fifth century, Europe, which had been loosely governed by Rome, broke into dozens of small principalities, with innumerable linguistic, cultural, and political divisions among them. The Catholic Church, which had been the official state religion of the Roman Empire at its collapse, however, retained a grip on power even after the fall. The church had its own army, and using the doctrine of "rule by divine right," provided the moral justification for the reign of the various monarchs of the Middle Ages.

During this time, most intellectual activity in Europe, including art, architecture, and writing, centered on the church. Economic activity was reduced partially because, without the stability provided by the Roman Empire, traders and merchants could not risk long voyages. Because of the low population density, and without large urban centers, Europe returned to what it had been before the Roman conquest – an area with low rates of literacy, with society mostly devoted to the production of agriculture. The economic system of this time is called *feudalism*, in which peasants lived on big estates owned by powerful local lords, working the land and paying with a substantial proportion of their harvest (see Chapter 21). This was largely an economy of barter, rather than money, and economic growth was minimal during this period. These were the so-called Dark Ages, and persisted for roughly 1,000 years.

This was also the period of a series of plagues across Europe, the Middle East, and Asia. Europe, especially, suffered from significant losses to its population. Politically, the fall of the Western Roman Empire and the establishment of multiple small principalities meant that wars were relatively common, and the destruction to the populations and the farmland reduced economic output. The Hundred Years' War (1337–1454), especially, laid waste to major regions of France and Belgium, but there were many other wars before and after.

Finally, the rise of the great Islamic empires meant that Europeans were often at war with the Muslim armies seeking to expand their control. In the seventh century the Rashidun Caliphate conquered the Middle East, leading to the Crusades, which was a major drain of money and resources. In 711 the Umayyad Caliphate conquered the Iberian Peninsula (Spain and Portugal), followed by an attempt to conquer France.

The Renaissance and the Age of Discovery

The year 1492 marks a shift in the fortunes of Europe. This was the year that Christian armies pushed the Muslims out of Iberia. This year was also when Christopher Columbus traveled to the Caribbean, and later Latin America, setting the stage for the colonization of the Americas. The influx of wealth from the Americas was a major contributor to the rapid growth of urban areas in Europe and the expansion of European economies. Other countries also sent explorers abroad, looking for trading or colonization opportunities. The Portuguese, who had advanced open-ocean shipbuilding design, were very active in Renaissance exploration, and sent Vasco de Gama (1460–1524) to sail south from Europe down the African coast, around the Cape of Good Hope, and to India. They also sent Ferdinand Magellan (1480–1521), who

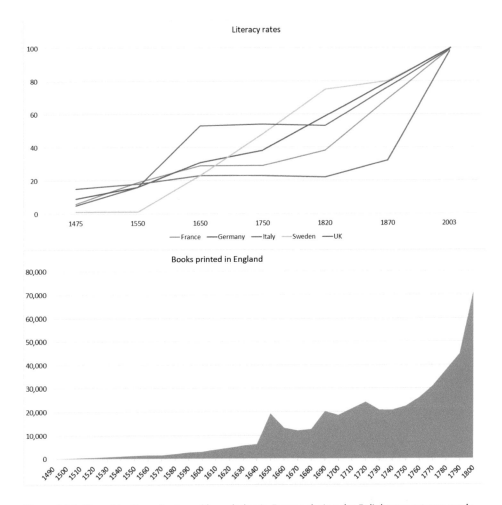

Figure 19.1 The wide dissemination of knowledge in Europe during the Enlightenment was made partially possible by the efficiency of the newly developed Gutenberg printing press and moveable type. The Gutenberg Bible (1450s), for example, was the first widely available version of the Bible, and its printing allowed much greater access to the Christian text. (The Gutenberg Bible was, however, printed in Latin, and it would not be translated into German and English until the next century.)

led the first expedition to circumnavigate the Earth, as well as others. Other countries, seeing the opportunities for economic benefits, followed suit, including England, France, Italy, and the Netherlands. Over the next centuries, all of these countries became significant traders and colonizers, with the wealth accruing across those regions.

With the increase of wealth, literacy rose, and with that came significant intellectual expansion in the region (see Figure 19.1). Europeans had traveled to the Islamic cities to learn about science, mathematics, navigation (especially the astrolabe), and the Greek philosophers, and brought translated texts back, notably Aristotle. The fall of Constantinople in 1453 made a significant contribution to the expansion of knowledge during this time, as scholars fleeing the fall of Constantinople (which had been a major center of knowledge since the fall of Western Rome) sought refuge in Christian Europe.

These various influences led to what is known as the Renaissance, first starting in Italy. Significantly, this marks the rise of Europe as a world power. From this period onwards, Europe, which had always been behind the other major centers of civilization, started to become a leader in technology, art, philosophy, scientific discovery, and political thought.

Challenges to the Catholic Church

Part of this intellectual revolution was the development of European science. Major mathematical and astronomical advances had been made in the Islamic world and India, and the spread of those ideas enabled European astronomers like Copernicus and Galileo to measure the orbits of the planets. The Catholic Church took notice of these new ideas, and when they felt it necessary, tried to suppress their spread, as in the case of Galileo (Figure 19.2) (Copernicus died before he could be similarly persecuted).

Figure 19.2 Galileo was persecuted by the Roman Inquisition for publication of the geocentric model of the Solar System (although it is not clear that he faced the threat of torture quite so boldly as is presented in this image). But he expressed the general principle of scientific investigation when he said, in his *Letter to the Grand Duchess* (1615): "I think that in the discussion of natural problems we ought to begin, not with the Scriptures, but with experiments, and demonstrations."

This is also the era in which people began to challenge the church on theological grounds. Martin Luther (1483–1546), the founder of Protestantism, challenged the Catholic Church over several church practices, and argued that common people should have the right to interpret scripture. He was the first person to translate the Bible into German, which allowed ordinary Germans to read and interpret the Bible. Luther's challenge to the church was met with excommunication by Pope Leo X. In England, William Tyndale (1494–1536) translated the Bible into English, for similar reasons, and was burned at the stake on the order of Catholic authorities. Luther avoided the stake, and his ideas found fertile ground after 1,000 years of rule by an increasingly wealthy and decadent Catholic Church. Protestantism was soon spreading throughout northern Europe.

One of the formative events of this period was the Thirty Years' War (1618–1648). This was primarily a conflict between Protestant and Catholic countries, and over the thirty years ranged across modern-day France, Germany, Austria, Belgium, Denmark, and Sweden. The high death toll (some 8 million) and destruction of the countryside, in a war between Christians, led some thinkers to challenge the very principle of organized religion.

The Age of Revolutions

Clearly, one of the major thrusts of this intellectual growth was a willingness to challenge established doctrine, and to independently evaluate arguments based on their merits. Writers reassessed the principles of government and began to reject the ideas on which feudalism and "rule by divine right" were based. Access to the literature of ancient Greece and Rome presented them with alternative models of government, as Athens was a participatory democracy and Rome was a republic.

As people across the world absorbed many of these new ideas, they sought to change the world in accordance with these them. This was the time of revolutions (North and South America, France, Haiti), and when the principles of monarchy were challenged, and in some cases overthrown (Figure 19.3). This was also the time when slavery became broadly challenged on ethical grounds, and when it was outlawed across much of the world.

These challenges to long-established modes of life and thought (to both religion and monarchy) marked the Enlightenment and established the trajectory toward our modern world. This was a time in which we see the appearance of ideas that we would recognize as very much like ours. The modern concepts of individual liberty, civil and human rights, democracy, separation of church and state, science, and freedom of the press come from this period.

Figure 19.3 The Enlightenment was known for its revolutions, depicted here. In France (1789, top left), Haiti (1791, top right), and the USA (1776, bottom). These revolutions overturned the long-established order in these countries and started the modern trajectory toward democracies becoming the norm. England had an earlier revolution (1688), and although it replaced one monarch with another, it also strengthened the power of the legislative body (Parliament), established the rights of citizens, and limited the power of the monarch.

In an important sense, the Enlightenment was the most significant intellectual break from humanity's past in our history. Today we take for granted many of these ideas, but they are relatively recent, and they are the product of a very particular set of historical circumstances.

Box 19.1 Where did civilization go after the fall of Rome?

Despite what you may have read, the fall of Western Rome did not mean that civilization paused. It simply went elsewhere. The Eastern Roman Empire (based in modern Turkey) continued for another 1,000 years, and this time period also saw the rise of other great civilizations. In India, the cycles of the great empires continued unaffected, with the rise and fall of the Gupta Empire, and the emergence of the Delhi Sultanate, all producing enormous works of art, literature, and architecture. This period is considered one of the "Golden Ages" in Chinese history, with the Tang Dynasty ruling peacefully over large areas of mainland China. It is famous for its production of poetry, art, and monumental architecture, and the cosmopolitan population of its capital, Chang'an.

For Westerners, however, the most important political event of the time was the emergence of the great Islamic empires. Following the establishment of Islam in the seventh century, Muslim armies conquered vast areas in North Africa (including Egypt), the Middle East, Iberia (modern Spain and Portugal), Iran, Iraq, and much of India. For almost 1,000 years, Muslim armies battled against Christian European armies for control of the Mediterranean. In general, the Muslim armies were victorious, capturing much of the old Western Roman Empire and toppling the last vestige of the Eastern Roman Empire, Constantinople, in 1453.

The Muslim empires were very sophisticated and their famous libraries in Granada, Cairo, and Baghdad preserved the ancient literature that had been lost with the fall of Rome. The Islamic "Golden Age" ran from roughly the eighth to fourteenth centuries, and was one of the few places where travelers from around the world could readily exchange ideas. At a time when Northern and Western Europeans had almost completely lost access to the knowledge of ancient Greece and Rome, and scientific learning was largely stifled, the mathematicians of Baghdad were inventing algebra and celestial navigation, biologists were mapping the human circulatory system, and scholars were reading, critiquing, and translating Aristotle. Furthermore, Islamic society was far more permissive of other cultures and religions. The Jews of the diaspora had settled in Moorish (Muslim) Spain, and when Christians reconquered the Iberian Peninsula (1492) the Jews were so vigorously persecuted by the Christians that they fled to Muslim North Africa, where they were welcomed and allowed to live in peace. To this day, in Granada, the former capital of the Moorish state, the spectacular Islamic architecture draws visitors from around the world.

Box 19.2 Major ideas of the Enlightenment

Today, much of the world is defined by the ideas of the Enlightenment, which shifted us away from the way the world had been ruled, organized, and conceptualized for almost 10,000 years.

Age of Reason

The Enlightenment is sometimes called the "Age of Reason" for its adoption of scientific explanations of the natural world. Isaac Newton is perhaps the preeminent example, having identified gravity, and explored, mathematically, its implications through astronomy and physics. In this period, we also see the appearance of modern geology and zoology, magnetism, electricity, and at the end of the era, evolutionary biology. The spread of these ideas, and their clear predictive power, gave science a legitimacy that caused people to question other ways of explaining the world.

Rejection of Religious Authority in Matters of Fact

Maybe the most important change during the Enlightenment was the exploration of ideas that challenged church doctrine. During the Renaissance, Galileo had argued that astronomical patterns could be best explained by directly contradicting the official doctrine of the Catholic Church – that the Earth did not move. Once the fallibility of the church became evident, philosophers and scientists felt more comfortable looking for explanations that were more mechanistic and consistent with the patterns of nature, rather than requiring explanations from the supernatural. One fairly characteristic expression of Enlightenment thought can be seen in Thomas Jefferson's edits to the New Testament, in which he removed all aspects of the supernatural (including the virgin birth of Jesus, the resurrection of Jesus, his walking on water, etc.), in order to capture the moral teachings without relying on anything outside normal naturalistic phenomena.

Separation of Church and State

The examples of history (the Thirty Years' War especially) had shown Enlightenment scholars that religion had to be divorced from influence over government policy. Martin Luther felt that the intermingling of political power and religious power led to corruption of both, and argued for the idea of the "Two Kingdoms" – one of spirituality and one of law. Luther was most concerned with preventing the political corruption of the church, but subsequent writers, particularly those trying to create new forms of government, were more worried about establishing a government absent of the conflicts of religions.

Box 19.2 (cont)

The first explicitly secular state was the USA, and the separation of church and state is established in the first line of the First Amendment. Other countries followed suit, notably France almost immediately afterwards (after the French Revolution) but also many others over the next 200 years.

Establishment of the Rights of Individuals

Prior to the Enlightenment, the hierarchies of societies were held to be, in some important way, natural. The monarchs ruled by divine right, and the various nobility were similarly inherently better than their social inferiors, all the way down to the lowest peasant class, who were naturally inferior (and therefore justifiably controlled). During the Enlightenment, this long-standing position was rejected. Philosophers like John Locke, Thomas Paine, and Georg Hagel argued that a government was not the result of the natural superiority or divinity of a particular class, rather it was the result of political power and history. Individuals, they argued, had inherent rights that should never be denied.

The Most Natural Form of Government

The historical examples of republican Rome and democratic Athens showed the writers of the Enlightenment that there were ways to establish long-lasting and successful governments that did not rely on the concept of the "divine right of monarchs." In his work called *Politics*, Aristotle (384–322 BCE) identified constitutional democracy as the least harmful form of government. Writers during the Enlightenment expanded on his ideas, and argued that it was the only truly legitimate form of government, because in a constitutional democracy individual rights are protected from governmental oppression, and perhaps more importantly, the citizens must consent to the various restrictions and coercions (Box 17.2) that are necessary for a government to function. Previous to this time, taxes, rules, and military conscription might be imposed by a monarch without any consultation of the people to be affected.

The Enlightenment is known for its many revolutions, starting with the American revolution, but followed by France, Haiti, and the expulsion of Spanish control from much of Spanish-speaking South America under Simón Bolívar. These revolutions were explicitly based on the principles of the Enlightenment, and the new governments were all (at least initially) founded on the principles of constitutional democracy (England had a "Glorious Revolution" in 1688, but it fell somewhat short of the standards of the time: the

Box 19.2 (cont)

monarchy was retained, and no written constitution was created, even if there was an expansion of citizens' rights and parliamentary power as a result).

However

It is worth noting that the Thomas Paine's ideas of the "Rights of Man" ran directly counter to the justifications for exploitation of Africans for slavery, and the founders of the USA were very aware of the internal contradiction posed by the text of the US Constitution. Generally speaking, Enlightenment thinkers were strongly opposed to slavery, and this is the period when we see explicit arguments against the principles of holding people in bondage for labor (although most of these thinkers might still be considered racist by modern standards). These arguments largely held sway, and the leaders of many countries were persuaded by the moral arguments. This is the period when slavery, as an international institution, became outlawed across the world. Within the USA, this contradiction would form the basis of the conflict over slavery that led to the US Civil War (1861–1865).

Although this period is known for the declaration of "The Rights of Man," there is no question that for many, rights were envisioned for males only, and by today's standards sexism was the norm. Nevertheless, this period laid the foundations for gender equality that would lead to the rise of the Suffragettes and the push for the right to vote for women in the early twentieth century, as well as the general expansion of the political and economic power of women.

Box 19.3 Why didn't the Enlightenment occur during the Islamic Golden Age?

One of the questions that has puzzled historians is why the Enlightenment occurred in Europe rather than in some of the other centers of literacy and education around the world. One of the best candidates for "an enlightenment" was in the Muslim world, during the Islamic "Golden Age" (roughly 800–1300). During this period, Europe was still mostly in the Middle Ages (also called medieval era, as well as the Dark Ages), with low rates of literacy in most of the region, and few large, densely populated cities where centers of learning could accumulate substantial bodies of knowledge. Those few centers of learning that did exist were controlled by the Catholic Church, which would suppress learning that contradicted its doctrine. For example, the

Box 19.3 (cont)

study of Aristotle was forbidden by the University of Paris starting in 1210, on penalty of excommunication.

However, in the Muslim world there were several great cultured cities with famous libraries and large teams of scholars. These were found in Muslim Spain, Egypt, and Iraq, among others. During the Islamic Golden Age, mathematicians developed algebra and trigonometry, and studied human anatomy and medicine, advanced astronomy, optics, chemistry, biology, cartography, navigation, and engineering. All of these scholarly activities are consistent with what happened in other wealthy empires, in which dense populations of literate citizens naturally generated new ideas.

The libraries of Islam also retained the writings of ancient Rome and Greece, often translated into Arabic, that medieval Europe could not yet access. The Muslim scholars were aware that ancient Athens was a participatory democracy and that Rome was a republic. They were also aware of the Greek skeptics, who challenged the notion of a divine creator.

But why, then, did the Islamic empire not generate a movement comparable to the Enlightenment, in which the ancient ideas of democracy and skepticism were combined with the science of the day? Why did the Islamic Golden Age end, rather than continue and influence the present as the Enlightenment has?

Historians have offered various ideas. Some historians attribute it to the invasion of the Mongol hordes. In 1258 the Mongols sacked Baghdad and destroyed the famous "House of Wisdom," which was probably the greatest library in the world at the time. It contained hundreds of thousands of manuscripts, including many sole copies of works from ancient Greece and Rome that are now lost to history (one anecdote describes the Tigris River as running black with the ink of the manuscripts thrown into the water by the invaders). The Mongols also destroyed the economic base of the region by burning the fields and scattering the inhabitants. There is little doubt that such destruction would have set back learning and intellectual development in the region.

However, there were other famous libraries in the Muslim world, and the Mongols conquered neither Egypt nor Spain. Rather, the explanation likely has more to do with the cultural relationship of religion to power. One of the major intellectual shifts of the European Enlightenment had to do with the decline in power of the Catholic Church. One major loss for the Catholic Church was the development of the Protestant Church. The other had to do with science. By the sixteenth century, it was clear that, despite the persecution of Galileo, ideas were circulating in society that contradicted Catholic Church doctrine. The

---

Box 19.3 (cont)

failure of the church to totally crush these ideas, and the willingness of other scientists to continue their work, meant that people perceived that new ideas could not be readily suppressed. Philosophers began questioning "rule by divine right," as well as other elements of church doctrine, including the essential question of divinity itself. Some philosophers began studying religion as we do now – as a phenomenon to be explained.

But in the Islamic Golden Age, as scientists and philosophers began to circle around some of these same ideas, religious scholars pushed back. One of the most famous of these was Al-Ghazali (1058–1111) who wrote a strong critique of such ideas in *The Incoherence of the Philosophers*. This work charged many of the Muslim philosophers of Aristotle (and other Greeks) with heresy and with rejecting Islam. Al-Ghazali's book was highly influential, and Muslim scientists and philosophers of the time understood that there were definitely limits to what they could write. Importantly, they could not challenge the basic ideas of the religion, lest they be persecuted. Since the foundation of the Enlightenment was, in many ways, based on the rejection of religious authority, a strong religious response to "enlightened" ideas ensured that the Enlightenment could not occur in the Muslim world.

It may be tempting to view the Islamic Golden Age critically, since it was unable to provide a sufficient challenge to the authority of religion, and therefore could not provide a true "enlightenment." However, the Islamic Golden Age provided many foundational tools essential for later European scientific inquiry, as well as the preservation of the critical works of ancient Rome and Greece, and was therefore essential to the Enlightenment. But more than that, the Islamic Golden Age followed the pattern seen throughout history up to that time, in which the learning of the society advanced the technological knowledge of the society, without providing a challenge to the established power structures. This was the pattern in every other great empire that had existed throughout history and medieval Islam deserves no more of a critique than ancient Rome, Persia, or China for failing to produce an enlightenment.

Why Europe?

The European Enlightenment happened to come at a time when religious power was fracturing in Europe, reducing the ability of any one church to suppress ideas (although suppression was common enough). Another factor was the fracturing of political power. In the Muslim world at the time, the power of the various empires across vast areas meant that central authority had considerable reach (this is also true of the Indian and Chinese Empires). Europe

---

Box 19.3 (cont)

was broken up into dozens of small states, which meant that the reach of any church authority was limited. The leaders of local regions could choose to enforce, or not enforce, the religious edicts from Rome. Martin Luther escaped being burned at the stake by moving around central Europe, protected by various leaders at different times. No such escape would have been possible had all of Europe been under central control. Many controversial thinkers used the borders as protection, although it was no guarantee: Spanish scientist Michael Servetus (1511–1553) fled Catholic Spain under threat of persecution, but was ultimately burned at the stake by Swiss Protestants, and William Tyndale, who translated the Bible from Latin to English against church wishes, fled England only to be burned at the stake in Brussels.

Figure 19.4 During the Renaissance and Enlightenment, the artistic focus of Europe was on classical Greek and Roman antiquity. By doing this, they intentionally overlooked the medieval period, and much of the art can be interpreted as a rejection of the religiously oriented medieval art. In this figure, the art of classical antiquity is on the left – the Roman pantheon (built 27 BCE–14 CE) and a statue of Caesar Augustus (first century BCE). In the middle we see Notre Dame (built116–1260) above a statue from the Collegiate Church of St. Stephen (eleventh century). On the right we have Enlightenment art clearly emulating the art of ancient Rome: the Baltimore Basilica (1806–1863) above *Napoleon as Mars the Peacemaker* (1802–1806).

> **Box 19.3** (cont)
>
> The failure of church authorities to successfully suppress threatening philosophical, scientific, and religious ideas forced them to change. Whereas Christianity had traditionally used the threat of violence to coerce obedience (from ancient Rome onwards), their loss of power led to a shift toward permissiveness. Across Europe, a call for religious tolerance suddenly found a new acceptance. This change might have been the critical shift that allowed for new ideas to spread. By the eighteenth and nineteenth centuries, the heads of the great European dynasties, not to mention religious leaders, were themselves calling for reforms to government and religion in line with the scientific and philosophical ideas of the time. In the absence of repression, with an "open marketplace of ideas," the Enlightenment was able to flourish for the first time in world history (see Figure 19.4 for the artistic expression of the period).

> **Box 19.4** The importance of literacy
>
> One of the most important trends that led to the Enlightenment in Europe was the dramatic increase in literacy, from the end of the Middle Ages (roughly 1400) through the nineteenth century. Throughout history, philosophers and other writers had put ideas down (on paper, papyrus, or clay tablets), and for most of history those ideas were little read and the majority are now lost. During the Enlightenment, the explosion of literacy ensured that there were many copies of the important philosophical or scientific works, and that these ideas were read and absorbed by the general population.
>
> Several factors led to the relative expansion of literacy. One was technical – the invention of the printing press with moveable type by Johannes Guttenberg in the fifteenth century. Although printing with woodblocks or ceramic had been available for hundreds of years (starting in China), each image for each page had to be individually carved. The invention of Guttenberg's moveable type (with a different block for each letter) meant that the costs of printing dramatically decreased. As the costs of books and newspapers dropped, it became practical for people with lower incomes to learn to read. Also, European languages are relatively easy to learn as written languages, since they are phonetic. To learn to read European languages only twenty-six to thirty-five letters need be learned. In contrast, Mandarin letters are partly conceptual, and to be literate in written Mandarin requires learning, at a minimum, hundreds of characters if not thousands. Additionally, there is a

Box 19.4 (cont)

cultural tradition of written scholarship in Europe, as the Catholic Church produced voluminous scholarly works that had to be read across Christendom. In ancient India, traditional scholarly learning was by rote memorization of ancient poems and stories rather than reading ancient texts.

In Europe there was some degree of variation in literacy, and this was a factor in the spread of the Enlightenment. In wealthy trading countries with growing middle classes, like France, England, and Germany, the ideas of the Enlightenment spread rapidly throughout the populations. But some countries were slower to move out of the medieval era. Russia, famously, never really went through the Enlightenment. The serfs were staggeringly poor and had historically been virtual slaves, with no rights to move away from poor working conditions, and who could be bought and sold by landowners. Killing a serf resulted in a modest fine until the nineteenth century. Among these people literacy was low, and with little disposable income, books were uncommon. These are not fertile grounds for the spread of ideas, and Russia did not contribute to the ideas of the Enlightenment in meaningful ways, even if the leaders did try to encourage it during the eighteenth and nineteenth centuries. But in countries with literate, middle-class populations, the Enlightenment made rapid progress.

After the invention of the printing press, more of Europe's average citizens were able to read. As more and more people became literate, the ideas of such notable thinkers as John Locke, Thomas Hobbes, and Jean-Jacques Rousseau, who were all proposing political reforms, became widely available and highly influential. This spread of knowledge outside official governmental or religious channels was a new phenomenon. There were no comparable ideas percolating through the societies of ancient Rome, Muslim Spain, Song-era China, or the Vijayanagara Empire of southern India. But it had a huge influence on the European societies of France, England, Germany, Belgium, and Italy, as people became aware of these ideas and were exposed to alternative models of government, religion, and society. The literacy of these populations had positive effects on their economies, as these societies were able to become part of the Industrial Revolution (Chapter 22). As we know today, educated people were able to contribute to their country's economies in ways other than simply providing labor or growing food, by creating new ideas, art, innovations, and inventions. As the Industrial Revolution created export markets, societies with educated citizens grew wealthy. In many ways, this is where the shape of the modern world was first formed, and today literacy is considered a top priority by leaders in virtually every country.

Box 19.5 Slavery and colonialism in the Enlightenment

The existence of slavery during the Enlightenment is a matter of record. During a time when political and moral philosophers were espousing the view of human rights as innate and inalienable, humans were captured in large numbers and sent to plantations and mines in North America, South America, and the Caribbean. This was a time of colonialism, and large amounts of the Americas, Africa, and Asia were under the control of distant European countries, which were extracting the natural wealth of the fields and mines from these countries and often enslaving or harshly ruling over the people doing this work. Because the same countries that were espousing freedom for humanity were engaging in the oppression of humanity in other countries, the accusation of hypocrisy has hung over the Enlightenment.

However, to reject the Enlightenment because of this would be to lose track of the fact that this was a period of diverse opinions and radical change. For the entire reach of history – some 10,000 years – humans had been engaging in these acts. The empires of antiquity and the Middle Ages across Asia, Africa, Europe, and the Americas enslaved other peoples and took what they wanted. The right of the strong over the weak was universal, and moral arguments held no sway.

It was not until the Enlightenment that these ideas were seriously challenged. There was considerable diversity of opinion – while some people, particularly those engaged in slavery and colonialism, defended those practices as natural and just, many others rejected them on moral grounds. While slave revolts had occurred many times throughout history, it was only in the United States that a country willingly went to war to free another group of people due of principles alone. This was the period in which the countries of Europe (and many others) outlawed slavery. At no other time in the past was there near universal moral agreement rejecting this long-standing institution.

Many of the countries of the Enlightenment (England, France, the Netherlands) were engaged in colonialism over the period of the nineteenth to twentieth centuries. However, the seeds of the liberation of these countries were planted during the Enlightenment, even if it took several hundred years to finally reach maturity. But even from the outset of the Enlightenment there were significant critics of colonialism (Adam Smith, Immanuel Kant, Denis Diderot, James Cook), and it is only because of the relative freedom of the press, and the free exchange of ideas, that these ideas were allowed to be spread.

It took World War II to finally end colonialism, but it was the arguments about the inherent rights of all humans, first widely articulated in the

---

Box 19.5 (cont)

Enlightenment, that were used to justify the freedom of the various colonies. This was a first – at no time in the past had a conquered people been freed simply because the moral arguments outweighed the economic ones. So, in an important sense, we are all descendants of the Enlightenment.

Box 19.6 Resistance to the Enlightenment

The period of the Enlightenment (roughly speaking, the sixteenth to nineteenth centuries) is sometimes portrayed as a period of rapid growth in new, humanistic viewpoints, unchecked by the old ways of thinking. But there was most definitely resistance from the institutions that relied on recognition of their traditional authority.

The Catholic Church famously pushed back against the appearance of the Protestant Church with the *Counter-Reformation*. This was a resurgence of Catholic power across the world, involving missionary work by Catholic priests in predominantly Protestant areas of Scandinavia and England, and which included the forced migration of Protestants from Catholic countries. Persecution by the Spanish and Italian Inquisitions for religious, philosophical, and scientific writings were part of this reaction. In addition to such religious activities, there was a series of wars between Catholic and Protestant countries to try to reestablish Catholicism by force. The Thirty Years' War was one of these conflicts, but there was a series of smaller wars that ran from the sixteenth to eighteenth centuries. In a victory for the principles of the Enlightenment, the *Patent of Toleration* in 1781 and the *Edict of Tolerance* in 1782 established a degree of religious freedom for Protestants and Jews living in the Holy Roman Empire.

The resistance of the Confederacy in the US Civil War is a similar rejection of the principles of the Enlightenment. Although several of the signers of the US Constitution were slave owners, they clearly anticipated the end of slavery, even if they could not figure out how to form the new country without the support of slave owners in the southern states. But by the 1850s, the southern states had pushed back against the ideas of the founders and against the Enlightenment generally. This is made patently clear in Alexander Stephen's "Cornerstone Speech" (Box 12.1), in which he explicitly rejects the ideas of the founders of the USA and the principles of the innate rights of all humans. The distribution of power and wealth in the southern states disproportionately

Box 19.6 (cont)

favored the few large landowners. Even free citizens of the south were often sharecroppers, a form of tenant farming inherited from medieval Europe. In many ways, the US Civil War was a war between Enlightenment ideals and pre-Enlightenment (medieval) ideas.

Humans will naturally pursue as much power as they can accumulate, and, in the absence of institutions to check individual power, pre-Enlightenment models of government reappear. In the twentieth century, this took the form of fascist totalitarianism (Chapter 23), and it took several wars to reassert the principles of democracy. In the future, this will inevitably reappear if the institutions created during the Enlightenment to assure the balance of power are weakened.

Chapter 20: The Industrial Revolution

Imagine you are living in 1720, whether in London, Istanbul, Delhi, New York, Paris, Peking, or any other large urban area on Earth. Let's say you need a pair of shoes. The places we get our shoes today do not exist – shopping malls, online retailers, mail-order warehouses. In fact, there aren't even shoe stores. If you want a pair of shoes, you have to find a person who makes shoes (a cobbler) and ask him to make you a pair. He traces your feet on a piece of paper and you have to wait, maybe a week, maybe longer, for the shoes to be built.

For most people at that time, having a decent pair of leather shoes would be a substantial investment, and you would hope that those shoes would last for years. The reason something as simple as shoes were so expensive is because of the skilled labor involved to make them. A cobbler would have to cut out the soles and insoles from heavy leather, then stitch down the uppers, while stretching the uppers around a last to make the shape of a foot inside. Then the heavy outsole had to be nailed down to the insole. Finally, the upper had to be trimmed and stitched at the cuff, with either grommets for laces, or straps and buckles attached. All of this had to be done by hand, without any power to assist – for just one buyer. Stitching heavy leather is hard work and time-consuming, and requires someone with skill – not just anyone can make a decent pair of shoes.

So, when you buy a pair of shoes in 1720, in addition to the cost of the materials (leather, etc.), you are buying the skilled labor of that one person for the length of time it takes to make the shoes. Let's say it takes him a full week – you have purchased the labor of a skilled person for 1/52 of the year. That cobbler probably has a shop, as well as a family to support, and if he is a good shoemaker he will want to charge enough so his family can have food and shelter and live comfortably. For comparison, let's look at what we think a shoemaker might earn today. Let's say a modern cobbler makes $50,000 a year. If you bought a pair of basic hand-built shoes that took a week to make, it would cost around $1,000. In the modern world, that is a lot of money, and might represent quite a bit of your monthly paycheck.

In 1720, whether you buy a pair of shoes, or a coat, or a tri-corner hat, or a harness for your horse, or a musket – everything is handmade and custom-built to order. So everything is expensive. Only the rich have lots of shoes and coats because only the rich can afford to pay for the labor. This was also the state of affairs in 1620, 1220, 220, year 0, 220 BCE, and 520 BCE. For most of history, if you wanted to have something complicated made, like formal clothes, or leather shoes, or tools, it would be expensive, and in many cases out of the financial reach of most people, or at least a major financial commitment.

Now, let's go to 1920. If you are in a major city and want a pair of shoes, you go to a shoe store where there are lots of shoes in lots of sizes. According to the Sears Roebuck Catalog, you can buy a pair of decent leather shoes in the USA for $2.50. At that time, the average income (according to the Internal Revenue Service) in the USA was $3,269. So the cost of a pair of shoes was 7/10 of 1 percent (0.007 percent) of the annual income for an average worker. In 1920, an average worker could probably afford to buy several pairs of shoes without making any real dent in their annual income. And this was the case for most things – by 1920, your average worker could afford several jackets, several hats, and even things that had been far out of the reach of a common worker in 1720 – a pocket-watch or a new hunting rifle.

So why could a worker afford so much more in 1920 than in 1720? It is because of an economic and technological revolution that occurred in the intervening 200 years. In many ways, this revolution overturned many of the foundations of economic power that had existed for almost 10,000 years. Today, we live in a post-Industrial Revolution world, where the results of this rapid period of change are still being felt and the consequences are still being worked out. This change has generated enormous benefit for billions of people, but at a steep cost – to traditional ways of life, to thousands of languages and cultural practices, and not least of all to the environment.

Changing the Equation

There are several key changes in the way products were built that made them affordable. The first change was the *assembly line*. In an assembly line, workers only build one or two parts of the final product. Imagine the shoemaker – they would have to know all the parts of the shoe, from the sole to the finishing of the leather, which meant it might take years to teach an apprentice to learn to build shoes well. In an assembly line, the process is broken down into steps, and a worker only has to know how to do a small part of the complete process. An assembly line worker can be taught how to do their few steps in a far shorter time than the training of a skilled craftsperson. It might take weeks, or even less, to teach an assembly line worker how to do their work well. In an assembly line, the requirements of extensive apprenticeships are eliminated.

The other aspect was the introduction of external energy into the system through machines. In 1720, there were only a few sources of power – the horse, the water wheel, and human hands. There was no way to harness a horse or water wheel to make shoes, so it was done by hand. But the introduction of the steam engine at the end of the eighteenth century meant that machines could be developed to help with manufacturing. Steam engines would power various tools that would make the manufacture of products like shoes much faster – for example, steam might power the tools to cut and stitch the leather, dramatically decreasing the time required.

This introduces the concept of *economies of scale*. A cobbler will charge for the labor and materials necessary to make a pair of shoes. Two pairs is twice as much labor and materials, and three pairs is three times as much. But in a factory, once the factory is built, making more shoes drives down the cost of the shoes. The cost of factory-made shoes represents a few things ("factors of production"): the materials, the labor, the power supply, and the cost of the factory (which is a fixed cost, no matter how many shoes are made). If the factory only ever makes a hundred pairs of shoes, they have to be expensive, because each pair has to be 1/100 of the fixed cost factor. But if the factory makes 1,000 pairs, each pair only has to be 1/1000 of the cost of the fixed factor of production (i.e., the factory costs). The labor, power supply, and materials might increase as they make more shoes, but the factory (building plus machines) is a fixed cost. So, as production increases the price of each pair will decrease.

As machines became more and more sophisticated, and assembly lines became more efficient, the prices of goods were driven lower and lower, so that by 1920 shoes were no longer an expensive financial commitment but a regularly purchased commodity. Across the consumer spectrum, almost every product imaginable suddenly became much more affordable, from clothes to food to housing.

Box 20.1 Are handmade products better than manufactured products?

Sometimes you will hear a product described as "handmade." This word carries the connotation that it is in some important way "better" than the factory-produced item. But is it true? Certainly, a handmade item will be much more expensive, because it represents such a large amount of work, typically by a skilled craftsperson. But in what way is it better (i.e., more functional)?

Let's return to 1720. Think about how a complex mechanism like a musket (Figure 20.1) is made. Each part has to be hand built, in a specific sequence, so that each of the parts fits together. While the gunsmith will certainly tend to make the various components the same way each time they make the musket, there will inevitably be minor variations because they are built by hand. Further, each different gunsmith will make the various components differently.

Box 20.1 (cont)

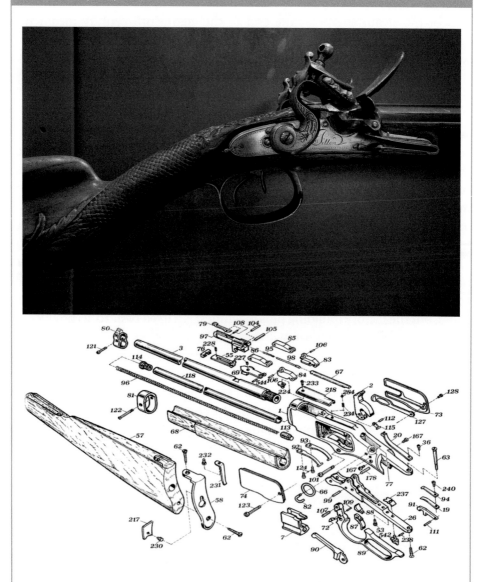

Figure 20.1 In a hand-built rifle, like this eighteenth-century flintlock (top), each piece of wood and metal is handmade. The gunsmith, a highly paid artisan, must be able to bore a barrel, forge the metal lockwork, and carve and finish the wooden stock. Properly made, this rifle will last lifetimes. However, if any part fails it must be replaced with a custom-made piece that must also be hand-fitted. The labor to make such a rifle is enormous, and the skills take years if not decades to master.

By comparison, the Winchester rifle (bottom), from the nineteenth century, is assembled from parts made on an assembly line, by relatively unskilled labor. Because of the assembly line process, the Winchester is a fraction of the cost of the handmade flintlock. However, the Winchester is not a bad rifle simply because it is mass produced – with correctly calibrated boring machinery, a factory-produced rifle is more accurate than a hand-bored rifle.

Box 20.1 (cont)

So, there will be differences in quality and performance, and the only way to know you are getting a quality product is word of mouth.

Further, should a part of your musket break, the replacement part will have to be custom-made by a gunsmith. If you are not in the same town where the musket was originally made, you have to try and find a gunsmith in your new town to match the original part which will certainly require hand-fitting. Such a repair will be time-consuming and expensive.

Compare this to a late nineteenth-century firearm like a Winchester repeating rifle. As you can see from Figure 20.1, each piece is a manufactured part that is designed to fit on any rifle that is the same model. This means that, if a part breaks, you just buy the (relatively inexpensive) part and install it, probably just using common tools, without any specialized knowledge.

And in what way would a hand-built rifle, even today, be better than a factory-built rifle? There is no reason to expect a hand-built rifle to be more accurate, especially if the barrel is bored by hand. Nor would it be more reliable – hand-built rifles inherently run the risk of small defects introduced by the hand-building process, which machines, if designed properly, would not have.

So a factory-built rifle is more accurate, reliable, and easier to fix, while also being a fraction of the cost. This is the case for most manufactured items. Today, a factory-built Toyota is more reliable and efficient (functional) than any hand-built car, while being a fraction of the cost. Some luxury cars are at least partially hand built (like Rolls Royce and Ferrari), but these cars are known to be expensive and have nowhere near the long-term reliability of the Toyota. And they would not necessarily outperform a high-performance Toyota, or have a smoother ride than the most stately Toyota (although car aficionados will argue both these points).

This, of course, highlights the difference in another way. People who see you in the Ferrari will marvel at it, and it does stand out as a symbol luxury. This in many ways demonstrates the value of handmade products in the modern world – as a display of wealth (social cachet).

Box 20.2 The invention of the consumer

Prior to the Industrial Revolution, people largely purchased what they needed from a local producer or produced it themselves. People certainly advertised their services locally, but because the scale of production was relatively limited

Box 20.2 (cont)

for any craftsperson, even in a shop full of apprentices, there was no incentive to try and acquire a substantial customer base. As the population increased, demand increased, but the ability of craft shops to supply that increasing demand was unlikely if demand was great. Once factory production methods were introduced (Figure 20.2), and factories could produce large numbers of products relatively cheaply, factory owners had a strong incentive to sell large numbers of their products.

Factories are expensive, so it was critical for a factory owner to be able to sell enough of their products to pay for their investments in the building, equipment, materials, and labor. Initially, the products made in factories were the same types of products that people had previously purchased from craftspeople – clothing, shoes, tools, furniture, etc. But production quickly expanded to luxury items that people had not really needed before, in the utilitarian sense. So, the trick was to convince them to purchase something unnecessary and thereby augment their concept of satisfaction, adding hierarchical brand identification to utility. To do this, they did not have to meet an existing demand, they had to *create* a demand. This was the advent of advertising, and by the nineteenth century marketing was ubiquitous.

Figure 20.2 The high product output of factories and mines in the eighteenth and nineteenth centuries was mostly due to the development of the steam engine. In a steam engine (left), a fire (wood or coal) boils water in a cylinder, and when the water boils it creates steam. This steam creates pressure, which pushes the top of the cylinder (the piston) up. The piston connects to a lever, which then turns a wheel.

In the cotton mill (right), the steam engine would power the spinning rods mounted to the ceiling of the room, and those rods would spin belts that led to the machinery below. In this image, hundreds of machines are being powered by the steam engine.

Note the lack of guards on those machines – any long hair, fingers, or limbs could easily be caught up in the machinery, which was essentially impossible to stop. Injuries and deaths were common early in the Industrial Revolution.

Box 20.2 (cont)

These are the origins of the modern consumer society, where people have a wide array of products from which to choose. Since multiple companies produce the same items, they must compete on factors like price, quality, style, etc. Marketers try to drive consumer demand by changing styles, since often, no actual technical improvements are available (Figure 20.3).

Automobiles are probably the most famous example. Year to year, cars show very few improvements in efficiency, safety, comfort, and performance, but auto manufacturers need to sell new cars. So manufacturers change the shape of the sheet metal, or the sizes of the rims, or the colors, to try to create demand for their cars. In general, consumers (being social animals) are suggestible. Car companies (and many other companies as well) take advantage of humans' need to display their success through products that they purchase.

Figure 20.3 The economics of consumption changed with the Industrial Revolution. A factory was a major investment that only made money if it was constantly producing its products. This meant that demand had to be created where there was none previously. The answer was advertising, in which marketers tried to convince people that they needed to purchase newly available items, whether there had been a demand previously or not. Prior to the Industrial Revolution, there was no marketplace full of "consumers" who were ready to purchase the latest products. Today, our economy is largely based on the consumer, and companies like Nike, Apple, and General Motors rely on consumers readily purchasing new fashions in shoes, electronics, and automobiles.

Box 20.2 (cont)

Other products reflect this same process. The consumer urge to have the latest iPhone (which, of course, largely functions as well as the previous model) is a result of the drive by the Apple, Inc. to generate demand for new iPhones. New styles of shoes are produced by Nike annually to generate sales, although most people have not worn out their previous pairs. Fashions in clothing follow the same pattern. And so it goes.

Box 20.3 Social changes caused by the Industrial Revolution

Although the acquisition of less expensive goods was most assuredly a significant improvement for most people, there were other large-scale social changes caused by the Industrial Revolution

The Rise of the Middle Class

Up until the eighteenth century, economic power was firmly in the hands of the large landowners. Outside the land-owning classes, most of the population was rural, earning their living by farming. This was the case across the world. Major urban areas, with educated populations, had small proportions of the population.

But large factories need workers, and to convince people to work in them they had to offer wages that would draw people away from their traditional lifestyles. In places like London, with thousands of factories offering jobs, wages in the early nineteenth century grew to roughly five times what a worker could have earned a hundred years before.

At the same time, the assembly line production process caused the prices of goods to drop. Suddenly, the average worker had substantially more economic power than in previous centuries. Families who had worked on assembly lines might be able to afford education for their children. Also, with economic power comes political clout. The Industrial Revolution is mostly responsible for the creation of the modern middle class and the movement to an increasingly urban oriented society.

Class Conflict

But even as the working class gained economic leverage, this was also the period in which a new source of wealth appears, creating a new and powerful upper class. These are the factory and mine owners, and later oil and railroad

---

Box 20.3 (cont)

owners (*industrialists*), who in a relatively short period of time, acquired enormous wealth and political power. They exploited their economic advantages, through business trusts and by other means, to maximize their profits, often at the expense of factory workers and lower-income people who had land or resources they desired.

Factory and mine conditions could be brutal, and the workers would sometimes try to force the owners to make the work safer, or to pay higher wages (Figure 20.4). They would petition the government to pass laws enforcing such reforms as outlawing child labor, limiting the work day to eight hours, and installing safety equipment. The factory and mine owners tended to resist any such changes using their wealth and political power to prevent the passing of labor reforms.

Sometimes workers would attempt to apply economic leverage on the factory owners by striking, leaving the factory or mine idle and sometimes forcibly preventing substitute workers from filling their positions. Factory owners often responded with violence, and there were some famous battles between workers and the armed guards hired by the owners. One of the most famous was the Haymarket riot (1866), in which workers protesting for an eight-hour work day were assaulted by Chicago police, and in the ensuing violence both protesters and police were killed, with many more wounded. But there were hundreds if not thousands of other such events, as workers tried to force concessions from factory and mine owners.

In general, the average worker had very little economic leverage to improve his or her circumstances at a mine or factory, since they were easily replaceable on the assembly line. Their response to this power imbalance was to create *unions*, in which the group negotiated as a single unit, thereby combining their bargaining power. Industrialists responded by attempting to have unions outlawed, and today, in some US states, union power is limited by statute.

However, political organization followed unionization. Over time, many of the workers' demands were met, and in industrialized nations across the world a series of laws were passed to protect the rights of workers. Today, the eight-hour work day, safety protections on the job, child labor laws, and a host of other reforms are the product of organized labor's efforts to improve conditions in factories and mines. The natural competition between these two factors of production labor and capital – eventually produced results that were both socially positive (e.g., improved work conditions) and negative (unions became power structures in their own right, with issues of hierarchical power and political corruption).

Box 20.3 (cont)

Figure 20.4 Although the countries that participated in the Industrial Revolution acquired substantial economic benefits for its citizens over the long run, this period could also be quite hard. Housing was cramped and uncomfortable, and since housing was often near the factories, it was dirty and polluted.

Early in the Industrial Revolution, child labor was common (top right). Children had no legal protections and could be paid lower wages than adults, since they had difficulty advocating for themselves.

The struggle for workers to work in safe conditions, and to obtain better wages, resulted in strikes and violent protest (bottom). Often, the factory owners were politically influential, so laws were passed to suppress the labor movements. However, struggles between corporations and labor has been a relatively constant factor in the history of industry to this day.

Box 20.3 (cont')

Figure 20.5 The dehumanizing aspects of the Industrial Revolution led some artists to look backward rather than forward, expressing a sentimental (and often unrealistic) attraction to the simpler pace of life before industrialization. This artistic movement is known as Romanticism, and was a broad movement across literature and the fine arts. Poets like Byron, Shelley, and Keats, and novelists like Mary Shelley and Jane Austen, who dealt in Romantic themes, are still popular. The Arts and Crafts movement, with its handmade wooden furniture and buildings, comes from this time.

Visual artists also expressed these themes. Here we see (left to right) *On the Beach,* Thomas Doughty (1828), *Fra Hardanger,* Hans Gude (1847), and *The Fighting Temeraire, tugged to her last berth to be broken up,* Joseph Turner (1838). In this last painting (one of my favorites), the past, represented by the elegant and ghostly sailing ship, is being brought to her destruction by the modern, fire-breathing steam tug.

Box 20.3 (cont)

Women in the Workplace

Although we often think of industrial workers as being exploited during the first century of the Industrial Revolution, it is clear that factory work created economic and political power for many of the workers. The demand for factory workers was strong, and this was the first time that women entered the public marketplace in large numbers. Previously, women had made substantial contributions to the economy with their work on family farms and family businesses. But in those cases, the income was tied to the family, and the women, although contributing to the income, were not part of the public marketplace. But once women were working in factories, they acquired income independent of any family farm or business. A woman could be single, living alone, and still make a reasonable salary.

Disentangled from the economic dependencies of a family farm or business, women became a significant economic and political force for the first time. This is the period in which the call for the women's rights became a significant public issue, and marks the appearance of the right to vote for the first time in most of the Western world. Female strikers were a common sight in the late nineteenth and early twentieth century, insisting on better working conditions at factories and demanding more political rights.

It is worth noting that, in the developing world today, factory work is still a route out of rural poverty for women seeking to avoid oppression by conservative and patriarchal traditions. In these places, women can obtain economic power and independence for the first time, and the empowerment of women is most often found in countries with industrial rather than rural economies. As the industrial economy has progressed, the classical benefit of having children (increasing the family's labor force and earning power) began to subside. The economic, religious, and general social impacts have become increasingly evident and profound (see Figure 20.5 for artistic reactions to the Industrial Revolution).

Box 20.4 Industrialized agriculture

For much of world history, the greatest concern for a person, and their families, was getting enough food to eat. Food was often expensive and anything beyond the basic grain staples (wheat, rice, potatoes, barley, corn) was hard to

Box 20.4 (cont)

get and harder to afford. But today, in much of the industrialized world, people struggle not with getting enough to eat but with eating too much. Food is, in effect, so cheap that people can afford to overeat.

The effects of the Industrial Revolution were not limited to the mine or factory – it had significant effects on farms and agricultural output. In the nineteenth century the steam engine was powering grain mills. Increased steel output in factories enabled the invention of a variety of horse-drawn machinery, such as harvesting machines. Once the steam engine became ubiquitous early tractors began pulling plows, and later disk machines that created multiple furrows in each pass. Threshing machines, towed behind the tractors, harvested the grain, and railroads delivered it to market. With this increased efficiency, the price of food was much lower than in previous centuries. Over the length of the Industrial Revolution, the volume of food was sufficient that some countries began exporting grain, and today the industrialized nations export significant volumes of food, whereas nonindustrialized nations, particularly those with large populations, must import food.

One consequence of the industrialization of agriculture, as well as the ready availability of jobs in the cities, is that the populations of many countries shifted from predominantly rural to predominantly urban. In the USA, at the end of the eighteenth century, only 5 percent Americans lived in cities, but by the end of the nineteenth century that had multiplied seven times over, so that 35 percent were living in urban areas. That trend continues, and today more than 75 percent of Americans live in or near a city. The lack of farm workers was no impediment to crop production – despite the migration away from rural areas, per-acre crop production grew more than sevenfold from the early nineteenth century to the mid-twentieth century.

In spite of often large national food surpluses, hunger in the poorer urban areas of industrialized nations still exists. However, famines, which were once common across the world, are virtually unheard of in industrialized nations. Today they occur in places where people are living in fragile environments, or when corrupt politicians need to apply leverage to their populations. But overall, the world produces enough to feed the world comfortably. The increase in food production (together with medical and public health advances) has been largely responsible for a significant increase in life expectancy in industrialized nations. Today, people live, on average, more than twice as long as they did in 1720, thanks in significant part to industrialized agriculture.

Box 20.5 The Great Divergence

As I said in Chapter 19, prior to the Renaissance and Enlightenment, Europe was a relatively unsophisticated place, and the standards of living were worse in Europe than in places like Delhi, Istanbul, Peking, and Cairo. But for about 150 years the economic development of Europe and North America far outpaced the rest of the world, and only now are some regions catching up. The embrace of the Industrial Revolution has largely been responsible for initiating this change in fortunes, which is rapidly accelerating with the breakneck advance of the information technology revolution.

Among countries that quickly adopted industry as a foundation of economic growth, wealth has increased steadily. But some countries resisted this change and they found themselves at a significant disadvantage.

One of the clearest examples of this divergence, and its consequences, is Japan. By the seventeenth century, Japan had been repeatedly visited by European explorers and traders (mostly Portuguese). They brought technological advances, such as firearms and advanced sailing ships. The Japanese found these foreign influences to be culturally destabilizing, particularly firearms. A poorly trained peasant with a matchlock musket was more than a match for a highly trained Samurai warrior, even one with a decade's worth of combat training. The nobility of Japan found this shift in power to be sufficiently corrupting that they passed a law in 1633 expelling all foreigners and forbidding any Japanese from going abroad. Western technologies, such as firearms, were similarly outlawed.

This effectively isolated Japan from outside influence for some 200 years. However, in 1853, an American expedition headed by Commodore Mathew Perry took a flotilla, including steam-powered side-wheel gunboats, to Japan. He used the threat of force to open up the country to foreign trade. The Japanese, who had outlawed firearms, had no ability to respond militarily, and this allowed Perry to force a treaty on the Japanese, opening up their country to trade with the United States on favorable terms to the Americans.

The Japanese isolationism, rather than strengthening them, had made them vulnerable. The immediate effect was the collapse of the Japanese government, to be replaced by one that took the country on a very different course. Over the next fifty years, Japan sent its citizens around the world to study at European and American universities, and to study Western industrial methods. By 1900, Japan was a completely industrialized nation, and in a complete surprise to the Western world soundly defeated Imperial Russia in the Russo-Japanese War (1904–1905). Notably, Japanese battleships had been demonstrably superior to the Russian warships in every way, including long-range armament,

Box 20.5 (cont)

gunsight optics, and radio communications. At the Battle of Tsushima, the Japanese utterly destroyed the Russian navy, the first time a non-Western military had completely beaten a Western military since the fifteenth century.

Japan had learned the importance of industrialization by the nineteenth century, and over the next hundred years most countries tried, with varying degrees of success, to follow suit. Today, the term "developing nation" largely refers to countries that have not yet successfully developed industrial-based economies. The most prominent example of a country that has recently changed its economic fortunes by adopting industrialization is China. Since the 1980s, China has rapidly adopted industrial manufacturing and export to international markets as the basis for its economic growth. By most economic measures, it has been extremely successful.

Economic Changes

As the factories churned out new, inexpensive products, new economic forces were at play. The factory owners became wealthy, in some cases vastly wealthy, and challenged older, established landowners for political power. At the other end of the spectrum, the workers acquired new economic and political power as well. To keep factories fully staffed, factory owners had to pay competitive wages, and hired men and women (and children, initially) in increasingly higher numbers. Often, poorer communities benefited substantially, and manufacturing employment became a path to the new middle class. But, in many cases, factory working conditions were dangerous and unhealthy, and workers had to struggle to ensure that they were safe and paid a fair rate. Many of the political struggles of the nineteenth and twentieth centuries were between the competing interests of the industrialist class and the working class (see Chapter 21 for a discussion of Marxism).

Populations in industrialized cities soared, as people fled rural life for the economic opportunities of factory work. At the beginning of the Industrial Revolution, more than 90 percent of people lived on farms in rural settings. By the end, in the twentieth century, most people in industrialized nations lived in or around cities.

The modern marketplace of goods competing for consumer interest also started with the Industrial Revolution. Factories only maximized profit when they operated at a big scale, so there was an economic incentive for industrialists to make sure their products were being purchased in large numbers. As part of this drive to generate consumer interest, advertisers attempted to persuade people to become regular consumers of newly developed luxury and convenience products.

Because of the need for engineers and scientists to drive the technology of production forward, education became a priority in industrialized countries. Many colleges, universities, and libraries were established during this time, often with the economic support of the new industrialists, as the world could now see the value of an educated populace. Illiteracy rates dropped and became relatively rare by the latter twentieth century in developed nations.

Although the Industrial Revolution started in England in the eighteenth century, various countries have moved through this transition at various times. Japan started in the late nineteenth century and rapidly caught up, becoming an industrial powerhouse by the mid-twentieth century (which enabled its military expansion prior to World War II). China is undergoing this transition right now, having started in the late 1980s. The economic power of China today comes in large part from its ability to manufacture products for export cheaply and efficiently. The Chinese economy has grown sevenfold since 1990 and has passed many other developed nations due to its adoption of industrial manufacturing, application of new technologies, and transfer of new technologies (sometimes illegally) from other industrialized nations. The citizens of China, many of whom work in factories, benefit economically.

One major downside to inexpensive production is, of course, the ecological cost. If it is cheap to burn fossil fuels, we will do so in large enough amounts to affect our climate (as we see today). And if the products themselves are not curated carefully, they can become pollutants themselves, plastic being a major source of ocean pollution. And, if the exploding population of potential workers is not attentive to these and other resultant risks, irreversible damage to society as we know it today is a distinct possibility.

The Industrial Revolution gave us our modern economic world, with the economic opportunities provided by an open marketplace full of products and choices, but also with the downsides of the struggle for workers to ensure fair treatment and the damage done to our environment.

Chapter 21: Economics

Of all of the topics in this book, this is the one that is most likely to pertain to your day-to-day life. If you listen to the radio, or watch TV, or read the news online, you will inevitably have heard journalists interviewing economists who discuss interest rates, the government debt, the stock markets, and the value of currencies. From all of this, you might have inferred that economics is all about money. It is true that money is part of economics, but economics is itself not about money.

This is because economics encompasses a vast array of topics, and is not even limited in scope to humans. In fact, scientists have taken economic approaches to almost all living things, from ants to chimpanzees. So what is economics? Economics, simply put, is the systematic study of choices. It is a way of understanding why things behave the way they do.

When biologists look at why elk are distributed across certain patches of a mountain range, an economic perspective would examine the costs of movement (in, say, calories), against the benefit of the food availability. Elk will normally take shorter routes to denser patches of food, and that is consistent with an economic perspective on behavior. Similarly, when a wolf decides whether or not to try and kill a young elk in a herd, she is judging the risk of being injured or killed by the mother elk against the benefit of the meat that young elk represents. Certain circumstances will change her calculations – the hungrier the wolf, the more important it is for her to eat, which may have an effect on her economic calculations of risk/reward.

Even ants have economics, and these are related to their genetics. Worker ants have no offspring, so they cannot directly pass on their genes; their genes are passed on indirectly by their queen (who is also their mother). When a worker sacrifices itself to make a bridge over water, it is an economically sound decision if it helps the queen pass on her genes, thereby indirectly passing on the genes of the worker.

In primates, particularly the social species such as baboons or chimpanzees, biologists use economic models to determine how decisions are made. When would a chimpanzee share a piece of meat? When it makes economic sense, in terms of increasing reproductive access for a male, or in terms of increasing the likelihood of survival of offspring for a female.

So, economics is the study of choices. Most specifically, economics attempts to explain the decisions made by an actor (animal, human, etc.) when faced with unlimited desires but limited choices. Economic models can be used to analyze an enormous variety of phenomena in the natural world, but by far the most common use is in understanding human behavior.

Scale

Economics can be studied at a variety of levels. At the level of government, we can see that countries are making decisions based on economic calculations of risk and reward. For example, in World War II, Nazi Germany invaded Poland and Czechoslovakia based on the theory that the reward was high (they acquired the assets of those countries) and the risk was low (no other countries would intervene). They were essentially correct. However, when they invaded the Union of Soviet Socialist Republics (USSR), North Africa, and France, their risk–reward calculations were off, and they consequently lost the war.

The same logic applies at the individual scale. People make decisions based on their perceptions of risks and rewards, and economists use their understandings of human behavior to try to figure out why they did what they did, and what they will do in the future.

Human Economics

Normally, modern economics attempts to understand the production, distribution, and consumption of goods and services. Some of the questions of economics: Why do we purchase and why do we save? Does a decision have a financial impact on ourselves and on broader society? What should a government do to ensure that people are healthy and able to meet their basic needs? What is the relationship between personal wealth and physical health? What causes economic recessions/depressions in societies? As you can see, the questions posed by economics range in scale from the gene (as in ants) to the behavior of billions of people across continents. This is why economics is so powerful.

One of the underlying premises of economics is that individuals are making choices. Normally, we tend to think of humans as acting rationally, making decisions that are appropriate to their specific circumstances. (Of course, what people may think of as a rational decision may in fact not actually be, potentially because of genetic predispositions to act on short-term gains at the cost of long-term losses – see Chapter 13 – or because of cultural beliefs.)

So economists try to understand why a rational actor would make a given decision in a given circumstance, and, in turn, how many millions of decisions, by millions of actors, affect the world.

A Brief Economic History of the World

Hunter-gatherers make economic decisions, as do shoppers in the mall. It is far too complex to try to understand every decision by every individual across the world. So in order to make sense of these economic decisions, we tend to frame them within systems, which have evolved over time.

Capitalism

Choice at a nongovernmental level is the key element of the *capitalist* system. In a modern economy, of course, we are not acting as hunter-gatherers simply exchanging fish for arrowheads. In the modern economic system, products must be manufactured from raw materials and then sold. So in a capitalist system, at every stage of the process, choice plays a key role. For example, in the manufacture of hats, the wool manufacturer offers wool at a particular price. The hat manufacturer may accept the price, or reject it and look elsewhere. Once wool is purchased, the workers in the factory may accept their jobs at the salary and working conditions offered, or they may quit and take jobs with better circumstances. Once the hats are manufactured, they must be trucked to the shops by drivers who either accept the trucking job or reject it based on the wages offered. At the shop, the sales people will also accept or reject their jobs. Finally, once the hats are on the shelf of the haberdashery, the consumer may choose to buy the hat (if it is offered at a good price and is made in the latest style), or they may reject it and buy hats from another manufacturer.

At every step along the way, the agents (people) are making decisions because they have choices. In a smoothly running capitalist system, these choices are freely made: the worker has plenty of jobs to choose from, and the manufacturer has plenty of workers to whom they can offer jobs. At the level of the consumer, there are many hats to choose from, and enough consumers that most manufacturers can thrive by offering good hats at reasonable prices.

Barter

Capitalism is the oldest economic system, and the simplest version is the *barter system*, in which individuals make exchanges based on what they have and what they need. This was the economic system that we inherited from our ape ancestors, and we still use it today. In this system, if you need something, you offer what you have or what you can do (your labor) in exchange. Capitalism, particularly in the form of the barter system, is considered the most natural economic system because it originated prior to the evolution of humans. You can see its continuance daily in, for example, children, who, when deciding to share, say, a bag of candy, will try to determine what they may gain, now or in the future.

Feudalism

Once humans formed complex agrarian societies, naturally the economic systems became more complex. The hierarchical nature of these societies, with a few powerful individuals at the top, and a much larger set of poorer and less powerful people below, is reflected in (and largely caused by) their economic system. This economic system is known as *feudalism*. Under this system, land became owned, and this owned land (private property) was consolidated by the powerful. But empty land is not productive unless there are people to work on it, so poor people were granted the right to live on the land and grow crops, but only under the condition that they give a proportion of their crops to the landowner.

This lifestyle could work out well for both the powerful landowners and the working farmers, if the harvest was good and the landowner didn't demand too much of the food grown. Alternatively, it could be very exploitative – the landowners typically controlled the armed forces and could enforce compliance to any decisions they made. And if the harvest was poor, and the landowners demanded a fixed amount of crops, there might not be enough left over for the farmers to feed themselves.

A critical element missing in this system was free choice at the individual level. Although an individual (peasant or serf) working the land could, in theory, make a choice to leave one landowner and look for a better deal elsewhere, in practice this was often not possible. Landowners would use various methods to prevent the peasants from having choices – colluding together on how much of the harvest to take as rent, or, in the case of pre-Soviet Russia, simply passing laws forbidding the serfs from moving away from one landowner without specifically granted permission. For thousands of years China had a similar system, where the peasants were legally tied to the land they worked and were not free to search for better conditions.

Feudalism is probably the longest-lasting complex economic system. It dates back to ancient Mesopotamia and Egypt, some 5,000 years ago, and persists in parts of the world today. Sharecropping was common in the American South until World War II, when industrialized agriculture made it no longer competitive. However, the power disparities made the system inherently exploitative, and there have been occasional peasant revolutions throughout history to protest the abuses of the peasantry by landowners.

None of these were successful in generating any long-term systematic changes because, over time, if new people were put in charge, exploitation would ultimately rear its head again. Broad-scale changes only occurred when large-scale transitions in politics, technology, and social organization occurred.

Mercantilism

From the perspective of a government, feudalism has pros and cons. In its favor, it tends to be relatively stable, assuming there are no large-scale climate changes (i.e., drought). Food will always be produced and people will always multiply. This provides a firm long-term economic foundation for a country. And, for thousands of years, barring an invasion by another country, the system was stable. Countries with bigger populations and bigger crop productions tended to be more powerful, as they could raise bigger armies. Competition tended to be between countries, who would occasionally invade their neighbors to expand their area, population, and crop production.

However, feudalism only ever shifts the wealth in a country from one part to the other. Generally, the wealth of the country would accumulate in the higher levels of the system – the landowners, who retained the right of choice. But there was a fixed amount of wealth in the system of a country, so in financial terms there was no growth in the economy. The leaders of some countries realized that they had excess production. A natural way to dispose of this excess is to trade it with other countries.

For the first time, there was an external source of wealth for a government. Some countries were able to generate enormous wealth by selling or trading away their agricultural products. The government controlled the exports, so, for the individual peasant, the result was much the same, although the introduction of additional wealth into the overall economic system might benefit the worker incidentally. In this system, the aspect of choice is still relatively constrained – crops are not sold on an open market because the government controls all trade and makes the choices.

This system is known as *mercantilism*, and it appears when governmental power is strong enough that it can control the landowners and regulate the production and trade of the country. Although mercantilism is associated with the European Renaissance and the expansion of European trade in the sixteenth century, it has appeared multiple times, whenever power is consolidated and central authority is strong. The Mughal Empire of India (1562–1857), the Hand Dynasty of China (202 BCE–220 CE), and the Byzantine Empire (330–1453) all established mercantile systems.

Under this system, in which governments control trade, there are economic incentives to control land that produces trade. The European period of mercantilism is when we see colonies established by many European countries. Advances in sailing technology had made long-distance ocean trade practical, so controlling far-off lands that produced resources for trade to other countries became realistic for the first time. This is the period when we see the appearance of special government-regulated monopolies that were, essentially, extensions of governmental policy. There were many: the Dutch East India Company, British East India Company, French East India Company, Swedish East India Company, South Sea Company, Sierra Leone Company, German New Guinea Company, etc. In all of these organizations, the government gave a monopoly to a company for the purposes of establishing a colony. The output of the company was to be regulated by, and directly benefit, the government of the country (notably Massachusetts and Virginia were founded this way), but the general manner and method of generating that output was left to the companies.

Mercantilism inherited from feudalism the idea that economics is a zero-sum game: for someone to win someone else had to lose. A country would become rich at the expense of another country becoming poor. This justified the top-down control of colonialism – for the colonizers to become wealthy the colonized would have to be kept poor. But also within each country, for a business to become successful the workers would have to be kept poor (and relatively choiceless), since there was a fixed amount of wealth. This philosophy and economic model was explicitly used to keep the lower classes from accruing wealth.

The Free Market

Ultimately, the exploitative aspects of mercantilism were its undoing. Perhaps the most famous rejection of the government monopoly on trade in the United States was the Boston Tea Party in 1773. The government-issued charters of the British colonies on the eastern side of North America imposed strict rules on the importation amounts and taxes imposed on tea (which had been produced in another British colony in India). The colonial citizens were expected to buy a certain amount of tea and to pay a certain tax. Mercantile rules meant that the citizens of Massachusetts were not permitted to purchase tea from any other source. This lack of free choice at the level of the individual market participant in the tea marketplace, and resentment at the taxes, is what inspired some of the rowdier citizens of Boston to storm aboard a tea-trading ship and toss the tea overboard.

Some of the impetus for the change in economic models came from the highly influential economic theorist Adam Smith (1723–1790), who argued that mercantilism was constraining economic growth. He proposed (Box 21.1)

Box 21.1 Incentives

Adam Smith's model of free market capitalism revolves around two main ideas. The first, discussed in the main text, is choice. Choice is critical, since it is by presenting alternatives that individuals can make the rational decision that best suits their circumstances. However, another of his ideas is the importance of incentives. In a free market economy a person is driven to act by incentives (see Box 21.2).

The role of choice in the free market is clear, but the idea of incentives is somewhat more controversial. Naturally, businesses are incentivized to make decisions that will make more profit, just as individual workers have the incentive to search for the job that gives them the best wages and working conditions.

However, incentives, particularly for individuals, can be quite harsh. Under some circumstances, the incentive to accept a job may mean the difference between eating and starving. Under those circumstances, unless there are as many acceptable jobs as there are employees to fill them, the incentive favors the employer, who can then offer employment terms favorable to the employer's priorities; this might mean very low wages and dangerous working conditions (mine towns, with a single, large employer, could be expected to provide these circumstances). Alternatively, if there is a strong social welfare system, people may not be incentivized to work at all. If you can get as much money from welfare as you would in a job, there is no incentive to take a job, even if there are sufficient jobs for full employment.

Societies have to perform a balancing act: maintain incentives and capacity for people to work (since the overall economy needs people to be working and building products and services), while protecting workers who are vulnerable to economic forces outside their control. For example, if a large factory lays off thousands of workers there may not be jobs for them anywhere in that town, and people need time to try to sell their homes, move their families, find new jobs, etc.

Some economists and politicians are content to let those people fend for themselves, arguing that, knowing of their vulnerability, they should have prepared in advance (e.g., obtaining higher education), and would anyway be highly incentivized to quickly find other work. The opposing viewpoint is that there are vulnerable segments of the population who are unable (through physical, mental, or economic disadvantage) to understand or undertake the advance preparation required to protect themselves. This debate has sometimes been framed in moral terms: Are workers on unemployment benefit *lazy*? Is cutting off unemployment benefit *cruel*? Politicians (and media handlers) are apt to try to inflame public opinion using this technique.

Box 21.1 (cont)

One idea that has been floated by economists is *universal basic income*. This is a government program that ensures that everybody receives a minimum amount of money, even if unemployed. This would, without a doubt, help people fend off personal economic catastrophe when unemployed. But it would also remove a critical incentive for people to find work. From 2015 to 2018 Finland ran a basic income experiment, but the results are still open for debate; currently a few other societies are experimenting with it.

Today, when you see political arguments over welfare, food stamps, and the social safety net, you are seeing this debate play out. You can be sure that your grandchildren will see the same debates.

Box 21.2 The mathematics of incentives

Economists often look for ways to quantify or graphically demonstrate economic patterns. This graph is known as a *Lorenz curve* (Figure 21.1). In it, there are actually three different economic patterns, but they are shown on one figure so you can compare them. The basic idea of the graph is to show how much of a country's wealth is earned by its citizens, ranging from the poorest to the richest. The green diagonal line, at a 45 degree angle, shows a healthy economy, where the poor (on the left) earn less money than the wealthy on the right. The reason this is considered healthy is because there is clear economic incentive. If you work your way into a better income bracket, you move up and have a substantially improved economic situation. Similarly, you don't want to slide to the left, so there is a negative incentive not to avoid working.

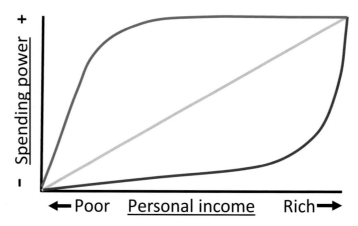

Figure 21.1 Lorenz curve.

Box 21.2 (cont)

But look at the red line. Here, moving to the right actually doesn't get you much more spending power. So you have a much harder time improving your economic circumstances. You have to get far to the right to make things better for yourself, and in some economies this is probably not possible. This is the curve you would see in a feudal or mercantile economy, where there is, in effect, no large middle class. In this economy, there is no opportunity for most to improve their lives materially, and very little incentive for harder work or innovation. In some cases, a high level of inequality can lead to political instability.

The blue line is the opposite. Here, almost everybody in the population has the same (or similar) economic circumstances. This is the theoretical distribution under Marxism. But, again, there is limited incentive for innovation and hard work. By moving yourself to the right, you don't get much in terms of material improvement. This economic distribution leads to economic stagnation and inefficiency, which was characteristic of communist governments.

One way of quantifying this curve is the *Gini coefficient*, and it can be used to make comparisons across countries or to track trends in inequality across time.

Box 21.3 Smith and Marx

It is impossible to intelligently discuss economics (and, indeed, modern history) without a knowledge of Adam Smith and Karl Marx. These two thinkers have had a far-outsized influence on the world, extending beyond discussions of economics into such realms as politics, warfare, human nature, mathematics, and morality. Debates over public policy incorporate their ideas, and politicians frequently borrow their concepts (often ignorant of their origins).

Both were concerned with the conditions of society, as well as the individual worker, and both thought their models would produce the best possible outcome for the economy at large. So, in an important sense, their ideas were moral arguments. The free market, completely without government interference, is sometimes characterized as a place where the cruel laws of nature apply and the weak are naturally consumed by the strong. But Adam Smith was very concerned with poverty. He saw the harsh conditions imposed by feudalism and mercantilism on the poor and was appalled. He observed that labor was forbidden, by law, to organize and negotiate for better wages,

Box 21.3 (cont)

whereas employers were permitted to collude. For him, this imbalance was immoral as well as economically short-sighted.

Marx had the advantage of living a hundred years after Smith and actually observing the ways in which the Industrial Revolution played out. Because of the role Soviet-style communism played in international affairs, Marx is frequently mischaracterized for political purposes, but his goals were much the same as Smith's. He wanted people to maximize their economic opportunities and achieve personal happiness. The problem is that satisfaction is defined very differently for different people, and what provides happiness for one may provide misery for another. Accommodating those differences is the great challenge of governments, legal systems, economists, and philosophers (including Smith and Marx).

Communism in the USSR and elsewhere was harsh and severely constrained personal freedoms (see Box 21.3), but Marx never would have envisioned such a system (he died before the Russian Revolution). He was a strong advocate of open democracy, voting rights for all, personal liberty, and freedom of the press, and he disdained regulations of personal affairs by the government.

Marx had two things that have worked against him. First, he tended to use bombastic language that could be easily (and often intentionally) misinterpreted. For example, he spoke of the "dictatorship of the proletariat [working class]." This doesn't sound very democratic, but what he meant was that, in a democracy, if the working class could vote, since they were the numerical majority, they would effectively drive governmental policy.

The second strike against Marx is that he turned out to be incorrect about his predictions for the future. He admired capitalism for its efficiency and dynamism, but he thought that it was unstable. He noted the boom-and-bust cycles of many economies, which were due to the attempt of investors (for him "capital") to exploit any imbalance in an economy to acquire wealth. For example, if there was a demand for housing, investors would greedily overinvest in housing production, such that the overproduction would often collapse a local economy. Further, automation (which started with the Industrial Revolution) would gradually eliminate the need for workers, resulting in mass unemployment and social instability. Finally, as corporations competed they would become so efficient at production, and be forced to push prices so low to retain the consumer, that they would operate on an ever-reducing profit margin. This final element, he argued, would result in the ultimate collapse of capitalism, which would have to give way to a subsequent economic form: communism.

Box 21.3 (cont)

This was, of course, incorrect (so far). Capitalists have figured out ways to innovate, and found new types of products, so that there is always a market for the products of industry, at prices that keep the industries profitable.

But the biggest problem with Marx is that his ideas were (ostensibly) used as the inspiration for the foundation of the Soviet Union, with all its repression, inefficiency, and ultimately, failure. So Marx is thrown into the dustbin of history with the communists. But many of his analyses of the power relationship between the workers and factory owners stand, as does his perception that industrialists generate demand (ask any advertiser). When some absurd new product is hyped, and you scoff, you can be sure that Marx saw it coming.

that economies could not reach their true potential until government monopolies (the mercantile companies) were revoked. He argued that a true "open market," in which all individuals were free to make choices about economic decisions, would result in an equilibrium that would maximize wealth and political power for all. One of the critical ideas was his idea of competition: when two companies were permitted to compete, each would have to improve, either by producing better products or selling them at a lower price. When companies had to compete for workers, they would have to offer better wages or working conditions. The ultimate outcome of all of this competition would be a marketplace with the best possible combination of quality goods at low prices, with well-compensated workers.

In some ways, the free market is the natural outcome of the Industrial Revolution. With many companies producing the same wares, and the factories competing for the best labor, there is a wide choice of products from which consumers can choose, and it is easier (theoretically) for workers to search out the best wages and working conditions. In a feudal or mercantile economy, economic circumstances within a country are not as varied, and opportunities for individuals to find the best products or working conditions are much more limited.

The economic theories of Adam Smith appeared in the middle of the Enlightenment, and the exploitation of the poor by the mercantile system was inconsistent with the humanist impulses expressed during this period. Unlike the mercantilists, who were primarily focused on maximizing their own wealth through control and limitation of the choices available to individual market participants, Smith advocated the free market because it would benefit all society. Today, his ideas are still those expressed most often in the developed nations of the world, and, in general (although with some important caveats), the freer systems tend to produce the best results for all citizens.

Marxism

Despite expanding the economic opportunities for many people, the free market could also be exploitative. The twelve-hour work day, child labor, and injury or death on the factory floor or mine were common. Workers could be fired for asking for better conditions or wages. Adam Smith's ideals of free choice did not appear to apply to many workers because there were so many people looking for employment that the individual workers had no leverage. If a worker protested their wages or was injured, they would be fired and replaced in hours. In many towns, only few factories or mines provided all the employment (especially in mine towns), so a worker could not readily look elsewhere for a better job. Further, factory owners would collude among themselves to keep wages down, rather than competing for workers by providing competitive salaries. Sometimes workers were paid in "company script" that was only good at the company-owned store, preventing workers from accruing any real capital of their own.

This imbalance of power led some thinkers to question Adam Smith's model of the free market. Karl Marx (1818–1883) argued that there was such an imbalance of power between the factory owner and the individual worker that, in actuality, there was no real element of choice for a worker. On the production side of the marketplace, Marx argued that workers would never be able to negotiate successfully with factory owners because in a purely capitalist system the factory owners, who viewed workers as a commodity much like coal or iron or wheat, could find replacement labor, assuming that they were always permitted to collude and were not forced to compete, as envisioned by Smith.

At the consumption end of the marketplace, Marx was equally critical. He viewed the competition for market share created by the production of new products as artificial. For Marx, the factory owners were not producing important products so much as creating demand for things that were not truly necessary. The factory owners (and advertisers) were manipulating buyers (often the workers themselves) into becoming consumers; they were, in effect, deceptively generating demand rather than simply producing things the people truly needed. So in his view, the choice of the marketplace was an illusion created by deceptive corporations, and that deception fueled profits.

For Marx, the issue of classes in conflict was central. The wealthy classes were in natural conflict with the working classes, and the struggle between the two classes was the defining element of capitalism. Because the working classes were less educated, and with less available capital, they were forced to work toward short-term economic self-interest. One of his main goals was to "enlighten" the working classes to their plight, and teach them how to best work for their own long-term interests.

The arguments Marx put forth found fertile ground in the industrial world of the late nineteenth century. The Industrial Revolution had created a new class of decadent super-wealthy industrialists (the "robber barons") who had enormous economic and political leverage. Simultaneously, the exploitation of workers by these powerful factory and mine owners was obviously generating enormous suffering. Groups of factory and mine workers attempted to organize into labor unions, so that they could exert some leverage on factory and mine owners to extract concessions on working conditions and wages. In many cases, the industrialists leveraged their political power to have laws against labor unions passed, and to suppress the economic and political power of industrial workers (such collusion was contrary to Smith's model).

The cause of organized labor was widely expressed across the industrialized nations of the world, and was often articulated as the type of class conflict Marx envisioned. In most nations, over time, the new-found power of organized labor was used to pressure governments into passing a series of labor reforms: eight-hour workdays, safety reforms, child labor laws, explicitly giving legal recognition to organized labor unions, minimum wages, two-day weekends, etc.

But these reforms fell well short of Marx's ideals, which called for an entirely new economic system (communism/socialism). The ultimate expression of these radical social changes occurred in Russia in 1917, when the monarchy of Russia was overthrown, the royal family killed, and, after a six-year-long war, a new government installed, theoretically founded on the principles of Marx's ideas.

This new country, known as the USSR, was based, economically, on a completely top-down economic system, in which no corporations existed and in which the individual had no economic or political choice. All factories, mines, and farms were owned by the government. The government decided how much to produce, how to distribute the results of production, and who would be employed at any given job.

The result was a highly authoritarian government with a highly inefficient economy, and ultimately communism collapsed. The reasons for the failure of Soviet communism are explored in more detail in Box 21.5, but it persisted long enough (for most of the twentieth century) for it to present a competing economic model to the rest of the world. Many of the post-World War II struggles were between the "Western" industrial countries that embraced capitalism (western Europe, the Americas, Australia, New Zealand, South Africa, and a few others) and the communist countries: initially only the

USSR, but later, China, North Korea, Cuba, Vietnam, Ethiopia, Somalia, Angola, Laos, Yugoslavia, Poland, Czechoslovakia, and half of Germany, among others. Of those, only a few cling to communism today: Cuba, Vietnam, and China, largely in name only, with North Korea the last country maintaining a true communist state.

Among historians and economists, communism is regarded as one of history's grand experiments, and the world has moved on. But many of the reforms and attitudes associated with the nineteenth-century reform movements have been retained. Today we expect employers to ensure worker safety and provide fair wages, and we expect the government to have the right to intervene when workers are mistreated. The creation of social safety nets for poorer workers, who were unable to save on their own, is also an expectation of governments: social security and health care for the poor or elderly. So, even while Marxism was a failed experiment, many of the reforms associated with the expansion of labor rights come from those same political forces.

Box 21.4 The economics of slavery

Slavery has been around as long as civilization itself – probably much longer. From an economic perspective, slavery makes sense: work is accomplished at the relatively low cost of simply keeping the slaves fed and housed. Many economies throughout history have been based on slavery. One of the more famous was the economy of ancient Rome, in which slaves were captured by military campaigns and taken to Rome to become economic assets. Ancient China also had a largely slave-based economy during the Yuan and Qing Dynasties. The Mongolians used large numbers of slaves during their conquest of Asia. Slaves were an important part of the Ottoman Empire.

In the United States, work by economic historians has demonstrated how slavery was a critical element in the expansion of American economic power during the eighteenth and nineteenth centuries. Slavery-based enterprises were the foundations of the fortunes that helped build Harvard, Brown, and Yale Universities, among others. Modern corporations, like Lehmen Brothers, JP Morgan Chase, Aetna, and Norfolk Southern Railroad are just a few examples from a long list of the thousands of corporations that invested in, insured, or used slaves in the eighteenth and nineteenth centuries. It is fair to say that slavery was one of the most significant factors that drove the newly powerful American economy. And although it is sometimes said that relatively few people and corporations actually owned slaves, it is more

---

Box 21.4 (cont)

accurate to note that thousands of corporations were involved in the slave trade, and that, in the Confederate states, the 3.7 million slaves (as of the 1860 census) represented 36 percent of the population. By any possible measure, slavery was pervasive throughout the American economy, both in the North and the South, and was a major factor in the wealth of the new nation.

However, it was a remarkable expression of Enlightenment principles that so many corporations and people were ultimately willing to work against their economic self-interest in the name of morality (albeit belatedly). The Confederate states (having more to lose than the northern states) were unable to put morality above economic self-interest and shared much with the pre-Enlightenment world. Perhaps it was too much to expect from an agrarian society that had otherwise resisted the advances of industry, modernity, and even the Enlightenment generally.

Box 21.5 The failure of Soviet communism

When I was growing up, and over the lifetimes of my parents and grandparents, communism loomed large in the public consciousness. It may be hard to imagine now, but for most of the twentieth century there was an alternative to the market economy we share today.

In 1917, the Russian monarchy was overthrown and the royal family was killed. Following a six-year civil war, a new country emerged, founded (supposedly) on the economic principles of Karl Marx: the USSR. From 1922 to 1991 the USSR was one of the most powerful countries in the world, and supported communist revolutions and governments around the globe. During World War II the USSR almost single-handedly drove the Nazi military machine from the gates of Moscow back to Berlin. This country was technologically advanced and economically powerful – it was the first to put a satellite into space, and the first to send a human into orbit around the Earth. The USSR was also the most visible and powerful opponent to the industrialized countries of western Europe and the Americas. Then, in 1991, the country suddenly collapsed, taking communism with it.

Why Did This System Fail?

Under Soviet communism, the government controlled all economic production and distribution (this is known as a *command economy*). This

Box 21.5 (cont)

meant that there were no privately owned factories, farms, or mines – it was all owned by the new Soviet government. The government also decided the allocation of production – wheat, corn, bacon, shoes, wristwatches, antibiotics, underwear, pencils, frying pans, automobiles, vacuum cleaners, houses, firewood, and virtually every other material or object a citizen would need. All planning was in the hands of a centralized planning authority, rather than in the hands of individual and corporate market participants.

On the employment end, the government set the priorities for the economy, and obviously needed a great many factory, mine, and farm workers. This meant that people might not have a choice of careers. You may have preferred to become, say, a poet, but if the central government already had enough poets, and needed a drill press operator, you might well find yourself pulling that lever on the factory floor all day. Naturally, this meant that there were quite a few dissatisfied workers.

But there were two economic aspects of the system that were critical shortcomings. One was the issue of incentives (Box 21.1). In a communist system, that element, identified by Smith as essential to motivate workers, was largely reduced or missing. Under communism, people all received food, shelter, medical care, and other material needs, irrespective of their ranks or positions. Workers might not feel the need to excel, because there would be no benefit. In a capitalist system, good workers receive promotions, bonuses, and other incentives, but in a communist system these are rare.

Under Soviet communism, workers were expected to try to excel for philosophical reasons – because they believed in the communist system and wanted to help the country and their fellow citizens grow in economic strength. But most workers, it turns out, are not thinking along philosophical lines day to day. They are more easily motivated by specific monetary incentives, but those were largely missing in the communist system. Further, the disincentive that is present in normal capitalist employment – the threat of being fired and subsequently poor – was largely missing as well, since the state met all the basic needs of every citizen. This lack of incentives meant that workers did the rational thing – the minimum work to avoid being punished. The result was that production of most goods was extremely inefficient – both in industry and in farming. Even though the USSR had some of the largest wheat fields in the world, it was compelled to import grain from other countries during most of its existence.

Box 21.5 (cont)

Second, in a command economy the overall system is poor at responding to the effects of things like drought or long winters. As outlined in Box 15.3, chaos theory predicts that in a nation's economy, with millions or billions of different elements, it is impossible to make an accurate prediction of the outcome just by manipulating a few factors. Repeatedly, over the history of the Soviet Union, the central government misallocated resources and made unrealistic demands of the agricultural system that resulted in famines, with tens of millions starving to death (Figure 21.2). And the authoritarian nature of the communist system meant that when production could not be met, the farm workers would be denied access to the very grain they were growing, partly as punishment for failing to grow more. There were also significant shortages of industrial products, repeatedly, although those shortages had nowhere near the impact of the famines.

Social Reasons

The constant failures of the economy to provide the basic needs of society meant that there was significant resentment within the population. To maintain control, the Soviet government had a powerful internal security apparatus. Free speech was nonexistent and dissenters were either killed or sent to prison camps (*gulags*) in distant and frigid Siberia. Leaving the Soviet Union without specifically granted permission was illegal, and people were imprisoned or shot trying to escape to western Europe. Part of the problem may have been that communism started in Russia. Russia, under the Czars, already had a strong tradition of suppressing dissent, and the living conditions of the serfs in pre-Soviet Russia were probably the worst in Europe. There was no strong tradition of personal liberty or freedom of expression in Russia (as was present, say, in France and England), and the population was the most poorly educated in Europe at the outset of the Russian Revolution.

Conditions in the Soviet Union did improve over the twentieth century (especially after the 1953 death of Joseph Stalin, who was famously iron-fisted), and education levels did rise to near-parity with western Europe and the USA. And, as noted above, the USSR did produce some incredible feats of science and engineering, such as the Soviet space program. However, the inefficiencies still meant that, for the average citizen, material conditions never matched those of western Europe. Coupled with the restraint of personal freedoms, the average Soviet citizen was well behind those in the West.

Box 21.5 (cont)

Figure 21.2 Russia swung from one end of the Lorenz curve to the other. Under the prerevolutionary economic model, there was dramatic imbalance between rich and poor. In the top two images we see Russian Burlaks pulling boats up the Volga River. On the bottom we see the product of the communist system, with the Soviet Famine of 1932–1933 causing the deaths of some 4 million Soviet citizens. This was a direct result of the Soviet government demanding specified amounts of grain from the farms, which simply couldn't produce enough. The grain was all shipped to the large cities, and the farm workers were denied access to the grain they had produced. In fact, several hundred thousand were prosecuted for picking through the harvested fields (gleaning) looking for scraps of edible grain. In the Ukraine, the famine is known as the Holomodor.

Box 21.5 (cont)

Figure 21.3 Between 1989 and 1991, communism collapsed in eastern Europe (East Germany, Poland, Czechoslovakia, Yugoslavia, Romania, and the Soviet Union, notably). In those countries, the militaries refused to fire on the protesting citizens. In the left-hand image, Russian citizens on this tank joined with the soldiers in celebrating the end of communism.

At roughly the same time, Chinese citizens in Beijing protested, calling for reforms to Chinese communist rule. Here a protester is standing in front of a column of tanks that have been called in to break up protests in Tiananmen Square, Beijing. There was hope the military might not support a crackdown on the civilians, and that the government might peacefully shift toward democratic reforms. Unfortunately, a few days later the military fired on protesters in the square, and all protests against the communist government were crushed. The fate of the man in this photo is not known.

By the end of the twentieth century, it was impossible to justify maintaining the Soviet-style system. Citizens started to rise up in the Soviet Union, as well as in countries under Soviet control. Given the alternative of a brutal military crackdown on civilians (which had been used multiple times since the founding of the USSR), the communist leaders, remarkably, allowed a relatively peaceful overthrow of communism (Figure 21.3).

Box 21.6 China

Although the Soviet Union collapsed in 1991, China, which had adopted communism after World War II, has not had an overthrow of the government. Looking at the differences between China and the Soviet Union provides a useful contrast that helps illustrate several of the main flaws with communism.

The founding of communist China is a somewhat complex story, but China, through World War II, was essentially preindustrial, and had been under partial colonial control of the British and then the Japanese. Following World War II, a civil war broke out, with the communists establishing control over the

Box 21.6 (cont)

mainland part of the country (but not Hong Kong or Taiwan) in 1949. The Soviet Union supported the foundation of communist China, and provided advisors and technical assistance, including the establishment of modern factories.

From the 1950s through the 1970s China suffered many of the same problems as the Soviet Union, with large-scale famines (and deaths in the tens of millions) and harsh repression of dissenting citizens (as well as anyone remotely associated with Western ideas). However, with the death of its harsh leader Mao Zedong in 1976, China slowly oriented itself toward the West. Unlike the Soviet Union (which, to the end, had firmly resisted entangling itself with the Western economy), in the 1980s China started opening its economy, internally and to the world. With its relatively cheap labor and reliable political stability, over the next thirty years the country became the preferred manufacturing center for Western corporations. Today, the Chinese economy is second in size only to the USA and citizens there have enjoyed a substantial improvement in material conditions and choices.

However, the role of personal liberties is still an open question. Chinese citizens who protest against the government are regularly imprisoned, and there is neither freedom of speech nor freedom from investigation. China leads the world in the application of facial-recognition technologies, which are used to enforce compliance. Social scientists had long associated economic freedoms with personal freedoms, and in general, countries with strong economies are those with the most civil liberties (strong economies require educated citizens, who may not accept government regulation of their liberties). However, China may be the exception, with an open and extremely powerful economy but strict constraints on personal liberties. Whether the Chinese people will accept this state of affairs over the long term is an open and interesting question.

Box 21.7 The politics of communism in America

During the nineteenth and early twentieth centuries, the Industrial Revolution became a major economic force in the United States. As in the UK, factories and mines became major employers of large numbers of relatively unskilled and poor workers. Many of the same problems with working conditions and poor wages were present in the USA, and the response was the same: organize labor for a better negotiating position.

Box 21.7 (cont)

However, the appearance of Marxism, with its call for the "dictatorship of the proletariat" and the elimination of private property, became a powerful symbol in the United States. There was no serious attempt to institute communism in the USA (although a few radicals did call for it), but labeling someone a communist became an easy way to turn them into a target for persecution. During the middle of the twentieth century, the Federal Bureau of Investigation (FBI), under J. Edgar Hoover, regularly persecuted groups it suspected of communism. After World War II, when the Soviet Union expressed clear international political opposition to the Western democracies, anti-communist sentiment swept the nation.

Since labor unions expressed some of the same sentiments as Marxists, they became ready targets. The most famous expression of this persecution was "McCarthyism," named after the Wisconsin senator Joseph McCarthy. McCarthy instituted a series of public hearings (1953–1954) to flush out people suspected of sympathy with communism. To achieve maximum visibility, he notably pursued public figures, including academics, writers, musicians, and movie actors, writers, and directors, but he also had teams pursue low-level workers involved in organized labor (he also pursued homosexuals, on the grounds that they were, like suspected communists, "un-American").

Although McCarthy finally overstepped the mark, and was censured by the US Senate, the persecution of communists by the FBI and other interests continued for more than twenty years. Soviet communist philosophy articulated opposition to racism (as part of its class struggle), and one way to discredit the civil rights movement in the USA was to accuse it of affiliation with Soviet communism. The FBI infiltrated groups like the National Association for the Advancement of Colored People and attempted to associate it with communism. Even religious groups entered the fray, especially Southern Baptists. In a 1965 sermon, Baptist minister Jerry Falwell accused Martin Luther King, Jr. of being a communist "subversive."

Accusations of communism are still regularly part of social discourse in the USA, even though it has not been an international political threat for almost thirty years. One of the most recent was probably the debate over the Affordable Care Act, in which support for the act was compared to communism and socialism. Whether the presence of universal health care truly does act as a disincentive to an efficient economy is an open question, but the role of the accusations was clear enough.

Box 21.8 Managing a modern economy

Today, since the world has rejected the top-down command-style economy of communism, governments use just a few tools to regulate the economy, and encourage it to grow, while also making sure it is operating in a way that is "fair."

Monopolies

Since competition and choice are key elements of the free market model, governments try to ensure that no corporation can so dominate the marketplace that it effectively prevents competition. There have been numerous times in US history that the government has broken up a corporation because it was too large and pervasive. One of the most famous was the US telephone system, which was effectively run by just a single corporation for almost a century: Bell Telephone. The US Justice Department broke up the corporation in 1984. Standard Oil, which controlled 80 percent of US oil production, was dismantled in 1911.

Today, there are multiple technology corporations that have been accused of monopolistic practices. Microsoft was sued by the European Union in 2007, resulting in a large financial judgment and enforced changes in software bundling (but not the dissolution of the corporation). Google, Amazon, and others have been, and are currently being, similarly scrutinized.

Import Taxes

If the industry of a country is not able to compete against imported goods, one way to protect that industry is to impose import taxes (duties) on the imported goods, effectively raising the price of the imported goods. This is a common practice, even though it goes against the principles of the free market. For instance, the USA imposes duties on imported steel, sugar, and vehicles to help US corporations compete. Other countries have done the same, and often tariffs are a response to tariffs imposed by other countries on their products.

One famous example is the so-called "chicken tax." In the 1960s, France and Germany imposed a tariff on imported chicken meat to protect their domestic producers against the competition of the efficient industrialized US chicken producers. In response, US President Lyndon Johnson imposed a 25 percent tax on imported light trucks, which, at the time, were coming into the USA from France and Germany. France and Germany are no longer exporters of light trucks to the USA, but the tax remains so that, nowadays, US consumers pay a 25 percent tax on light trucks coming from Japan.

In that case, as in others, a tariff really constitutes a tax on domestic consumers, through the increased cost of the goods subject to the tariff.

> **Box 21.8** (cont)
>
> Governmental protection can also take the form of subsidies, loans, tax benefits, and other generally more beneficial means of favoring domestic industries, with less direct impact on the consumer.
>
> Monetary Policy
>
> One way that modern governments can stimulate economic growth is to put money into the economy to stimulate consumption. Normally, a country wants to see an economy with steady but slow growth – too slow leads to stagnation, with business slowing or closing, and jobs lost; too fast leads to the "boom-and-bust" cycle, with prices on things like homes skyrocketing then collapsing.
>
> One way a government can control the pace of economic growth is to make sure that people can access loans for big things like business growth or home mortgages. The government can do this by setting the interest rate on loans and making loans itself. A high interest rate tends to discourage people from borrowing money, while a low interest rate encourages borrowing. This interest rate acts as a throttle on the economy, and it is always a balancing act to try to keep the economy moving, even as external factors such as technological innovation, droughts, foreign competition, and wars can have positive and negative effects on the economy.

Today

Today, capitalism is widespread. This is not because capitalism is not without flaws – the cycle of boom and bust is a constant threat, and the competitive tension between labor and ownership for control of choice in the marketplace continues. The role of monopolies is debated and is a threat to free competition. Political questions about how much to intervene in the free market are common: Should tariffs be imposed on imported goods? How much should the minimum wage be? Are mine owners liable for black lung disease in workers? Should public transportation become mandatory in densely populated regions? Who is responsible for environmental clean-up in mining towns? Is it legal for employers to collude in setting salaries? Should state governments have to negotiate with unions over salaries?

Because so many of these questions are outstanding, we acknowledge that a free market capitalist system is not perfect, and that it must be constantly tweaked to be fair. However, the only real alternative available (the Chinese mix of communism and capitalism) is so recent a development that capitalism is, in a sense, the last system standing. And one thing is clear: since the Industrial Revolution, the free market system has generated the most material wealth for the most people. An average worker today is, in material terms, wealthier and healthier than at any time in history. That is a significant accomplishment, considering that the change has largely occurred in the last 250 years, after almost five millennia of nearly unchanging worldwide feudalism.

Chapter 22: Globalism

Money and Power

You have probably heard the expression: "money is power." This is a fairly direct expression of a truth that applies at all levels of an economy. It applies to individuals, since wealthy people live as long as two decades longer than poor people, and are able to afford things like better housing, clothing, and medical care, as well as lawyers, accountants, and security guards to protect them from legal, financial, or physical threats. But it also applies to countries – wealthy countries are more powerful than poor countries, since they can afford stronger militaries with better weapons and larger numbers of soldiers. Much like individuals, countries strive to acquire greater wealth because of the prestige and power it gives to their leaders, as well as the advanced public goods provided to its citizenry.

As explained in Chapter 21, countries cannot truly increase their net wealth internally. If a country is a closed economic system, the wealth potential is fixed by dint of its population and natural resources. To truly grow economically, a country must reach out externally and engage in commerce across international borders. There are two basic ways in which this happens, and most countries today engage in a combination of both.

Production and Export

To acquire money (financial capital), a country can export something it produces. This can be something it grows, such as wheat, tea, or opium, something it mines from the ground, as in oil, coal, or diamonds, or something it manufactures, for example cars, watches, or cellular telephones. This was the basis of the mercantile system, in which countries would sell a product and increase the net worth of the country by acquiring money worth more than cost of producing the product. This was common across history – the ancient Chinese, Roman, Indians, and Ethiopians, among many others, engaged in the export of various goods and this has continued through history, since various resources and products are grown, mined, or manufactured in specific geographic locales around the world. Archaeologists and historians have evidence of long-distance trade between the Han Dynasty and the Roman Empire, a distance of some 10,000 km/6,400 miles over land, some 2,000 years ago.

Trade

However, another source of wealth is the trading itself. Some nations have become enormously wealthy by facilitating the movement of goods across borders and over long distances. There have been complex trade networks throughout history, but one of the most famous early examples of this type of economy is found on the famous "Silk Road." Trade between China and Europe passed (going east to west) south of the Gobi Desert, north of the Tibetan plateau, around the Taklamakan Desert, through modern-day Afghanistan, then south of the Caspian Sea by way of Iran, Iraq, and Turkey. Large, powerful city-states grew up along this route, providing food and rest as well as military protection against the desert raiders of Central Asia. In return, they imposed a tax on goods passing through. The Silk Road route persisted for more than 2,000 years, generating enormous wealth for the region, and some of the cities were legendary for their opulence. The trade routes only declined after the sixteenth century, when a variety of factors like political instability and the advent of large ocean-going trading vessels made land transportation dangerous and uneconomical.

Several technological improvements, such as the astrolabe (developed to practical utility during the Islamic Golden Age) and the magnetic compass (imported from China sometime in the twelfth century), were combined with European shipbuilding advances to make long-distance open-ocean trading practical for the first time by the fifteenth and sixteenth centuries. Some European countries took to trading with vigor, notably the British and Dutch, and much of their later wealth and power derived from their large trading and naval fleets. Early European trade was part of the mercantile system of the European powers: colonies of these countries would feed the trading routes, and, in many cases, wealth was generated by having colonies purchase (at fixed prices) the products of other colonies of that same country (see Chapter 21 for more details of the mercantile system).

Is Trading Always an Exchange of "Choice"?

The imbalance of military power between countries means that trading is not always done by the consent of both parties. This is particularly the case when one country effectively controls another. In ancient Persia under Darius, the Persian Empire controlled Egypt, Syria, Turkey, Mesopotamia, and parts of India. Darius expected these new areas to produce wealth for his empire, and built a 1,600-mile "Royal Road" for trade. His government extracted taxes that generated enormous wealth for the Persian Empire. This was the model followed by most empires throughout history.

One of the most notorious impositions of imperial trade in recent history was that of the opium trade (1838–1860) under the British Empire. Under imperial control of Great Britain, India produced most of the world's opium in its large poppy fields in the north of the country. The British, looking for markets for the opium, turned to China, where they had it smuggled in (in ships) and sold illegally (often trading it for tea, which was shipped back to England). The Chinese government, alarmed at the damage to their society caused by opium addiction, tried to stop the import of the narcotic. Having diplomatic pleas ignored (including a letter to Queen Victoria), the Chinese finally had opium seized and destroyed.

British merchants protested to their government over the loss of the valuable opium, so in 1839 the British government sent gunboats and soldiers to Chinese territory. The Chinese were forced to sign an agreement allowing the import-ation of British opium and pay reparations for the value of the lost narcotics. This display of weakness destabilized the Chinese government, and after a brief civil war a succeeding Chinese leader outlawed opium importation. Again the British sent a military force, defeating the Chinese, legalizing the trade in opium, allowing Western businesses and missionaries to operate, and further weakening Chinese governmental authority. After 600 years of stability following the recovery from the invasions by the Mongols under Genghis Khan, the British incursions started a pattern of destabilization in which wars, invasions, and uprisings shook the country for almost a hundred years. But, from the British perspective, the weakening of an international power like China was all to the good, since they could now extend their military and economic trading power far into the eastern hemisphere, and it was not until 1997 that Great Britain finally gave up its last colony in China (Hong Kong). Many of the motivations of the Chinese government over the last fifty years have been, for them, framed in the history of their struggles against powerful Western militaries.

Trade and Power

In terms of Europe itself, this part of the world only became powerful once it had obtained the wealth and power derived from its international trade. Until modern times, they never had any products the rest of the world really needed – no great deposits of gold, nor spices, nor oil, nor surplus grain that could simply be sold. It was only with the advent of colonialism under the mercantile system that money started to flow into Europe (Spain first, then the rest of Europe). Once the Industrial Revolution caught hold in Europe, factory-produced goods were in demand around the world and exports became vital.

However, today the rest of the world is starting to catch up (Figure 22.1). Japan became an industrialized exporter more than a century ago (and has the

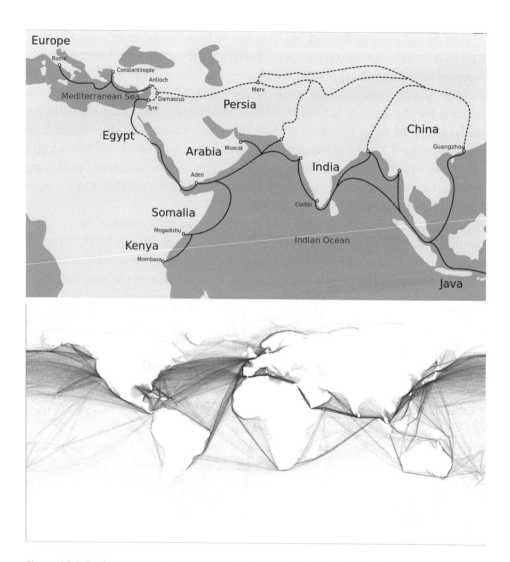

Figure 22.1 Trade routes have been sources of wealth throughout history. There were significant routes across all the continents, but the first large-scale trade route to generate wealth across multiple continents was the Silk Road (top image), which ran for more than 2,000 years from China, across Asia and the Middle East, and into Europe and Africa. Cities on the Silk Road were famous for their wealth and splendor.

Today, large-scale commerce is most often ship-borne, as transportation by water is more efficient than by land (or air). The lower image shows GPS data for thousands of ships transporting products around the globe. Note that, in this image, a disproportionate amount of shipping originates, or lands, in North America, Europe, Japan, and China. This may be the clearest visual demonstration of the economic power of those regions, because those shipments generate economic activity – investment, employment, and accrued wealth. The most direct way for a country to raise its standard of living is to become a place where those blue lines converge.

third largest economy in the world); today it is not uncommon to see Chinese-manufactured goods around the world, and India is also now producing cars, motorcycles, computers, and other industrialized goods that the world readily purchases. As a result, China and India now have greater international prestige and wealth, and their militaries have started to become powerful and important (both countries possess modern aircraft carriers, airplanes, tanks, and nuclear weapons).

Box 22.1 The power of an incoming dollar

To understand the value of money coming from outside a country or community, consider a single dollar (or pound, or whatever currency) earned through exporting something like corn. That dollar first goes to the farmer who farmed and sold the corn for export. But the dollar then has a path through the community or nation, building value on its path. It might next pass to the tractor repair shop, whose owner might use it to buy lunch at the café. The café owner might then use it to buy shoes at the shoe store. Then the shoe store owner might spend it at the movie house. And so on, throughout the economy of that community, literally making the economy happen. But a dollar spent outside the community does the opposite. This is why exports are so important – they literally generate wealth for the entire community or nation.

Box 22.2 The history of money

Why do we use money? The first real economic system was the barter system, and it has been around since before humans even evolved (chimpanzees regularly engage in barter for food or reproductive access). Under this system people trade something they had (say, a fox pelt) for something they wanted (say, some spearheads). Under this system there is no need for money, because the objects themselves serve as the currency. However, this system has an obvious problem: it requires *a coincidence of wants*. It works well if you have a fox pelt and want some spearheads. But what if you want some fruit but the fruit-grower doesn't want a fox pelt? Well, you could go around and try to figure out what the fruit-seller wants, and trade for that, but you can see how inefficient that is, plus you might never find the right combination of trades.

The solution is a medium of exchange. Money serves as an abstracted representation of goods and/or services and allows exchanges of different commodities. The advantages of using money are that it:

Box 22.2 (cont)

1. Enables the exchange of commodities, even if there is no coincidence of wants.
2. Allows storage of value for perishable commodities (think of fruit or meat).
3. Gives a clear "metric" of exchange, so that people can understand the value of what they are exchanging.

There have been various forms of money throughout history. Salt has long been used (hence the expression "worth his salt"), as have various other useful objects or materials – food, tea, nails, iron ingots, beaver pelts, etc. This type of money, which is composed of something that has inherent value, is known as *commodity money*. In prisons in the United States, cigarettes serve as commodity money among prisoners.

Precious metals, which can be turned into jewelry, have long been used as commodity money. Gold is especially valued because it is stable (it does not rust), highly malleable, and relatively rare. Silver and copper are valuable for similar reasons (and nowadays copper has utility in electrical/electronic applications). Precious metals have value by weight, and so are organized into ingots or coins, which are worth an amount determined by their weight. Coins made of precious metals are seen by 800 BCE in India, China, and the Mediterranean cultures.

However, coins are heavy, and it can be difficult to transport large amounts of money in metal form. Further, coins can suffer from the problem of debasement (coin clipping), in which the edges are trimmed away, with the intent of fraudulently passing the coin off as whole. Since it is inherently valuable, the clipped metal can be sold off. Another technique was to drill out the center of the coin and pound the remaining metal to close the hole, plugging it (hence the expression "plugged nickel").

One additional problem with coins made of precious metals is that the coins themselves would be subject to the variations in the value of the precious metals. A new discovery of silver might flood the market with the metal, driving down the price, and making the currency of the country less valuable.

Fiat Currency

In the eleventh century, the Chinese government started issuing paper notes, in which the piece of paper represented the value of a certain amount of precious metal. This was the origin of what we call "fiat currency" today, where the value of the note is determined "by fiat" – that is to say, the government. China, under the Song Dynasty, was a geographically

Box 22.2 (cont)

widespread country with a large, powerful, and stable government. This is critical to the success of paper money. People have to believe that it is actually worth what it says it is worth.

Initially, paper money could be literally exchanged for its value in precious metals. So, for example, the British pound is called a "pound" because it was held to be equivalent to a pound of silver (hence the expression "pound sterling"). In the United States, the US dollar was fixed at the exchange rate of $20 per ounce of gold from 1789 to 1934. This fixing of the dollar to gold is known as using the *gold standard*, and is the reason why the US government stored gold in Fort Knox – to back the value of the US dollars in circulation (if you have seen the James Bond movie *Goldfinger*, you might remember that the villain's plot was to irradiate the gold in Fort Knox to make it unusable, thereby increasing the relative value of the villain's horde of the metal).

But paper money has its own weakness. Sometimes a government will print more money than is valued by the gold in its reserve. In fact, this is the case for most currencies today. In 1934, during the middle of the Great Depression, US President Franklin Roosevelt declared that each ounce of gold was worth $34. This suddenly increased the value of the money in the US Reserve, and allowed it to spend more on the many projects designed to help the country out of the economic depression, such as building dams, roads, and bridges. Hopefully, it is apparent from this act that the actual value of the dollar relative to gold is arbitrary. A government could fix its currency anywhere, and at any value. In 1971, President Richard Nixon stopped linking the value of the US dollar to gold, and suddenly the US dollar could buy whatever people thought it was worth.

This means that paper money is truly only as valuable as people think it is. This is known as a *floating currency*, because the value of the currency can rise and fall. If a government is strong, and the economy is strong, a currency is typically stable. However, if a country has a poor economy its currency can lose value rapidly. After World War I, when the German government had to pay the French and British governments reparations for the war, its economy was in a shambles, and the value of the deutsche mark fell dramatically – at its lowest point dropping to one trillionth of its value (this economic disarray contributed to the rise of Hitler and the Nazis, who promised to stabilize the economy and did).

This has happened numerous times throughout history. In the United States, during the American Civil War, the Confederacy issued banknotes,

Box 22.2 (cont)

but these were not based on a gold standard, and by the end of the war they were worthless. In the 1990s, in Zimbabwe, the government was forced to issue banknotes worth one hundred trillion Zimbabwe dollars. Venezuela, in the grip of an economic and political crisis from the late 2000s, has seen its currency, the Bolivar, drop to less than one millionth of its early 2000s value.

Box 22.3 The dangers of a specialized economy

When you are the leader of an exporting country (or region), sometimes the world needs what you produce, and sometimes it does not. When you happen to have a highly desired product, your exports can generate significant wealth, but you are still vulnerable to shifts in any one of a variety of market forces.

One of the starkest examples occurred in Brazil at the turn of the twentieth century. The Amazon rainforest is the home of the rubber tree (*Hevea brasiliensis*), which naturally produces a gummy sap called latex. Collected and processed, latex forms the basis for natural rubber. Although rubber had long been known, it was relatively rare and expensive. The period from the middle of the nineteenth century to the early twentieth century saw enormous new demand for rubber products, from boots to overcoats, to vehicle tires, to gaskets and seals for machinery. This demand created a rubber boom in Brazil.

Production of rubber centered around the city of Manaus, deep in the Amazon Basin, and on a major tributary of the Amazon River. Exported rubber formed the basis of the economy of the city and the region. Manaus had an effective monopoly on rubber production (rubber was also available in the Congo, but there was no large-scale production). Owners of rubber export companies became fabulously wealthy and decadent, with the "rubber barons" building lavish estates, buying yachts, and constructing the world-famous Amazon Theater: a spectacular opera house, with Italian marble, Murano glass chandeliers, French roofing tiles, and Scottish steel walls. The city itself had electric streetcars and lights long before other Brazilian cities.

But in 1876, an Englishman, Henry Wickham, smuggled out tens of thousands of rubber plant seeds, and the English started growing small seedling plants in England. Over the next forty years, the British nurtured and then transplanted these seedlings in their Asian colonies, so that by the early twentieth century the English, in their colonies in Southeast Asia, had

Box 22.3 (cont)

developed rubber plantations that could produce rubber at a much lower cost than the Brazilians. With the British, then the Dutch, flooding the market with cheap rubber, the international price of rubber dropped from 800 English pounds a ton in 1910 to fewer than 75 pounds a ton by 1922, despite world demand skyrocketing with the spread of automobiles. With the monopoly on rubber broken, the economy of Manaus collapsed and did not recover until the 1960s. Today, Manaus is full of nineteenth-century buildings that are faded and often empty, but still elegant and beautiful.

Currently, the clearest example of such a narrow economy is seen in Saudi Arabia. The economy of this country is almost entirely based on oil production, and lacks the diversification that could support the current economy. The Saudi government has made attempts at diversification, but the country has limited alternative resources and little industry outside the petroleum sector. Because of the wealth and relatively small population of the country, Saudi citizens can receive a government oil stipend. Most workers employed in the petroleum industry and elsewhere are not Saudi citizens but visiting guest workers. Today, Saudi Arabia is one of the wealthiest countries per capita (per citizen) in the world, but a dramatic dip in the international oil market, either from the discovery of new petroleum sources or by the development of alternative technologies (solar, wind, nuclear power), could change the country's economic fortunes dramatically. It remains to be seen whether the lessons of Manaus are lost on the country's leaders.

Box 22.4 Two sides to the international economy

Negative Consequences

The interconnection of the international economy has some powerful consequences. When someone in England orders a cell phone, they are engaging many parts of the world economy in profound ways. The natural resources for the phone come from around the world, but some, notably tantalum, are extracted from war-torn regions of Central Africa. The reason a person in the Democratic Republic of Congo is willing to risk their life mining tantalum for a cell phone that is a trivial purchase in England is because of the dramatic disparity in wealth between those two countries. Similarly, when someone pumps gasoline that comes from the oil drawn from the Niger Delta, they are almost certainly funding what most citizens of the world would likely

Box 22.4 (cont)

consider local corruption and environmental destruction far from the automobile drivers filling their tanks in Western countries.

Perhaps the grimmest form of this disparity was seen in the nineteenth-century transatlantic slave trade. A wealthy plantation owner in Georgia would have had enough resources to pay for a slave who had been captured on the other side of the world. The effect of internationalizing commerce is that people can reach out across the world. The fact that consumers in England cannot see the desperate mining conditions makes it easy to ignore the implications of the purchase.

The so-called transnational corporations, such as Exxon, Royal Dutch Shell, and British Petroleum (oil), DeBeers (diamonds), Apple (technology), Nike (clothing), and Archer Daniels Midland (agricultural products), to name just a few, have come under criticism for leveraging the wealth disparity between Western industrialized nations and the developing world through highly mobile capital deployment, to employ workers who are willing to accept lower wages than domestic workers, avoid taxes, and avoid domestic environmental controls in countries far from their consumer base. It should be noted that the large slave-trading corporations of the eighteenth and nineteenth centuries were also transnational, and were criticized for using their economic leverage in similar ways for exploitative reasons.

Another criticism of the transnational trade has been the loss of some aspects of local culture, due to a cheaper or better product being suddenly imported, or simply by exposure to a competing culture. The appearance of McDonald's restaurants is regularly criticized as an oppressive American cultural product that is damaging local food culture. Perhaps no McDonald's was more controversial than the first one to appear in Paris (although it seems to have had no effect on Parisian culture).

It is true that, over the last few hundred years, cultural traditions have been lost because of exposure to foreign styles, ideas, and art. Today, for example, few Navajo will wear traditional Navajo clothing because of the powerful influence of American culture. Similarly, across the globe, fewer and fewer people wear the traditional dress of their ancestors or, perhaps more universally, speak the language of their predecessors. Almost 200 documented languages went extinct in the twentieth century, as people stopped using their traditional languages and culture and shifted toward the languages of the cities and towns, where they have moved in the search for for economic opportunities. Local religions throughout the world have also suffered, as missionaries have, for millennia, continually attempted to convert, by force or

Box 22.4 (cont)

pursuasion people from their local (often shamanistic folk religions) to one of the large proselytizing religions or the faith of powerful outsiders.

When a young woman from rural Vietnam moves to Hanoi for the opportunity to work in a garment factory (a common occurrence for young women in Vietnam), she will be unlikely to pass on her village traditions to her children; instead, she will raise them like the other children in the city. Her folk traditions may be lost, and she will dress, speak, and eat like the other urban peoples of Hanoi.

However …

As Western economic influence has spread, not all consequences have been negative. Modern Western thinking is largely the result of the Enlightenment, and even as some large corporations have acted immorally, exploiting their economic advantages, some have spread Enlightenment ideas. This is seen in the spread of democratic ideals (even if unevenly applied), but perhaps most starkly visible in the relations between men and women. In many traditional cultures, women are discouraged from acquiring education, and encouraged to submit to male authority. Exposure, in the twentieth and twenty-first centuries, to Western culture (especially media, television, and movies), with strong female role models, has caused some documented changes in attitudes toward educating girls in rural Indian villages, for example. Similarly, attitudes on female circumcision (a traditional rite of passage in many African countries) has altered, and the practice has been outlawed for the first time in most African countries, with rates of circumcision dropping significantly in the twenty-first century.

The power of the international industrial and technology economy has demonstrated the importance of education, for both girls and boys, and rates of formal education are up across the developing world. It is rare to find illiterate people under forty almost anywhere in the world, whereas just twenty-five years ago it was common. Finally, the health implications of this change have been generally positive. Between 1960 and 2010, life expectancy in countries like India, Brazil, China, and Indonesia has increased by more than twenty years, far outpacing the relatively modest increases in Western nations.

Consider that young Vietnamese woman, who has moved from her rural village to Hanoi to work in a garment factory. One may lament the loss of her traditional culture and relationships. However, she moved to Hanoi of her own choice, because the opportunities offered in the city are preferable to life in a

---

Box 22.4 (cont)

rural rice field. In Hanoi, she will have access to better health care and educational opportunities. The men she meets will be more likely to treat her as an equal, without the more traditional expectations of rural life. In Hanoi, where commerce and communication with all parts of the world are commonplace, her children will have easier access to education, and are unlikely to have to work in a rice field – in fact they will have the possibility of a job in the technology field. She will become a "Westernized" person, to some extent, with the attendant loss of culture, but also with the clear material and educational advantages that the technologically developed West offers.

Box 22.5 The Belgian Congo

International mercantilism blurred the lines between the economic interests of a corporation and those of a country. People in a country with a powerful and exploitative mercantile economy were often insulated from the most inhumane aspects of colonial exploitation. Perhaps the most egregious example of this system was found in the Congo (Figure 22.2, highlighted in red in the image on left). During the "scramble for Africa," when European countries were establishing colonies in Africa – Kenya (Great Britain), Tanzania (Germany), Algeria (France), Tunisia (Italy), etc. – the King of Belgium, Leopold II, claimed a large portion of the basin of the Congo River as his *personal* property, in 1885. He established a colony that claimed 900,000 square miles (2.3 million square kilometers) and created a colonial administration to extract wealth from the land and people.

He pursued ivory and rubber, especially, and used his private army (the Force Publique) to enforce quotas on the Congolese population. By decree, he established a serfdom for these people, who had to produce certain quantities of ivory (incidentally killing hundreds of thousands of forest elephants) and rubber. If the ivory and rubber were not produced, the Congolese were punished. The Force Publique was infamously barbaric, and the chopping of hands was one notorious punishment for many. Leopold (who never actually traveled to Africa) was responsible for the deaths, by punishment and starvation and disease, of roughly half of the population of the Congo basin.

Unfortunately for Leopold, missionaries, journalists, and diplomats in the region quickly spread word of the atrocities. (Perhaps the most famous portrayal in literature was in Conrad's *Heart of Darkness*, whose villain, Kurtz, is

---

Box 22.5 (cont)

a cruel ivory trader deep in the Congo basin.) Photographs and personal accounts of the mutilations and executions by the Force Publique became part of an international outcry and by 1908 Leopold was forced to surrender his personal colony to the Belgian government. The Congo was a colony of Belgium until 1959, but it has yet to reach long-term political stability.

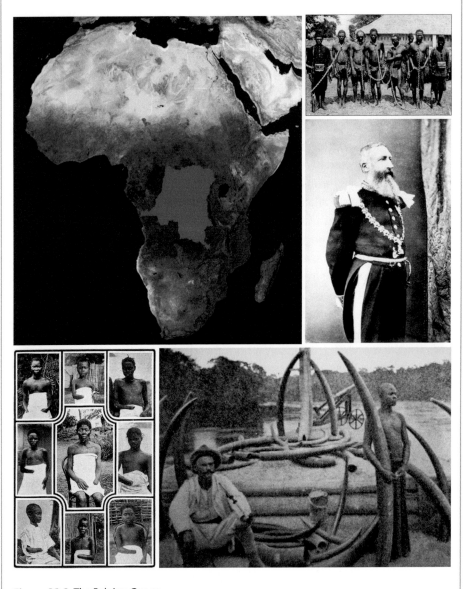

Figure 22.2 The Belgian Congo.

Box 22.6 Unintended consequences

As economic interests often intersect with political interests, governments have sometimes stepped in to protect their businesses, even in distant countries, and sometimes with dubious moral justification. One of the greatest errors in such economic and political intervention, with the worst long-term effects, came in the 1950s. Iran, which had once been one of the great international powers, even until the mid-nineteenth century, was undergoing a series of political upheavals, due to a famine in the late nineteenth century and occupation under the British and Russians during World War I. A secular constitutional monarchy, generally oriented toward the West, was established in 1921 and maintained through World War II.

During the first half of the twentieth century, British Petroleum (the Anglo-Persian Oil Company at the time) had developed and controlled the Iranian oil fields. With the spread of the automobile and the military transition from coal to oil, these fields were a critical strategic interest of Great Britain. In 1951, democratically elected Iranian Prime Minister Mohammad Mosaddegh decided that British control of Iranian oil fields gave the British too much influence in their internal political affairs. He "nationalized" the oil fields – that is, he cancelled the oil leases that the Anglo-Persian Oil Company had negotiated, with all oil rights reverting back to the Iranian government.

Britain responded by overthrowing the Iranian government, with considerable assistance from the CIA. The British convinced US President Eisenhower and US Secretary of State John Foster Dulles that the nationalization of Iranian oil was a communist threat. Using a variety of political, social, and military levers, the CIA and MI6 engineered the overthrow of Mosaddegh, and installed the monarch (previously a figurehead) as the ruler.

The Shah ruled Iran for the next twenty-five years with an iron fist – secular but brutal – crushing internal dissent and preventing any democratic reforms. However, his orientation toward the West, with the continued concession of Western companies developing Iranian oil fields, ensured support from Western countries. But the citizens of Iran, a generally well-educated population, finally rebelled against the Shah in the famous 1979 revolution. At first, the revolution was partially democratic and secular, but revolutions are hard to control, and a charismatic Islamist, Ruhollah Khomeini, had built a network of support parties, some of which were armed, and quickly took control of the revolution.

Dying of cancer, the Shah fled to the USA, partly for medical treatment but also for refuge. Khomeini's supporters demanded the USA return the Shah for execution, and when they were refused they attacked the US embassy in

Box 22.6 (cont)

Tehran, taking fifty-two US citizens as prisoners. The prisoners were held for 444 days, well after the Shah's death, as retribution against the USA.

Today, Iran is formally Islamist. The government is unpopular, and its citizens are said to yearn for Western-style freedoms, but any dissent is brutally suppressed. The governments of the USA and Iran have a strongly antagonistic relationship, with the Iranian government deeply resentful of past US interference in internal politics, and the USA still angry over the 1979–1980 hostage crisis. Despite this, the Iranian people do not appear to be critical of US citizenry, as the USA has a large population of immigrants from Iran. Iran has a highly educated and culturally sophisticated population and a tradition of secular democracy. As a non-Arabic population, Iran might be a natural ally to the West in the region. Yet, the West is unable to escape its history of past interference on behalf of a large multinational oil company, as well as other political missteps.

These "historical hangovers" are not uncommon. Many people in Central America are still resentful of US intervention during the twentieth century in their internal politics (see Box 22.7). Other former colonizers face the some repercussions in North Africa (France), South America (Spain), and eastern Africa (Great Britain). But this is not limited to the colonizers of recent centuries. Cultural memories, and the grudges they carry, can be quite deep. In 2011, I visited Marrakesh (in Morocco), and a young Berber man expressed resentment at the Muslim invasion, and the imposition of the Arabic language, which had occurred some 1,300 years previous. To him, he was a Berber first and foremost, and spoke Berber as his native language; anything else (first Arabic, then French) was the language of an oppressor. One only need review the political histories of Northern Ireland and the former Yugoslavia for additional examples of long-simmering historical hangovers.

Box 22.7 Banana republics

When you see a bunch of bananas in the fruit aisle, you are probably just thinking about your stomach, and whether you are in the mood to eat one. But it is worth knowing something about the power and history of those bananas.

One of the most politically powerful corporations of the early twentieth century was United Fruit (now Chiquita). United Fruit controlled production of bananas in Central American countries such as Costa Rica, Honduras, Guatemala, Cuba, Panama, Nicaragua, and Mexico. It owned millions of acres

Box 22.7 (cont)

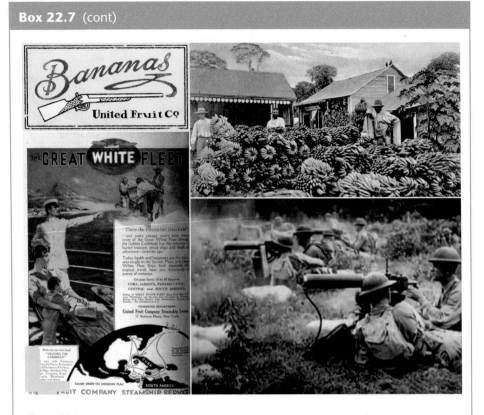

Figure 22.3

of land and was the largest employer in Central America. To move the bananas, United Fruit built railways and ports, as well as telegraph, telephone, and mail systems. It had its own fleet of modern steamships that was also used by tourists traveling to and from Central America.

All of this investment in these countries shows the enormous economic resources available to the company, which gave them political leverage. Many of these countries were run as extensions of the corporation, which had monopolistic control over most aspects of the economies. These were not democracies, but run as small dictatorships of the wealthy (the American writer O. Henry coined the term "Banana Republics" for these countries). Workers had few or no rights, and were often exploited in the name of profit. United Fruit also had leverage over the militaries of these countries, and employed its own armed security forces to enforce control over workers and land.

The autocratic control of most of the viable agricultural land in these countries, the lack of democratic rights, and the exploitation of the poor

Box 22.7 (cont)

workers naturally led to periods of extreme political instability. Revolutions would periodically occur, and the political power of United Fruit was strong enough in the USA that the US Marines would sometimes be sent to suppress the revolutions and support the autocratic governments. This occurred many times (notably in Honduras) from the early 1900s through the 1920s (the so-called "Banana Wars"). Highly decorated US Marine Corps General Smedley Butler later renounced his role in these interventions, and wrote *War is a Racket* in 1935.

The last intervention related to United Fruit was in 1955, when United Fruit persuaded the Eisenhower administration that the new democratically elected president of Guatemala, Jacobo Arbenz (who was instituting a series of labor and land reforms), was a communist agent. With direct CIA support, President Arbenz's administration was overthrown in a coup. Four decades of civil war followed, and this incident (as well as several other similar interventions) created a long-lasting distrust of the US government in Latin America that persists to this day.

The US government no longer directly intervenes on behalf of this company, but United Fruit (now Chiquita Brands International) has not avoided getting itself into trouble. In 2007, in a US Federal Court, Chiquita pleaded guilty to smuggling arms to a terrorist organization. They agreed to pay a fine of $25 million for smuggling assault rifles to a Colombian paramilitary organization that was forcing Colombian banana farmers to sell to Chiquita and killing or intimidating farm worker labor representatives in that country. Chiquita was also forced to pay reparations to the Colombian farmers and their families.

Chapter 23: Modernity

As Europe and the rest of the industrialized world passed through the eighteenth and nineteenth centuries, the trends of the Enlightenment continued. But in 1859 a publication appeared that caused a political, scientific, religious, and popular upheaval that continues to this day.

This publication was, of course, *On the Origin of Species by Means of Natural Selection, or the Preservation of Favoured Races in the Struggle for Life*, by Charles Darwin. Darwin knew it would be controversial, having essentially sat on the idea for more than twenty years for fear of the reaction. The book itself was a heavy read, and most of his pages present detailed evidence from the natural variation in wild animals, embryology, geography, the geological record, and animal breeding. The book mention humans only in passing in one line: "Light will be thrown on the origin of man and his history," but the central implication was clear: humans are animals, and arrived on Earth in exactly the same way as all other living things – through evolution.

The book sold out in one day and has never been out of print since. At the same time, it was condemned by a variety of groups and people. Some scientists challenged him, partly out of natural competitiveness, although the scientific community quickly recognized the power of the argument and presented near-unified support. The more aggressive challenges came from religious leaders, who, much as the Roman Inquisition did when persecuting Galileo, recognized that Darwin's idea was a direct confrontation to the authority of Judeo-Christian scripture (although it was the Protestant, rather than Catholic, clergy who protested – a pattern that has continued to today).

Darwin's book, presented as a popular rather than scholarly work, thrust science into the public sphere in a completely new way. Newspaper and magazine editorials argued for and against, popularizing the debate. Most importantly, for the first time the role of the scientist as a figure of public authority (or denunciation) had appeared. Suddenly, science was front and center – and unavoidable.

Many people found this shift in the philosophical position of humans extremely disorienting, since it more or less completely eliminated the last vestige of explanatory power in the Old Testament. What had started with

Galileo and Copernicus was completed by Darwin. As a result, religious leaders have continuously, from the time of its first publication to now, pushed back against Darwin.

The role of the scientist in the public sphere continued after Darwin's initial publication (he finally tackled human evolution in 1871 with *The Descent of Man, and Selection in Relation to Sex*). Newly discovered fossil dinosaurs and giant Pleistocene mammals filled museums around the world, confirming the expectations whetted by Darwin. Scientific explorers brought back exotic species from remote jungles and deserts, and probed the north and south poles.

A second great discovery that shifted world thinking again occurred in 1905. A Swiss patent clerk, Albert Einstein, published four of the most historically important papers in physics (and science generally). Two of these papers – "On the Electrodynamics of Moving Bodies" and "Does the Inertia of a Body Depend upon Its Energy Content?" – established the theories of special relativity and general relativity and, by doing so, changed the way we view reality itself.

Einstein's discoveries had an immediate and profound effect on science, as it formed the foundations of virtually all twentieth-century physics. One of the deeper implications of this work came about with the invention of atomic weapons and nuclear power. But, perhaps more importantly for society generally, his conclusions meant that our perceptions of space and time were illusory. When his hypothesis on the bending of light by gravity was confirmed in 1919 to wide acclaim (by observing starlight bend under gravitational pull, when passing by the Sun in total eclipse), his role as the preeminent intellectual of his day was established. The idea of relativity saturated popular culture, and was explored over the next fifty years in art, music, and literature (often inaccurately).

Darwin and Einstein were at the end of a transition away from intuitive and supernatural explanations that had started with the ancient Greeks, continued in the Islamic Golden Age, and then spread worldwide with Renaissance and Enlightenment. In a sense, they serve as philosophical bookends to a period of intense philosophical change. Today, we are still in this period, as society turns to scientists for answers in most crises, even as some more traditional thinkers push back. (As of the writing of this manuscript, the world is in the grip of the COVID-19 pandemic, and the struggle between science, and those challenged by its conclusions, is visible daily.)

Trends in Modernity

Although the changes were precipitated by the widespread acceptance of science, there were other important social and economic trends that moved in parallel. Any society that intended to become wealthy and powerful (Chapters 21 and 22) in the modern world also needed to be educated. The days of the illiterate but all-conquering Mongol hordes were gone, and small professional armies backed by industry and technology (such as the telegraph, railways, repeating rifles, cannons, and, later, machine guns) regularly routed much larger forces equipped with simpler weapons. In the world marketplace, economic power was driven by industry, and countries with more engineers were more able to compete in the world economy than those without.

This period is when we see the expansion of education. No longer was school reserved for the wealthy; countries invested in large public school systems, from which they could draw intellectual talent. Between 1852 and 1917, every US state passed laws requiring children to attend school. This was also the time when many public universities were founded. In the UK, University College London was founded as the first secular university in England in 1826, and in 1870, the Elementary Education Act provided for free public education in England and Wales.

But an educated populace is not necessarily deferential to authority. Educated citizens are more likely to challenge the decisions of leaders, and to push back against traditional norms that might be justified by "tradition." This expressed itself in many ways, including the rejection of religion, monarchy, and, notably, male superiority and white racial superiority (see Figure 23.1). By the middle of the twentieth century, these trends helped enable the end of colonialism. The power of women, especially, expanded dramatically, and the right to birth control and divorce helped women pursue careers outside the home.

The trend toward racial egalitarianism became a hallmark of the period. Although eugenics and racism were common at the turn of the twentieth century, by mid-century laws forbidding racial discrimination in public accommodations, housing, employment, education, and marriage were present in most industrialized nations (although, to be sure, racism itself has persisted). Although still imperfect, the current state of affairs is a far cry from mid-nineteenth century, when slavery was legal in many places around the world, and racial discrimination was not just allowed but obligatory in many places.

Modernism is also associated with dramatic changes in the arts – it is in this period when impressionism, expressionism, modernism, cubism, and modern art in general appears, rejecting the explicit realism and representation of virtually all previous artistic movements (Figure 23.2). Architecture shifted away from the neo-classicist orientation toward ancient Greece and Rome seen during the Renaissance and Enlightenment, and created entirely new forms, such as Art Deco, Art Nouveau, Bauhaus, and the International Style. Modernism in music include Schoenberg and Stravinsky, and in popular culture it led to jazz, blues, and, ultimately, rock and roll.

Figure 23.1 The modern period was a time of expanded opportunities for groups that had been subjected to repression for hundreds or thousands of years. The 1920s, especially, had women gaining the right to vote in many Western countries, and acquiring celebrity status in a variety of fields. Starting at the top left, Bessie Coleman (1892–1926) was an African American pilot famous for her daring flying in barnstorming exhibitions; Amelia Earhart (1897–1936) was the first woman to fly across the Atlantic Ocean; and Hellé Nice (1900–1984) was the first woman to drive in Grand Prix races.

World War II was a turning point for African Americans, partly because of their performance in World War II. Prior to the war, some military leaders had expressed reservations over the abilities of African Americans to successfully use complex machinery like tanks and airplanes. Fighting in segregated units, those such as the 758th Tank Battalion (bottom left) and the 332nd Fighter Group (bottom right) demonstrated their abilities in the face of considerable prejudice. The decades following World War II saw the fastest advancement in civil rights in the United States, before or since.

Resistance 1: Religion

But, as the world shifted, inevitably there was pushback. In the United States, fundamentalist Protestant groups organized resistance to the teaching of evolutionary theory. Famously, in 1925 in Dayton a Tennessee high school teacher, John Scopes, was prosecuted under a Tennessee state law that forbade teaching "any theory that denies the Story of the Divine Creation of man as taught in the Bible." Although he was convicted (and fined $100), the conviction was overturned on appeal, and the international media attention was enough to point out the absurdity of the law to the general public.

Figure 23.2 The radical changes in society over the relatively brief modern period produced a number of artistic movements. Instead of narrative art, portraying a well-known story (often from Christian traditions, or Greco-Roman mythology), art became focused on ideas and emotions. The early modern

Religious leaders across multiple faiths have resisted modernism. Within Christianity, Protestants have protested most vigorously, but Islam has also strongly resisted modernism, and in some Muslim countries people can be prosecuted, and even executed, for rejecting Islamic scriptural rules. Hinduism, with its long tradition of the hierarchical caste system, has also pushed back against the trend toward secularism. Although India is, by constitution, a secular nation, the current prime minister, Narendra Modi, is a believer in establishing Hinduism as the state religion and is supported by large numbers of the Indian populace.

As women acquired more political power (Box 23.1), some religious leaders articulated the view that women should remain under the authority of their fathers or husbands. A frequently espoused view has been that the role of women is reproduction, family maintenance, and general domestic responsibilities. Birth control was outlawed in many countries as greater access to information and technology made it practical for the first time. This struggle over the role of women is still visible in many countries with a fundamentalist religious tradition, such as Saudi Arabia, Afghanistan, India, and Chad, and in some parts of the United States.

Resistance 2: Fascism

This period saw the expansion of democracy across the world. For the first time, voting rights were extended to women and minorities (although in some cases voting was illegally suppressed, especially in states in the US South). The colonies of foreign powers started to demand voting rights or outright independence. Today, there are more democratic nations than at any time in history.

However, an alternative form of government reared its head in the first half of the twentieth century that recalled the days before democracy. This form of government, known as *fascism*, appeared in a group of powerful and aggressive governments across Europe, Latin America, and Asia (Box 23.2). Fascism is a form of government in which power is centralized in a lone charismatic figure and his party or group. These countries are, typically, militarily aggressive and frequently engaged in conflict, and so have an outsized influence on world affairs. The philosophy of strength over the weak was explicitly expressed by modern fascists, but would have been readily recognized by Alexander the Great, Darius of Persia, and Genghis Khan.

Figure 23.2 (*cont.*) period saw impressionists (Monet, top left) bringing attention to the perception of light, as did Van Gogh (lower left). The Art Deco movement (second column), in graphic art, architecture, and vehicle design, emphasized movement and clean lines, and was a completely new form focused on the future rather than looking to the past (as in the neoclassicism of the Renaissance and Enlightenment). Modern art could also be grotesque and evocative, as in the German expressionism of the 1920s (right two columns). German expression was deemed "degenerate" by the Nazis and banned when they came to power in the 1930s.

The most famous countries are those that formed the Axis powers in World War II – Germany, Italy, and Japan – but other fascist countries of that era included Austria, Spain, Portugal, Argentina, and Paraguay. These countries viewed democracy as a philosophy of weakness and expected fascism to be the world's governing philosophy. World War II was, in many ways, a struggle between democracy and fascism (Box 23.3), and the eventual loss by Japan and Germany has been seen, at least partially, as a general repudiation of fascism.

Now

Today, we regard modernism as having won the day, despite the occasional appearance of a dictatorial leader or a loud but small racist group, or a call for women to resume more "traditional" roles. For most people, the present is better than it would have been for them at almost any time in the past, but more importantly, people have accepted the principle of change in the direction of modernist principles.

Box 23.1 The expansion of democracy in the USA

We often think of the United States as the first secular democracy – disentangling the government from the traditional sources of power like royal lineages and institutionalized religion. And in that sense, it was the first modern nation. But within the United States, and most other countries, the democratic rights of individuals were actually relatively limited, because the right to vote was constrained to a few. In the late eighteenth and early nineteenth centuries, voting was normally only allowed by property-owning males, and in southern states only white property-owning males.

The restrictions on voting were lifted in stages over the next 150 years. By the 1828 federal election, most states had eliminated the requirement that voters own property. The extension of the vote to former slaves came with the ratification of the Fifteenth Amendment (1870), although it was not well enforced until the mid-twentieth century. In 1920, the Nineteenth Amendment gave women the right to vote, and in 1924 federal law gave suffrage to all Native Americans. In 1943, Chinese immigrants were given voting rights, and in 1964 all poll taxes were outlawed by federal law (poll taxes were used in the US South to prevent poor people, notably black citizens, from voting). The last major extension of voting rights came in 1971, when the voting age was dropped to eighteen, although expansion continues – the majority of US states have started to make it easier for ex-felons to vote.

Box 23.1 (cont)

Although some US states pushed back against the expansion of voting rights by holding "white primaries," redistricting to split black voters, and imposing literacy tests, these techniques, over time, were outlawed, and their justifications discarded. Over a period of roughly 150 years, despite resistance, suffrage became universal. While there are occasionally challenges to voting (politicians sometimes try to tip the scales by making it hard for some groups to vote), these challenges pale in comparison to the constraints of the nineteenth century. There are no important politicians who argue, for example, against women or blacks voting, or want to impose a minimum income requirement. And this same pattern applies around the world. Despite occasional challenges, the modern world is a more democratic world.

Box 23.2 Women's rights, women's power

The Industrial Revolution gave women unprecedented economic power outside the traditional bounds of the family. The nineteenth-century push for compulsory youth education gave girls and women new educational opportunities. Under these conditions, it was expected that women would gain access to power in the public sphere in new ways. And that is exactly what happened.

Voting Rights

Probably the most well-known struggle for women's rights and power was the fight for *suffrage*, that is, the right to vote. In a few places around the world, women (normally only property-owning women) had the right to vote in municipal elections (and some US states gave women complete suffrage), but the first country to grant women the right to vote in all elections, including federal elections, was New Zealand, in 1903. Over the next half century, voting rights were extended to women across the world, especially in industrialized nations. The right to vote was hard fought, and in many nations women suffered public criticism, arrest, and physical attack for advocating for their rights (Figure 23.3).

Property

Suffrage is often highlighted, but a suite of other rights was obtained by women over the nineteenth and twentieth centuries. One of the most

Box 23.2 (cont)

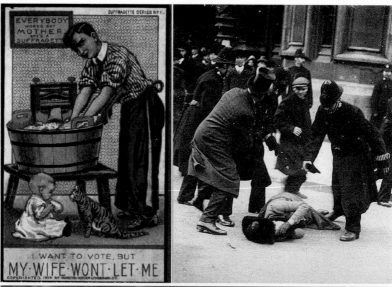

Figure 23.3 The granting of the right to vote to women was controversial, as well as expensive for the women fighting to acquire it. Propaganda aimed at men challenged the masculinity of those encouraging women to pursue suffrage (left). Women were often violently attacked (middle), and jailed, for marching for the right to vote. Imprisoned women would sometimes go on hunger strikes, and violent force-feeding was common.

Box 23.2 (cont)

important was the right to hold property. Property is often the foundation of wealth, but in the eighteenth century, many jurisdictions in the world constrained women's property ownership, particularly if they were married. In 1882, the British Parliament passed the "Married Women's Property Act," which granted married women the right to own and control their own property. Previously, under English common law, a married couple was, by law, a single entity under the control of the man. Anything owned by one was owned by them both but controlled by the man.

Naturally, this severely (and intentionally) limited the ability of a married woman to operate in her own economic self-interest, and constrained her ability to leave her husband. Similar laws were passed in the US states during the middle of the nineteenth century, granting women "separate economy." During this same period, the struggle over professional rights for women continued. In 1873, an existing Illinois law barring women from practicing law was upheld by the US Supreme Court, but only six years later, in 1879, the US Congress made it legal for any woman to practice law and argue in a US Federal Court. Laws granting women other professional licenses (as doctors, accountants, etc.) followed.

Divorce

The right of a woman to leave marriage on fair terms was often constrained. Divorce laws varied by country, and often by region, but typically a woman had to prove a variety of faults or actions committed by her husband, and typically the threshold for women was higher than for men – a wife's infidelity was sufficient for a man to divorce his wife, but infidelity on his part was often insufficient for her to be granted a divorce. In Islamic law, a man may divorce a woman simply through utterance, whereas a woman has to sue in court for a divorce (this, incidentally, is consistent with sociobiological models of male control of female reproduction – see Chapter 13). Laws on divorce changed roughly in parallel with the granting of property rights, and today most countries have enacted "no fault" divorce in which neither partner has to prove any specific actions. It is merely a legal contract out of which either partner may decide to opt out. Some religions still object to divorce, and today a second marriage following a divorce is not recognized by the Catholic Church.

Birth Control

One of the most important struggles was the fight for access to birth control. It is difficult to overstate the importance of allowing women to determine their

Box 23.2 (cont)

own fertility patterns for their social and economic progress in the modern world. We take for granted today that women have the right to determine when and if they become pregnant, but for much of history this was not the case, and many laws and power structures were created to ensure that men were in a position to control women's reproduction (it is worth noting that, for hunter-gatherers, who were/are relatively egalitarian, women's control of their own reproduction took various forms, including infanticide). Such legal structures as harems (the concept of consent did not exist for concubines), male polygyny, and restrictive divorce laws, as well as practices like female circumcision (also known as female genital mutilation) and laws on female dress, existed to prevent women from exercising their own reproductive decisions. In many societies, the role of women was to reproduce, and any attempt to control their own reproductive patterns was a challenge to the authority controlling their reproduction (note that male control of female reproduction is consistent with predictions of sociobiological theory, see Chapter 13).

In the modern era, the strictest controls on female reproduction waned: harems started to disappear and monogamy became the norm, even in more traditional countries (Turkey, for example, outlawed polygyny in 1926; see Figure 23.4). Behaviors to prevent conception through normal sexual practices (such as the rhythm method) could not be outlawed, but they were discouraged as "unnatural" by some groups on religious or philosophical grounds. Technology moved faster than society and devices that could be used to prevent conception (contraceptives) were outlawed in many countries on grounds of public morality or religious rule. In Ireland, the sale of contraceptives was outlawed in 1935. In the USA, the Comstock Act (1875) effectively outlawed contraception or the publication of birth control methods. In France, a similar 1920 law outlawed contraceptives and birth control literature.

Women in the late nineteenth and early twentieth centuries recognized that one of the keys to equality was to ensure women's ability to control their own reproduction. Birth control advocates, such as Margaret Sanger in the USA, argued that women should have the right to control the reproduction of their own bodies. The right to birth control in the USA was established by the Supreme Court in 1964, on the grounds that heterosexual married couples had an expectation of privacy in their personal (i.e., sexual) affairs. Other countries legalized contraception in the decades around that time (e.g., Ireland in 1985, France in 1967, and Italy in 1970). In Catholic countries today,

Box 23.2 (cont)

Figure 23.4 Turkey is an example of a country that transformed itself, almost overnight, from a traditional kingdom into a modern country. Prior to World War I the Ottoman Empire, ruled by Caliph Abdülmecid II (left), controlled much of the Middle East and North Africa. The Ottoman Empire allied itself with Germany in World War I, and as a result, lost all of its territorial holdings when the Germans and Ottomans lost the war.

At the beginning of the twentieth century, a group of secular revolutionaries known as the Young Turks overthrew the Muslim government. Although they were short-lived, their successor, Kemal Attaturk (middle), said: "The state will be ruled by positivism (science), not superstition." A new constitution was created that established a modern, secular nation. Rejecting the old religious government, the 1924 Constitution stated: "Turkey is republican, nationalist, attached to the people, interventionist, secular, and revolutionary" (Clause 1).Ataturk pushed through a series of reforms, changing many aspects of society from the script (adopting the Latin script) to the units of measurement, but more importantly granting women the right to vote (1930) and ensuring women served in the Turkish parliament. He is considered the father of modern Turkey, and Turkey joined the ranks of modern nations. However, this change was imposed from above, and many elements of society have never reconciled themselves to the secular government. Today, the current president of Turkey, Recep Erdoğan, has reoriented the country toward Islam, moving away from secularism.

access to contraception is sometimes difficult, even if technically legal. The debates over abortion further reflect this debate, and although abortion has become legal in most Western democracies since the latter half of the 20th century, the recent US Supreme Court decision (Dobbs v. Jackson) overturned US Constitutional protections for the right to abortion in the United States.

Employment

Laws guaranteeing women the right to equal opportunity and treatment in employment came relatively late. In the USA, the Civil Rights Act of 1964 made broad guarantees of equal opportunity, but the individual rights themselves were acquired piecemeal, often through court decisions. Other countries proceeded at roughly the same pace. So-called "marriage bars" prevented married women from working in many occupations around the world, on the

Box 23.2 (cont)

principle that their husband was responsible for providing the family income, and these were retained until the 1960s and 1970s in many European countries. Similarly, women in some European countries required the explicit permission of their husbands to work, a practice that was outlawed around the same time.

Recently, the "role" of women in combat has followed this same trajectory. Traditionally, in Western countries women were allowed to serve in support roles in the military. However, combat roles in the military, and the potential for decoration, are major factors in promotion, so by denying women access to combat roles they are, in effect, slowing their progression through the ranks. Justifications historically focused on social stereotypes: women were not sufficiently aggressive, or they could not handle the psychological rigors of combat. Some roles were denied on the basis of physical strength. Finally, women captured or killed in combat might be bad for the morale of the unit and country.

However, these justifications have, one by one, been proven false or insufficient. Today, in most Western nations, women can serve in almost any combat role. Even in non-Western countries, women are often in combat roles – in Kurdistan, women are regular combat infantry, although these roles are still denied to women in more traditional countries.

Resistance to Change

Although this narrative portrays the expansion of women's rights as a steady progression, in most cases these individual rights were hard fought. Many women fighting for equal rights faced jail, abuse, being fired from their jobs, or simply being shunned (in some countries they face all of these today). The extension of equal rights was controversial – newspapers and magazines argued vigorously on both sides of issues, and it formed a central part of the public debate, especially in the first half of the twentieth century.

Although arguments against women's rights and power are frequently framed in terms of religion or "tradition," it is worth emphasizing that arguments against women's rights were most often made by men, and for completely selfish reasons. Within any political or economic system there is a finite amount of power. If women were to obtain more power, the men would necessarily lose some (this applies to minority political power as well). It was in many men's self-interest to try to ensure that they did not have to compete against highly qualified women for education and employment, or that they, as men, would determine political outcomes. If a woman has no property rights in a marriage, then the man has much more power; but if she acquires "separate economy," then he loses his economic leverage over her. But the acquisition of

Box 23.2 (cont)

women's economic power and education, as part of the modern societal trends, ultimately made acquisition of political power inevitable, and most arguments against such rights have fallen by the wayside (with a few notable exceptions).

Women must still struggle for their rights, especially in places where traditional power structures limit their opportunities. Saudi Arabia is probably the starkest example. Strict interpretations of Islam have meant that women are controlled and restricted in almost all aspects of their lives: their clothing is determined by law (they must wear an *abaya* or some similar covering over their hair and body), they are forbidden to travel or be employed without permission of a male guardian (husband, father, uncle, etc.), and they do not have legal custody of their children. Fathers may marry off their daughters at any age. In some well-known cases, fathers have married off their preteen daughters to much older men to pay off debts, and the courts have upheld the marriages. Men may divorce their wives without even informing them, whereas women must sue for divorce in the courts.

The principle of controlling females is so strong that, in 2002, fifteen girls died in a school fire in Mecca because the religious police forbade the girls from fleeing the burning building, for fear they would be unsupervised by male guardians. Currently, Saudi Arabia is undergoing a transformation, and some limited rights have recently been extended to women (e.g., spousal abuse was made a crime for the first time, and women will now receive a text message if they have been divorced), but men still view women's rights as a threat – women's rights activists are imprisoned and tortured by the Saudi government.

Whether we regard a country as "modern" is, to a large extent, based on the degree to which we perceive women in that country to have access to economic and political power. Today, countries like Saudi Arabia are the exception (fortunately), and in the twenty-first century it is not uncommon to see a woman as the head of state in countries large and small.

Box 23.3 Fascism

In the nineteenth and twentieth centuries, one country after another shifted its government toward democracy (often imperfectly, it should be noted). Despite the many flaws of these governments, most countries explicitly

Box 23.3 (cont)

Figure 23.5 Fascism employs a variety of techniques to persuade populations. Fascists control information (as in book burning, upper left), have charismatic leaders, such as Adolph Hitler (top center) and Benito Mussolini (top right), and indoctrinate the youth (as in the Hitler Youth, lower left).

But the most destructive method is to adopt in-group/out-group psychology, in which the nation is framed in opposition to an identifiable group. In the lower right, the Nazis are imposing a boycott on Jewish-owned businesses, preliminary to destroying them during *Kristallnacht* (November 1938), in which rioters looted the businesses, followed by the deportation and genocide of most German Jews.

articulated the Enlightenment values of public participation in governmental decisions and representation. However, there was a group of countries that offered an alternative form of government: *fascism* (Figure 23.5). This form of government was a throwback to the days of the authoritarian states of Genghis Khan, Alexander the Great, Darius, Sargon II, Qin Shi Huang, Ashoka, or any of the hundreds of warlike and imperious emperors and kings throughout history.

Historians have often viewed the Spanish Civil War (Figure 23.6) and World War II as the ultimate expression of fascism's rejection of democracy. Once fascism was established in Europe, a war was probably inevitable. Fortunately for us, the fascists were roundly defeated, but there was a time when this conclusion was not foreordained. Because this was a war of philosophies, and

Box 23.3 (cont)

Figure 23.6 The Spanish Civil War was an early struggle between the ideals of modernity and their opposite, fascism, under Francisco Franco (lower left, reviewing German troops with Hitler). The Republicans fighting fascism were explicitly diverse, with men and women from around the world joining their ranks (top left and middle). However, Franco's troops were supported by the German and Italian air forces (Italian bombers and fighters shown on lower right), and the fascist troops ultimately defeated the Republicans. The bombing of civilian populations in republican Guernica was an early atrocity committed by the Luftwaffe (upper right).

in many ways fascism was a rejection of modernity, it is worth exploring the ideas and expressions of this philosophy.

Fascism is a very particular form of authoritarian government in which the government engages in a group of behaviors to obtain and maintain power:

1. *Extreme nationalism*. Fascist governments seek to appeal to the unity of their cultural group by establishing themselves with clearly expressed identity traits – things like a particular style of dress, language, religion, geographic region, "race," or economic status can all serve as markers of an in-group, and separate them from other out-groups (see Chapter 13). Further, in a fascist state individuals exist only to serve the state. So, in an important sense,

Box 23.3 (cont)

anything that is individualistic may be seen as a challenge to state authority and suppressed.

2. *Opposition*. Fascist leaders frequently frame their group identity in opposition to another group. Using group identifiers (language and religion are common) the existence of the in-group is presented as under threat from out-groups. (For example, Hitler often presented the German people as under threat from Jews.)

3. *Consolidated power*. Fascism is inconsistent with democratic principles, so fascist governments attempt to consolidate all political power, often by outlawing competing political groups or manipulating the political system to reduce their power to insignificance. Economic power is often controlled, not via a communist control of industry but by manipulating government spending to favor particular corporations.

4. *Constraint of freedom*. Under a fascist government there is no freedom of speech or freedom to petition the government for redress (protest). Other freedoms, such as movement, religion, rights of association, and marriage may also be constrained in the name of protecting the in-group. Fascist governments control the flow of information in order to further propagandistic goals (e.g., book burnings were common in Nazi Germany, but also seen in Franco's Spain). Secret police are typical of these societies, to maintain control over citizens.

5. *Charismatic leadership*. Fascist governments are generally dependent on the charismatic appeal of a single leader around whom much of the population rallies. Benito Mussolini, Adolf Hitler, and Francisco Franco were all powerfully charismatic personalities with whom the in-group identified. These are often excellent public speakers and skilled in propaganda and manipulation.

Fascist countries were obviously not democratic, so decisions tended to be rapidly made and implemented, unlike in democratic countries where ideas might be studied and debated in the legislatures for long periods of time, and only passed into law with consensus and compromise. A fascist government also accomplishes political goals without having to worry about laws designed to protect individual rights – if a journalist is writing something the government finds objectionable they may end up imprisoned without a trial.

Fascists saw this type of decisiveness as a strength, and they viewed democracies with contempt: as soft, indecisive, and slow moving. Hitler notably viewed the United States as weak because of its overall diversity, both in its politics and its culture. For him, the German people, who he felt would think and act as a single unit, were strong and quick through their uniformity.

Box 23.3 (cont)

The Power of the In-group

Some of the characteristics of fascism listed here are shared with authoritarian governments generally. But one especially important element of fascism is the way it creates in-groups and out-groups. This kind of us-versus-them philosophy is easily exploitable, probably for evolutionary reasons (see Chapter 13). Once an out-group is established as an enemy, it is very easy to justify oppressive, even genocidal behavior against them. In this light, it is worth examining the way the Nazis and Japanese treated their conquered peoples. If a conquered people was identified as a true out-group, they might be enslaved or exterminated. When the Nazi war machine moved into France, the French government was controlled, but Paris was relatively unmolested, and the (non-Jewish) French people were relatively free to go about their daily business. However, as the Nazis moved east, into Poland, the Ukraine, and Russia, the Nazis categorized the local people as "inhuman Slavs," and regarded them as an extreme out-group. Cities in these countries were frequently destroyed, and the people turned into slaves or exterminated (the Nazis, under their "Hunger Plan," intended to starve out all Polish, Ukrainian, and western Russian citizens to make room for the expansion of Germanic peoples – the famous *Lebensraum*).

The Japanese were similarly brutal. The Japanese had a strongly nationalistic ideology in which anyone not Japanese was regarded as inherently inferior, and therefore appropriate only as slaves or to be exterminated. The destruction of Nanking (the "Rape of Nanking") in 1937–1938 and Manila (1945) are two of the most well-known destructions of cities, but estimates of civilian deaths from Japanese occupation range from 10–20 million, with an additional half-million women captured and used as sexual slaves by the Japanese military across China, South Asia, and the Pacific Islands.

This should be contrasted with the Western Allies, which were largely modern democracies. Although the wartime Allied propaganda was certainly jingoistic, and used in-group identity to unify the country (and was sometimes racist), the Allies succeeded because the alliances were meaningful and based on shared humanistic values; because of this, the countries had a reasonable degree of trust in one another. After the war, the civilians in defeated Axis countries were neither enslaved nor exterminated. These occupations were characterized by humane treatment of civilians and rule of law, all in accordance with the principles of the Enlightenment and modernity

Box 23.3 (cont)

(the Soviet occupation was a notable exception – explored in Box 23.5). The very nature of modern democracy, in which diverse citizens are able to express political opinions and alliances across cultural groups are pursued for political ends, tends to work against the very worst types of in-group behaviors.

During World War II there were notable exceptions, and these were important, both at the time and for the subsequent history. One was the internment of Japanese Americans by the US government following the Japanese bombing of Pearl Harbor. Japanese citizens on the west coast of the USA were rounded up and sent to concentration camps. They were only allowed to bring one suitcase each, and leaving property meant that businesses were lost, mortgages went unpaid and were foreclosed, and vehicles were abandoned. The justification was the danger of these American citizens spying for the Japanese government, which was completely unfounded (no incident of such spying was ever recorded, and US General John DeWitt, a virulent racist, was later found to have falsified evidence).

Also, during World War II the US military was racially segregated. African Americans were put in separate units – at first only in support units (truck driving, unloading munitions from ships, etc.), but later in separate combat units – infantry, tanks, fighter planes, etc. Although this segregation was unquestionably wrong, from both a moral and military standpoint the clear success of these all-black units in combat was one of the major factors in President Truman's decision to integrate the military in 1948. (Military segregation was not reserved for African Americans – Japanese Americans formed the 422nd combat regiment, which fought in Europe and was the most highly decorated combat regiment in the war.) Both of these events, and the clear lack of military or moral justification were, in the United States, part of the backdrop of the drive for racial equality. In other parts of the world, the same forces spelled the end of European colonialism.

Ultimately, the philosophy of fascism has inherent flaws that made failure probable (if not inevitable), especially in the form we saw during and before World War II. The in-group philosophy of Nazism and Japanese fascism sets each of those groups against all the other peoples in the world. When the Nazis captured an area, their brutality immediately set local people in opposition to almost everything the Nazis did. A useful contrast might be seen in the ancient empires of Rome and Persia. When those empires captured a new area, the people were offered the opportunity to ally with the identity of the empire.

Box 23.3 (cont)

Serving a fixed term in the Roman army afforded full citizenship rights to conquered peoples. Although the Nazis frequently forced captured peoples to serve in militaries, this was not a consensual decision, and it certainly never gave a captured Ukrainian, for example, any rights as a German citizen. As a result, areas controlled by the Roman and Persian Empires were stable for long periods of time, whereas the Nazis had to contend with resistance movements at every turn.

In fact, even though the Germans, Italians, and Japanese were allies, each viewed the other with suspicion. There was no underlying philosophy uniting them, since the ultimate goal for each was to conquer the world. Eventually (as they well saw), if they had beaten the Allies, they expected to be at war with each other. The Allies, united by general democratic principles of self-governance, had no such struggles with each other.

After World War II

The defeat of the Axis put other fascist (or fascist-leaning) countries on notice, and Spain, Portugal, and Argentina avoided international intervention by limiting their oppression to internal dissidents. These governments did not last, and although there are still authoritarian governments across the world, none claim the mantle of "fascist." The term "fascist" has become an insult, readily thrown around at unpleasant bosses and blustering politicians, but in the middle of the twentieth century the tank columns and bomber squadrons of fascism were a threat to the very freedom and survival of our parents, grandparents and great-grandparents.

Box 23.4 World War II

In the 1930s, a group of countries, mostly in Europe, adopted fascism as its official governmental model. Two wars were fought, literally over the philosophy of governance. The first was the Spanish Civil War, in which a fascist dictator overthrew the democratically elected government. This was, in some sense, a warm-up to World War II, and the Spanish fascists were supported by the German Nazi government and the Italian fascist government, whereas the "Spanish Republicans" were supported by

Box 23.4 (cont)

international brigades of people opposed to fascism from around the world, as well as the Soviets.

Franco's fascists, with the support of the German Luftwaffe (airforce) especially, defeated the Republicans, which was an early win for the fascists. But soon, Hitler (Germany) and Mussolini (Italy) looked to start expanding their own borders militarily. In the late 1930s and early 1940s, Germany conquered France, Belgium, Denmark, the Netherlands, Norway, Czechoslovakia, Poland, Hungary, Greece, Yugoslavia, Lithuania, and large parts of the Soviet Union. Italy invaded and controlled much of North Africa – Libya, parts of Egypt, and Tunisia, as well as Somalia and Ethiopia – and Albania.

Japan, in another variant of authoritarianism, had already established itself as a colonial power over Korea and several islands in the South Pacific after World War I. Despite a few decades experimenting with democracy, by the mid-1930s the Great Depression (among other factors) ended the experiment, and Japan became an authoritarian regime run by militarily aggressive leaders. From the early 1930s through World War II, Japan expanded into large portions of mainland China, Taiwan, Indonesia, Malaysia, Myanmar (Burma), the Philippines, Vietnam, Laos, Cambodia (all three part of French Indochina), Hong Kong, Singapore, and a variety of South Pacific Islands (Guam, Wake Island, etc.). For propaganda purposes, this was known as the "Greater East Asia Co-Prosperity Sphere."

Philosophy on the Ground

When the German army invaded Poland, it was doing so in pursuit of *Lebensraum*. Under this program, people in countries near Germany were to be driven out, or murdered, to make way for the growth and expansion of the Germanic peoples. This was explicitly based on two philosophical foundations: (1) the German people were genetically superior, and other peoples, especially Jews and Slavs, were lesser humans and fit only for destruction or enslavement; and (2) the philosophy of strength justified the powerful ruling over the weak. Using these two ideas, the Nazis were willing to go to war in ways that Genghis Khan would have recognized – intentionally killing large numbers of civilians for the purposes of removing them from the land. Japan, and to a lesser extent Italy, had similar philosophies and methods.

The Axis philosophy was one of control or destruction of others for the growth of the in-group. The Allied powers pursued the war largely in self-defense, but also for the defense of allied countries generally. Although it had famously failed to defend Czechoslovakia, a change in government meant that

Box 23.4 (cont)

Great Britain finally did go to war to defend France and Belgium. The United States was slower to become involved, partially because of the diversity of opinions over intervening in what was considered a "European War." Although the bombing of Pearl Harbor settled the issue, the USA was still slow to arm, a likely result of the more deliberative processes involved in democracies. By contrast, the Axis powers made rapid, unexpected invasions.

Many historians view World War II as a contest between democracy and fascism. Democratic nations were, in a philosophical sense, carrying forward the principles of the Enlightenment, with diverse populations and political views, a humanistic value placed on human rights, and the principles of limited government. The fascist countries were, contrarily, a reaction to the trends of the Enlightenment. For fascists, democracies were weak and slow to act (especially in the context of the Great Depression). Leaders in fascist countries did not believe in Enlightenment values and saw them as a threat.

Prosecuting the War

To some extent, the actions taken by both sides reflect their philosophies. The Axis, as part of *Lebensraum*, would exterminate or enslave captured civilian populations. This made occupation difficult and expensive, as partisan units always cropped up to resist the Nazi occupation.

The Axis often took civilians from captured countries and forced them into military units. This is partially a reflection of the fact that fascism is inherently exclusionary. People in captured countries would never willingly join the German military, since, unlike in ancient Rome, it would never give them the rights of German citizens. The German army had many units composed of conscripts from foreign countries, but these units were quick to surrender to Allied armies.

Further, the Axis was the first to engage in the indiscriminate bombing of civilian centers. If a military opponent is "less than human," then there is little justification for humane treatment. The Nazis would also go out of their way to destroy cultural landmarks and burn libraries in captured countries to try to remove symbols that might encourage resistance.

These were generally not the methods used by the Allied armies, which tended to focus on military targets. It is unquestionably true that the large-scale bombing of Axis cities was employed during the war (including the only wartime use of nuclear weapons in history), with the intention of breaking civilian support for the war, but this was partially in response to similar Axis bombing, and, even

Box 23.4 (cont)

then, was well understood to be morally questionable within the Allied governments. On land, Allied armies tried to avoid killing civilians, and the Axis-style oppression of civilians never occurred on any large scale.

World War II was the most destructive war in history, and tens of millions of civilians were killed, but, by far, the Axis bears the bulk of the responsibility for the civilian deaths. The post-war treatment of the defeated Axis powers may provide the greatest contrast. In Germany, Japan, and Italy, the Allies held orderly war crimes trials, and over the next twenty years helped the countries establish functioning democracies and productive economies. Today, those three countries are highly productive, democratic, wealthy countries. If the Axis had won the war, you may be sure that the countries conquered by those militaries would not have enjoyed the same treatment.

Was Fascism Like Colonialism?

Some historians have argued that the German, Japanese, and Italian invasion of countries in Africa and Asia produced, in effect, little net change. These peoples, already oppressed by the European colonial powers (primarily England, France, and the Netherlands), had simply exchanged one controlling power for another. But this ignores two important things. The first is that the people in these countries generally resisted, rather than cooperated with the invasions of the fascists, and they did this because the new invaders were not liberators, but rather imposed even harsher rulers than they previously had. In just one example, Vietnam had been a French colony, but the Vietnamese strongly resisted Japanese rule, helping the Office of Strategic Services (the precursor to the CIA) with sabotage, and had a guerilla force to fight the Japanese occupation.

Second, the European colonial powers gave up almost all of their foreign holdings in the decades after the war. The end of colonialism was not always clean, and there was some resistance by the Europeans, but, ultimately, the European powers could not justify controlling other countries after having fought a war for liberation from the Axis (see Box 23.6). The contrast between colonial control (which was sometimes quite harsh) and the genocidal horrors of the Axis occupations could not be clearer. The end of colonialism, in line with other Enlightenment trends (such as outlawing slavery), was probably inevitable, although World War II likely accelerated the change. In contrast, the invasion and control by fascist militaries in World War II was a throwback to the kind of occupation seen under the harshest invading armies of the past, such as Genghis Khan and Alexander the Great.

Box 23.5 Was the Soviet Union "Modern"

One of the dominant questions in the twentieth century for Western democracies was how to regard the Soviet Union. Because Marxism and communism were new historical phenomena, there was no way to put them into some existing conceptual box. They weren't an ancient-style authoritarian dictatorship, because their philosophy of communism explicitly expressed an opposition to the power hierarchies of the past, and the economic system aimed to raise up the working classes.

The question partially comes from the fact that the Soviet government, often for propagandistic purposes, expressed modernistic sentiments. The official position of the USSR was in opposition to racial and gender-based oppression. Their propaganda would, for example, emphasize the multiracial aspects of their army. The USSR was, notably, the first country to have a female astronaut in space (in the USSR called a "cosmonaut"). They would frequently emphasize the modern industrial and scientific achievements of their country. This, plus their claim to be working to improve the conditions of the common workers, meant that some Westerners (especially in the 1920s) saw the Soviet Union as a kind of new, modern, scientific country that might be emulated by the rest of the world. A large group of intellectuals in Western nations were initially attracted to the Soviet Union for these reasons.

But they were clearly not a democracy, despite the fact that they claimed to espouse democratic values and seemed to hold elections (the last word of the name of the country, USSR, is "Republics," with the implication of democracy). Their rulers were, in effect, dictators for life (Lenin and Stalin), with fascist-style "cults of personality." Further, their governments never truly allowed the Enlightenment principles of individual expression and freedom. People were regularly rounded up and executed or imprisoned if they expressed dissenting opinions, and people weren't even free to just leave.

In their foreign policy they expressed opposition to the colonialism and military aggression of Western countries (and they supported the Spanish Republicans against Franco's fascists). But their own cynical expansionism might be best revealed by the then-secret Molotov–Ribbentrop Pact (1939), in which Nazi Germany and the Soviet Union secretly met and agreed to attack Poland and divide the country between the two of them. Just a few weeks after the pact was signed, the Soviet Union attacked Poland from the east, while Nazi Germany attacked from the west. While the German army later attacked

---

Box 23.5 (cont)

the USSR, and the Soviet Union was compelled to turn to the Allies, it was initially an aggressor nation.

There is still debate to this day over the "modernism" of the Soviet Union, although most writers put the Soviet Union in category of "authoritarian regimes," emphasizing the oppressive nature of the political realities in the country (and the failures of the command economy) over the officially expressed egalitarianism. Although the Soviet Union espoused some of the principles of modernism, it never lived up to them, and today it is regarded as one of the failed experiments of history.

Box 23.6 The end of colonialism

At the outset of the modern period, in the nineteenth century, large portions of Africa and Asia were under the control of European countries. Whereas in previous eras conquered territories tended to be relatively adjacent to powerful kingdoms and empires, the seafaring technologies developed in the fifteenth to seventeenth centuries made long-distance control, and the transport of goods to and from colonies, practical.

Almost all of Africa (Ethiopia being a notable exception) was under the control of Great Britain, France, Belgium, Portugal, Germany, Italy, and Spain. Parts of China were under British control, and Southeast Asia was under the control of France, the Netherlands, Great Britain, and Spain. India and Burma (modern Myanmar) were under British control.

Colonial control of these regions was an extension of the type of imperial practices seen in the days of the New Assyrian Empire, the Persian Empire, the Roman Empire, and all of the early conquering kingdoms – control of a region for the purposes of extracting its wealth. No philosophical justification was ever deemed necessary, other than the philosophy of raw power: the strong have the right to rule over the weak.

But the era of European colonialism overlapped with the Enlightenment, and many people in these European powers saw the contradiction. How can colonial rule be justified by countries articulating democratic rights? One way was by using the principles of eugenics: the colonized people were argued to be genetically inferior and incapable of ruling themselves. In this view, the colonial powers were doing the people in Africa and Asia a favor by ruling them.

Box 23.6 (cont)

But these ideas became difficult to justify over time – the flawed racism of eugenics became clear, and a large number of people in the colonies acquired Western university educations, completely falsifying the racist idea of natural inferiority. Many of these people, having read the writers of the Enlightenment, began to agitate for independence from the colonial powers.

The end of colonialism was probably inevitable, but the event that accelerated its demise was World War II. The European countries with the greatest colonial holdings – Great Britain, France, and the Netherlands – found themselves attacked and invaded by the fascists. They turned to their colonies for large numbers of soldiers, who, by and large, responded by taking up the uniform of their colonizers and fighting with grit and determination against the Axis. Countries like India, Sudan, and Nigeria provided large numbers of troops for the British army, and Senegal, Tunisia, and Algeria for the French.

At the end of the war, these colonies, who had been fighting for the freedom of their colonizers from the Axis, in turn started to demand their own independence. This took a variety of forms, and sometimes the transition was relatively smooth. In Indonesia, the Dutch were essentially gone, having been expelled by the Japanese, and to reestablish control would have required a military invasion, which the Dutch were not prepared to undertake.

But in countries controlled by France and Great Britain, the colonial powers never lost physical control of most of their colonies, so independence was not as easily acquired. In Algeria, a guerilla war against the French ran from 1954 to 1962 before the French realized they could no longer hold the country without destroying it, and so relented. The French president, Charles De Gaulle, was outraged by the humiliation of France in World War II and was determined to reestablish the country as a world power, partially by reestablishing control over its colonies. Both sides used killings of civilians and torture as methods, and the war was a very public source of embarrassment for the French.

Many countries, such as Burma and Kenya (against the British) and Vietnam and Algeria (against the French), waged guerrilla wars to gain independence. These wars were costly and could be embarrassing, as suppression typically required the colonial powers to engage civilian populations militarily. In India, however, Mahatma Gandhi found success by using pacifist methods, which highlighted the brutality of the British responses to the demands for independence (including the famous Jallianwala Bagh massacre of civilians earlier in the century). Belgium, on the other hand, extracted itself from Congo

Box 23.6 (cont)

too rapidly, without helping establish a transitional government, leading to a series of civil wars and dictators, and that country has never had a truly successful government since.

Cold War

Since decolonialization largely took place after World War II, it was often framed in the context of the Cold War. The Soviet Union tended to support insurgencies against colonial powers because of the possibility that the country might orient itself toward communism and become an ally. The political effect of this is that wars of independence were frequently presented (by the colonial powers trying to maintain control) as wars against communism.

There was a series of small wars across Asia, Africa, and Latin America in which the Western powers supported one group (with funds largely provided by the USA), while the Soviets supported another side. For the United States, the most consequential was in Vietnam, which, from the nineteenth century through World War II, had been a colony of France. The country had long advocated for independence (even sending representatives to negotiate freedom after World War I). After World War II, France, under De Gaulle, attempted to maintain control, but the Soviet Union offered military support for the Vietnamese independence movement (under Ho Chi Minh).

Initially, the war was fought by the French, but they lacked the military resources to maintain control, so they persuaded the USA to assist, selling the war as a fight against communism. Ultimately, the French withdrew, leaving the Americans to continue the fight, which lasted until 1975. The US government was unable to perceive the war as anything but a fight against communism, and never realized that they were fighting for colonialism, and against liberation. However, other parts of the former colonial world did, and the war was a huge blow to American credibility.

A Decolonized World

Today, the world is essentially decolonized. Africa and Asia are self-governing, and no countries since World War II have advocated the type of imperial or colonial control that the world had seen since antiquity. Although many countries in Africa and Asia remain poor, and some governments do not meet modern standards of democracy, no one is advocating external control. The world is far from perfect, but the political power of any one individual in these countries is greater than it has ever been.

Box 23.7 Resistance to modernity 2: tradition

Across the world, as new social ideas (such as the rejection of colonialism, women's suffrage and rights, racial egalitarianism, or LGBTQ rights) became more accepted, some people, who viewed these changes as threatening or objectionable, pushed back. This resistance took many forms, and here we explore some of them.

Traditionalism

In almost all countries until the modern era, political and economic power was retained by men. Occasionally, a country would have a female monarch, but these were the exception and only happened in very specific circumstances. For the rest of a society, men were typically given legal and political rights that were not extended to women: property ownership, voting rights, rights to own a business, licenses to be lawyers, doctors, etc.

This changed over the modern period, and today women are more powerful than they have ever been. Naturally, there have been strong reactions by men who view this change as threatening. It is mathematically true that expanding women's power reduces men's power, since there is a finite amount of power in any political system. Women voters, for example, are able to determine the outcomes of elections, and if women fill certain political positions, there are that many fewer available to men. Of course, this is at it should be, but some men have pushed back against this loss of power, particularly those who found themselves unable to compete on an even footing with women (Figure 23.7).

Figure 23.7 The trajectory toward modernity is not a straight line. In Afghanistan in the 1970s, women enjoyed the freedom to wear Western clothes and pursue professional careers (Kabul 1975, left). In the late 1990s, the Taliban established a harsh, reactionary government based on a fundamentalist interpretation of Islam, and women lost much of the progress of previous decades (right). In this image, a women in Kabul is being beaten by the Talibani religious police for removing her burka. This photograph was taken with a hidden camera.

Box 23.7 (cont)

The resistance has taken many forms. Sometimes, it is expressed as nostalgia for a time when women filled more "traditional" roles in society. Another variant is the idea that a woman keeping herself out of the public sphere to raise a family and household is more "natural." This concept is sometimes also used to justify repression of LGBTQ rights, on the grounds that anything other than heterosexuality is a corruption of what is "natural."

Religion

Traditionalism often overlaps with religion as a reactionary force when faced with the changes associated with modernity (as in Figure 23.8). The Judeo-Christian-Islamic tradition has generally opposed such trends as the empowerment of women and the acknowledgment of LGBTQ rights (although there are important exceptions). Only in this century, for example, has the official state religion of Great Britain (the Anglican Church) allowed women to serve as bishops, and the Catholic Church has yet to allow it. Within Islam, female imams are extremely rare, and are generally forbidden for mixed-sex congregations. Women were only ordained as rabbis in the late twentieth century (and only in certain sects within Judaism).

Figure 23.8 Reactions against modernism include social and religious resistance to modernist ideas. This resistance has proven to be near universal, and is found across the world, from Uganda (top left), to the USA, to Afghanistan (lower right). Phyllis Schlafly (center top) opposed the ratification of a constitutional protection for equality for US women, calling for protection of "traditional values." One of the worst reactions was the reestablishment of the Ku Klux Klan in the 1920s (top right), as a response to some of the empowerment seen in Figure 23.1.

Box 23.7 (cont)

Some religious leaders have spoken out quite strongly against female empowerment. Southern Baptist minister Pat Robertson, in a 1992 fundraising letter, wrote: "The feminist agenda is not about equal rights for women. It is about a socialist, anti-family political movement that encourages women to leave their husbands, kill their children, practice witchcraft, destroy capitalism and become lesbians." But this kind of reaction is not limited to Christianity. One of the most visible examples in recent history has been the broad-scale oppression of women in Afghanistan by the Taliban. They prevent girls from getting an education, and repress any attempt by women to obtain economic or political power, often through violent means.

Religious reaction is not limited to female empowerment, as there have also been strong reactions against trends toward secularism, LGBTQ rights, and the teaching of evolution. The explicitly stated objections are typically presented in terms of scriptural positions, but any rejection of the authority of an institution (religious or not) whittles down its power.

Return to the Primitive

Two of the most extraordinary reactions to modernity have been seen in Cambodia and Peru. In the late 1960s and early 1970s the Vietnam War spilled over into neighboring Cambodia, as the border between the two countries was relatively porous. Communist-backed fighters supporting North Vietnam and Ho Chi Minh flowed back and forth across the border, leading the USA to conduct a massive bombing campaign of Cambodia. The country was sufficiently destabilized by the war that in 1975 the king (installed by the French in 1954) was overthrown by the Khmer Rouge, an ostensibly communist regime.

Unlike the Soviets, who presented modernist tendencies, the Khmer Rouge espoused a philosophy of premodernist primitivism, and attempted to return the country to a precolonial agrarian peasant society. This had been attempted in China, and had resulted in the death by famine of some 30–50 million Chinese, but it was taken to an even greater extreme in Cambodia.

In a complete rejection of the modern world, anyone who could be considered a professional, or had an "intellectual" job, was summarily executed. Simply wearing eyeglasses was sufficient, or wearing the clothes of a city professional or small-business owner, or having a degree. Foreigners, representing a corrupting influence on the purity of Cambodia's peasant population, including those from the West but also Chinese and Vietnamese, were targeted. Torture was common, to force people to implicate friends,

Box 23.7 (cont)

family, or acquaintances. The killing was rampant, and ultimately resulted in the deaths of a third of the population of the country (roughly two million people). Because journalists were among those killed, almost no word of the genocide escaped to the outside world. The Khmer Rouge, under its leader Pol Pot, had destroyed the country's ability to function and communicate as a modern country.

The Khmer Rouge was also massacring Vietnamese populations on the border with Vietnam, and in 1978 the Vietnamese army invaded and toppled the Khmer Rouge. Journalists entering the country with the Vietnamese army discovered the genocide (and the infamous "killing fields"), which quickly became a focus of worldwide attention.

In Peru, a group became inspired by the Khmer Rouge, and attempted a similar reaction to modernity. The Shining Path (El Sendero Luminoso) advocated returning the country to a pre-Columbian agrarian society modeled on Cambodia. Although this group was a dangerous terrorist presence in the 1970s–1990s, it never controlled the country, and was finally suppressed by the capture of its leader Abimael Guzmán in 1992.

Chapter 24: Prospects for the Future

Today, in the twenty-first century, we have unparalleled access to information about our collective and individual pasts. Unlike historians of long ago, we can examine the health demographics of ancient hunter-gatherers, the writings of ancient scholars, the archaeological ruins of long-lost cities, the economic shifts from the transition to agriculture to industry, and population movements over hundreds of years. With modern quantitative methods and tools we are in a relatively unique position to try to understand the broad-scale arcs of history, and how to make sense of where we sit today and where we might end up in the future.

There are some clear general trends that are driving our world.

Education

I am not that old, but when I was a boy it was not unusual to meet people who had never learned to read. That was less than half a century ago, but today people who genuinely cannot read or write are relatively rare. This increase in literacy in the USA is part of the general worldwide trend, and as older generations disappear, illiteracy will fade.

One thing is clear – education is the key to success in the modern world. In almost every country in the world, illiteracy rates are dropping off as young people get access to schooling. This will have many downstream effects because educated people tend to make better personal and political decisions. Every country wants to have an educated workforce so that it can compete in the world marketplace.

One of the most important predictors of success for a country is the education of its girls and young women. Educated girls are less likely to be married off as children. The children of educated women are much more likely to survive childhood diseases and live to be healthy adults, reducing the stress on a country's health care system; these healthy children, once grown, become an economic asset rather than a drain on the country's economy. Researchers studying the broad benefits of education have found that educating girls increases a developing country's economic output by as much as an additional

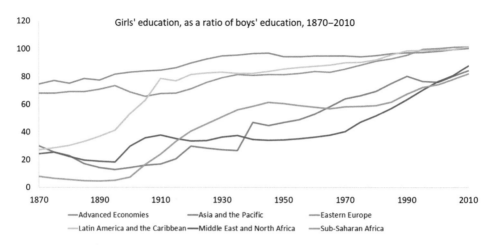

Figure 24.1 One clear trend across the modern world is the increase in the education of girls. This graph shows, on the vertical axis, the relative education levels of girls when compared to boys. On this graph, a lower number means boys receive more education whereas 100 equals educational parity and a number exceeding 100 means girls get more education. In all plotted cases, over the last 150 years, the trend has been toward increasing education for girls. In Latin America and eastern Europe, girls (on average) currently receive more education than boys. Since female education is one of the major indicators of social and economic progress for a region or country, this trend is evidence of significant improvement worldwide.

35 percent for each additional average year of education for the girls (Figure 24.1).

Freedom of Information

This trend, plus access to internet resources, has allowed more people to access more information than ever before. Paralleling this trend in literacy has been a change in the ways governments hold and keep information. The USA was the first major Western power to have a law institutionalizing access to information (the Freedom of Information Act of 1966), but other countries have passed similar laws.

However, these trends do face resistance and progress will probably be halting. One of the last countries in western Europe to enact an equivalent law was the United Kingdom, which does not have a strong tradition of governmental accountability. Tony Blair, the UK prime minister who signed the 2000 legislation, later said that it was the "biggest mistake of my career," because fear of accountability made decision-making in controversial situations more difficult.

The Chinese government has instituted a series of filters on internet traffic to prevent certain types of information from being accessible in China. Known as the "Great Firewall," this system tracks and monitors internet users in China and prevents Chinese citizens from accessing information critical of the Chinese government. Some sophisticated Chinese users have found ways around the filters and censors. Historically, well-educated people in thriving economies have insisted on political freedoms, and China is, in general, a well-educated country, so this censorship is part of the Chinese government's attempt to break the historical pattern. But China has only been open to the world since the 1980s, so we do not yet know if the Chinese people will ultimately reject the current authoritarian government.

Freedom of the Press

Journalists are frequently on the front lines of the fight for freedom of information (Figure 24.2). Governments such as those in Turkey, China, Saudi Arabia, and Iran regularly jail journalists who publish embarrassing or controversial information. Jason Rezaian, the Tehran (Iran) bureau chief for the *Washington Post* and an American citizen, was imprisoned – after a secret trial – for 544 days in 2015–2016 on false espionage charges. Unfortunately, this type of event is not uncommon; currently, Turkey has the highest number of jailed

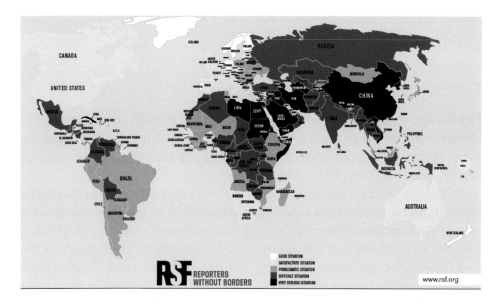

Figure 24.2 Freedom of the press is characteristic of countries governed by secular democracies (compare to Figure 24.5). Countries with authoritarian regimes typically suppress press freedom and try to control the flow of information. As is visible in this map, countries that are (or were) communist, or are run by religious governments, do not permit press freedom. Countries like China, North Korea, Iran, Saudi Arabia, Russia, and Egypt regularly jail journalists, although in some countries (such as Mexico and Colombia) journalists are not threatened by the government so much as criminal elements.

journalists, with between 60 and 120 imprisoned at the time of writing of this chapter.

Propaganda

Countering the trend toward access to information are new forms of disinformation. Traditionally known as propaganda, this is an attempt to discredit accurate information and provide a false alternative. The internet is the main source of this today, but it has always been a part of communication, extending back to ancient inscriptions exaggerating military victories and eighteenth-century pamphlets designed to stir up the American colonists against British rule. In the early twentieth century it often took the form of radio broadcasts. Charles Coughlin was an early radio propagandist in the USA who targeted Jews and supported fascism in the 1930s. Wartime propaganda during World War I and World War II, by all sides, was a clear example of deliberate disinformation designed to rally citizens for their respective sides.

Today, disinformation often travels through anonymous social media accounts, in much the way anonymously published political pamphlets were distributed in the eighteenth and nineteenth centuries. This false information can be difficult to sift out, given the virtual firehose of information that is available online.

Demography

Historically, one of the most important drivers of human social change has always been the size of the population. Demography drove the shift from the foraging lifestyle to agriculture, and the changes in economic, governmental, and religious structures throughout history.

Today, the world population is approaching eight billion. This has had a major impact on a variety of aspects of modern life. The most obvious effect is on the environment. There has been a steady degradation of tropical forest over the last 200 years, as consumption of hardwoods by urban populations, and demand for agricultural products, has driven deforestation. The modern use of disposable plastics by a huge world population has led to significant ocean pollution, and the infamous "Great Pacific garbage patch" stretches over more than a million miles of the northern Pacific Ocean. And burning of fossil fuels has led to enormous air pollution that is a source of lung disease in crowded cities and climate change worldwide.

Some of these environmental problems may be addressed with technology: alternative energy forms may make fossil fuel combustion less common, and scientists are working on ways to clean up plastic pollution. However, it is clear that an ever-increasing world population presents some significant problems. This is especially the case as more countries aim to acquire affluence, because wealthy populations consume far more resources than poor ones. If, for example, every family on Earth consumed at the rate of the average US citizen, the pollution and resource use would be enormous; it has been calculated that it would take four Earths to provide sufficient resources for all 7.8 billion people.

Wars and Famines

In still-developing regions dependent on small-scale agriculture, high rates of population density can be deadly. One of the starkest examples of this occurred in 1994 in Rwanda, a small African country in the mountainous regions west of Lake Victoria. The population of the country almost quadrupled in the twentieth century, and by the early 1990s Rwanda had one of the highest population densities in the world. Competition for arable land was fierce and overpopulation contributed to overfarming.

Because the country was broken up into two ethnic groups (Hutus and Tutsis), it was relatively easy for politicians to generate in-group/out-group sentiments. In 1994, the president was assassinated and Hutu leaders provoked the Hutu citizens into attempting to exterminate the Tutsis. Over the course of the genocide, some 75 percent of the country's Tutsis were killed, and several hundred thousand women were raped. One of the main rewards offered to Hutus by their leaders (via radio) was the redistribution of Tutsi land and cattle.

Overpopulation can also play a role in hunger and famine. In general, the world produces more than enough food for every human on the planet, and the number of famines has decreased over the last 200 years. However, in some circumstances, high population levels in ecologically fragile locations can lead to famine. In almost all circumstances, political events such as civil wars or government policies have been the direct cause of the famine, but population density certainly creates the appropriate preconditions. The Great Famine of Ireland is a fairly characteristic example, where a parasite destroyed roughly half of the potato harvest in Ireland in 1845–1846. At that time, Ireland was a colony of Great Britain, and British landowners controlled almost all the land. The British government refused to acknowledge the effect on the Irish economy by reducing rents. As a result, hundreds of thousands of Irish were kicked out of their homes and off the farmland to starve. Roughly a million died, and another million emigrated (many to the USA). To this day, the population of Ireland is less than it was in 1830.

Birth Control

The drive to reproduce is biologically determined, so it is natural that, as modern medicine spread around the world and industrialized agricultural equipment generated greater amounts of food, the normal constraints on human population growth would be released and the world population would increase. In the past, lack of food has been the primary constraint on growth, but this was mitigated by the advent of agriculture some 5,000–10,000 years ago and the efficiencies of industrialized agriculture over the last 200 years.

The development of modern birth control is one of the most consequential technological advances in history. Modern birth control has had two significant effects. The first is in enabling women to make decisions about the balance of work and family. In the past, most women typically spent some twenty years reproducing and acting as the primary source of childcare. In preindustrial agricultural societies children were a valuable source of labor, and rates of child death were high, so this amount of time spent in reproduction was deemed economically necessary. However, in modern societies there is less of a need for large families, and children are expensive to raise and educate to modern standards (i.e., high school and college). The development of birth control allows women to make their own decisions about reproduction. Birth control has, therefore, enabled women to move into the public economic and political arena in new ways (see Chapters 19, 20, and 23).

One consequence of women deciding when and if to reproduce has been a general decrease in population growth worldwide (Figure 24.3). In some Western industrialized countries there has been such a dramatic decrease in reproduction that populations have actually shrunk. Spain, Italy, Poland, and Japan have negative growth rates, as more and more women have decided against reproduction. Over the last seventy-five years, more and more countries have experienced a decrease in growth rates, with industrialized countries having lower rates but an increasing number of developing countries lowering their rates as well. This change from a high reproduction rate, in countries with low education rates, to educated, more industrialized nations with lower reproduction rates is called the *demographic transition*.

In general, when a country increases its rate of education for girls and women, the rates of reproduction decrease. The reasons for this are obvious – more educated women are more likely to be exposed to medical information, and educated women are more empowered to make reproductive decisions that benefit them over the long run. Countries that long had high rates of reproduction, such as Brazil, Columbia, Tunisia, Namibia, and Iran, now have dramatically reduced population growth. Modern countries with industry can generally function well without large families (Figure 24.4).

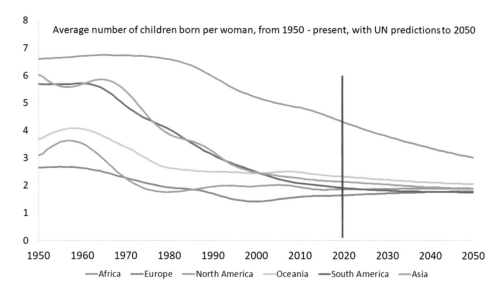

Figure 24.3 Globally, rates of reproduction are decreasing. This is consistent with the general trend toward reduction in world poverty and increases in education levels for both girls and boys. Currently, the United Nations predicts that the world population will continue to increase from current levels (about 7.8 billion) to a peak of around ten billion in 2050, after which the world population will start to contract. The more population levels increase, the greater the stress on the resources of the planet, in terms of resource extraction (such as deforestation or overfishing) as well as pollution. Also, the population levels are predicted to rise in places where resources are already stressed, so current predictions are for greater numbers of wars and civil conflicts over scarce resources.

In industrially developed countries, a decrease in population can relax a strain on natural resources and pollution, but a dramatic population decrease presents other problems. Since the working population of a country (through taxes and production) tends to support the elderly, as well as the young and indigent, population loss can lead to a decrease in a country's standard of living.

Japan presents an interesting example. As one of the most industrialized countries in the world, the population growth rate has been in a steady decline since World War II, and since the turn of the millennium the total population number has been decreasing. Unlike the United States, which maintains population growth through immigration, Japan is a very insular society with strict laws on immigration, and has low rates of foreign workers and residents.

Japan has one of the longest life expectancies in the world, so a decreasing population imposes greater burdens on each worker to support a rapidly expanding population of older citizens. Further, since the 1980s there has been a shortage of workers to fill jobs in the workplace, decreasing overall industrial and technological output. In Japan, as people die, homes become empty, and there are some eight million abandoned homes in the country (Tokyo is the only region of the country experiencing population growth).

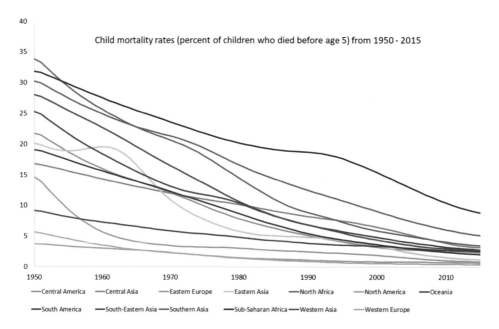

Figure 24.4 As developing nations move out of extreme poverty, they start to have healthier populations. Worldwide, *child mortality rates* have decreased. A decrease in mortality is coincident with a decrease in reproduction levels, as parents no longer feel the need to have a lot of children to maximize the chance that some survive. This particular decision-making process has already occurred in more economically developed nations, and is likely to have a basis in biological instincts. Note the increase in child mortality in East Asia (China) in the late 1950s and early 1960s – this shows the effects of the famine caused by the Great Leap Forward (see Figure 21.3 caption).

The Japanese government has responded by proposing changes to immigration laws and offering financial incentives for couples to reproduce, as well as government-sponsored maternity leave and childcare. Whether these changes will be enough to keep Japan economically dominant remains to be seen.

Disease

As of the writing of this book, the world is gripped by the spread of a highly contagious disease. This disease, known as COVID-19, is a respiratory disease caused by a coronavirus (SARS-CoV-2). Between January and March 2020 this disease spread around the world and at the time of writing has caused over six million deaths. It can be spread both by contact and by inhalation of small particles breathed or coughed out by contagious individuals. Unlike blood-borne diseases, it is able to readily spread between people in the course of normal daily contact.

The spread of this disease is directly related to the density of the populations in which it spread. Highly dense cities, like New York City, were especially hard hit. Because this disease is so easily spread among people, all it takes is for people to be near each other, and in dense cities people are constantly in close contact.

This disease would never have spread before agriculture, when the population of the Earth was thin. But since that time, pandemics of infectious diseases have been a constant presence, often killing large numbers of people and having important effects on history. Europe was notably hard hit by the bubonic plague, and was only able to become a world power once the plague had receded. Smallpox killed enough Native Americans (up to 90 percent) that the Europeans, who retained an acquired herd immunity, could easily defeat them militarily.

Although modern medicine has mitigated the effects of many diseases, the high population densities of our species means that we will regularly have to face the dangers of contagious infectious diseases. We will almost certainly develop technologies to help, but the basic biology of disease means that the economic, social, and political disruption caused by the COVID-19 pandemic will be the rule rather than the exception, much as it was in the past.

Democracy

In 1750 there were no true democracies in the world. Today, more than half of the countries and the majority of the world's population exist under some form of government that follows democratic principles (even if some of the democracies are not completely representative). The trend has been relatively steady since the end of the eighteenth century, and the principles of democracy are so self-evidently defensible that, even when governments are highly authoritarian, they will often incorporate the word (either "Democratic" or "Republic") in the country's name. For example: Democratic Republic of Congo, Union of Soviet Socialist Republics, Central African Republic, Socialist Republic of Vietnam, Republic of Kazakhstan, Democratic People's Republic of Korea (North Korea), Republic of Djibouti, etc.

However, the trend is not linear. There have been periods when democracy has increased, such as the post-World War I period when the large Austro-Hungarian Empire and the Ottoman Empire were broken up, the postcolonial period at the end of World War II, and the appearance of democracy in the former Soviet republics after the fall of the USSR (see Figure 24.5). However, there have also been periods when democracy has receded. Prior to World War II, several democratic nations became fascist. Some countries have recently slid into a nondemocratic government. Russia, for example, was relatively democratic in the 1990s, after the fall of the Soviet Union, but more recently it has become authoritarian. Turkey, once a democracy, has also drifted away toward authoritarianism.

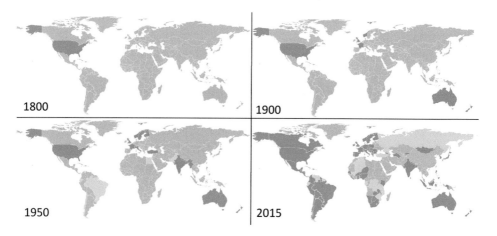

Figure 24.5 In 1800 there was only one democracy in the world. Today, the majority of the world's population has some form of representative government. In this figure, deep blue represents a full democracy, whereas light blue represents a partially representative government. As is visible from this figure, the world is clearly more democratic than it was only 200 years ago. There are still many governments that are not representative, but the trend appears to moving in the right direction.

In some developing countries, democracy is fragile, and can be readily toppled by powerful military leaders who can exploit in-group/out-group instincts. Democracy appears to function best in educated populations who are sufficiently well informed to vote in their long-term best interests, based on governing principles rather than group identity. Passing through a period of "enlightenment" appears to help predict the success of democracy in a country. Russia, for example, never had any period in which the population could truly participate in choosing leaders, from the Middle Ages, through the Czarist period, on into the USSR. Although they had a democratic power structure established after the fall of the USSR, it only lasted a few years, when it was co-opted into authoritarianism. The 2022 Russian invasion of Ukraine, under the military direction of a single leader, Russian President Vladimir Putin, was completely in-character for an authoritarian regime. One piece of evidence was their attempt to generate in-group/out-group support for the invasion by calling their opponents in Ukraine 'Nazis', despite Ukraine having recently elected a Jewish president in an open election.

Many developing nations have trouble maintaining democracy without having a tradition of Enlightenment principles. In countries where tribalism is strong, people tend to vote along tribal lines rather than for the broader self-interest of the country. However, as more and more people are educated, they are better informed about long-term decisions, and, even in the developing world, democracy is steadily, if sometimes haltingly, expanding.

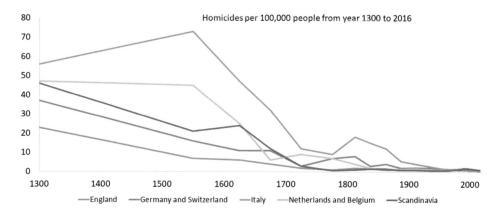

Figure 24.6 As countries become democratized and acquire wealth, they become safer. Countries with modern market economies tend to have more balanced distributions of income (see Figure 21.2). This means there are fewer people who are poor and hungry. Under these conditions, there are far fewer incentives for people to engage in violent crime. Countries with more poverty and uneven distributions of income tend to be less politically stable, and also have higher rates of violent crime.

Violence

Although the twentieth century saw two of the most violent and destructive wars in human history, there has not been a large-scale war between major world powers since, and to many the idea of such a war in the modern political environment is absurd. In fact, long-term trends in violence, both international and domestic, are dramatically down (Figure 24.6). There are various reason for these trends (the danger of nuclear war plays a major role in the larger powers avoiding direct conflict), but one is likely to be the decrease in poverty worldwide. Poverty tends to create political instability whereas material comfort and economic success tends to generate long-term political stability. Since the Industrial Revolution, there has been a worldwide increase in wealth. Worldwide poverty, while still a problem, has dramatically decreased, especially since World War II.

Within countries, trends in intrapersonal violence are down as well. Conditions of internal political stability, representative government, material wealth, and expanded personal liberties and protections have created circumstances in which individuals (on average) have fewer motivations to violence. There will always, of course, be some degree of violence (we are mammals, after all), but certain conditions predispose a society to violence, and understanding and controlling these conditions can dramatically reduce violence. In most of the world, a citizen is safer in their own country than at almost any time in history.

Box 24.1 The dangers of technology

Technology has, over the long arc of history, been a tremendous boon to humanity. Starting with the earliest stone tools that facilitated the consumption of meat some 2.6 million years ago, humans have become more and more reliant on technology. The great transitions of history, to agriculture some 5,000-8,000 years ago, and the Industrial Revolution in the 18th century both radically transformed the patterns in which humans lived on the landscape, ate, governed, reproduced, communicated, and worshipped, among others – almost every aspect of life.

Currently, we are in the very beginning of another revolution in technology – the so-called 'Digital Revolution'. This revolution has only taken shape over the last 50 years, but is likely to transform human existence into the foreseeable future. The internet, especially, has been a radical social and economic force – creating entirely new ecosystems where people can communicate, socialize, and exchange money for services across the globe with little obstruction.

However, with every technological advance, there are dangers. The most significant of these in the 20th century was the invention of nuclear weapons during World War 2. Although they were only used twice, there were several times during the Cold War when the US and the Soviet Union came very close to launching nuclear strikes, which would have resulted in the deaths of large proportions of the world's population.

In the 21st century, several scientists and computer researchers have warned about the dangers of artificial intelligence (AI). We are just seeing the initial uses of AI by governments to identify people via facial recognition software. In China, it is used to identify people who are pushing back against government authority, and it gives the Chinese government a powerful tool for repression. Recently, the Detroit Police Department falsely arrested a man based on flawed facial recognition technology. The militaries of several countries have experimented with decision-making algorithms that may replace humans in military actions, notably for drone strikes.

But perhaps the greatest danger would be if an AI acquired consciousness. If an AI were to become conscious, we should have no expectation that it would share our morality. The decisions made by the AI would be shaped by its hardware and training environment, but chaos theory predicts that, at truly high levels of complexity, we should not expect to be able to predict how decisions would be made, or that if it would share our ethical judgements. Once a decision-making machine has sufficient processors, we might well expect consciousness to emerge for much the same reasons that consciousness emerged in humans, dolphins, elephants, and crows.

If a conscious AI system were connected to the internet, it might well have the ability to affect much of the world, and in unpredictable ways. To many, this seems like preposterous science fiction, but for many computer scientists, it is an inevitability, and a genuine threat. One of the key difficulties is allowing for technological innovation while still avoiding the specific conditions under which this might occur, and whether it might be in the distant or near future.

On into the Future

It can be difficult to predict what the future will bring. Circumstances could dramatically change: if there were to be a sudden change in climatic conditions, creating a worldwide famine, violence would almost certainly increase dramatically. However, the trends over the last few hundred years are encouraging. Today, most people are safer and healthier than was possible at any time in our collective past. Governments are more representative and responsive to their citizens and more people are able to read about world events and make informed personal, social, and economic decisions.

We have to remember that humans are still biological creatures who are relying on instincts inherited hundreds of thousands to millions of years ago. We are still apes who make poor long-term decisions and have a strong in-group/out-group instinct. We still have the urge to reproduce and protect our young, whatever the cost. In certain circumstances, the biology of our decision-making process can still lead to some very bad outcomes.

However, we are now able to understand the factors that lead to trouble. A nation with hungry people is less stable than a nation where people do not have food insecurity. Previous economic models that relied on the exploitation of a particular group (the slave-based economies, feudalism, the mercantile system) have been abandoned, as they are not stable over the long term (and are immoral). A modern market economy that tries to ensure relatively equal access for all citizens produces healthy populations and wealthy and powerful countries. Egalitarian populations are more stable than populations with dramatic differences in civil rights or wealth.

The trends of the Enlightenment and modernity have moved us toward maximizing the long-term health and welfare of citizens. This does not mean that the world has met these standards – there are still plenty of people who are threatened by a more egalitarian society and who push back against these trends, much as in previous eras. But the trajectory is moving in the right direction in most places in the world. It will certainly be a bumpy ride, but the future is yet likely to be a better place even than today.

Additional Readings

Chapter 2

Burbidge EM, Burbidge GR, Fowler WA, Hoyle F. 1957. Synthesis of the elements in stars. *Reviews of Modern Physics* **29**(4): 547.

Eggen OJ, Lynden-Bell D, Sandage AR. 1962. Evidence from the motions of old stars that the galaxy collapsed. *Astrophys J* **136**: 748.

Guenther DB. 1989. Age of the sun. *Astrophys J* **339**: 1156–1159.

Hubble E. 1929. A relation between distance and radial velocity among extra-galactic nebulae. *Proceedings of the National Academy of Sciences* **15**(3): 168–173.

Laughlin G, Bodenheimer P, Adams FC. 1997. The end of the main sequence. *Astrophys J* **482**(1): 420.

Chapter 3

Alvarez LW, Alvarez W, Asaro F, Michel HV. 1980. Extraterrestrial cause for the cretaceous-tertiary extinction. *Science* **208**(4448): 1095–1108.

Bullard E, Everett JE, Gilbert Smith A. 1965. The fit of the continents around the Atlantic. *Philosophical Transactions of the Royal Society of London: Series A, Mathematical and Physical Sciences* **258**(1088): 41–51.

McKenzie DP, Parker RL. 1967. The north Pacific: An example of tectonics on a sphere. *Nature* **216** (5122): 1276–1280.

Patterson C, Tilton G, Inghram M. 1955. Age of the earth. *Science* **121**(3134): 69–75.

Zachos J, Pagani M, Sloan L, Thomas E, Billups K. 2001. Trends, rhythms, and aberrations in global climate 65 ma to present. *Science* **292**(5517): 686–693.

Chapter 4

Cleland CE. 2013. Conceptual challenges for contemporary theories of the origin(s) of life. *Current Organic Chemistry* **17**(16): 1704–1709.

Goldman S. 1980. A unified theory of biology and physics. *J Soc Biol Struct* **3**(4): 331–360.

Johnson AP et al. 2008. The Miller volcanic spark discharge experiment. *Science* **322**(5900): 404.

Lincoln TA, Joyce GF. 2009. Self-sustained replication of an RNA enzyme. *Science* **323**(5918): 1229–1232.

Martins Z et al. 2008. Extraterrestrial nucleobases in the Murchison meteorite. *Earth Planet Sci Lett* **270**(1–2): 130–136.

Shrodinger E. 1944. Order, disorder and entropy: What is life? Dublin Institute for Advanced Studies at Trinity.

Chapter 5

Darwin C. 1859. *The origin of species*. John Murray.

Dawkins R. 1996. *The blind watchmaker: Why the evidence of evolution reveals a universe without design*. Norton.

Dobzhansky T. 1973. Nothing in biology makes sense except in the light of evolution. *The American Biology Teacher* **35**(3): 125–129.

Forrest B. 2009. The religious essence of intelligent design. *Cold Spring Harbor Symposia on Quantitative Biology*, December 21, 2009, **74**: 455–462.

Ruse M, Travis J. 2009. *Evolution: The first four billion years*. The Belknap Press of Harvard University Press.

Chapter 6

Crawford MH. 2007. *Anthropological genetics: Theory, methods and applications*. Cambridge University Press.

Hall BK. 2012. Evolutionary developmental biology (evo-devo): Past, present, and future. *Evolution: Education and Outreach* **5**(2): 184–193.

Holland PW. 2015. Did homeobox gene duplications contribute to the Cambrian explosion? *Zoological Letters* **1**(1): 1–8.

Ijdo JW, Baldini A, Ward DC, Reeders ST, Wells RA. 1991. Origin of human chromosome 2: An ancestral telomere–telomere fusion. *Proceedings of the National Academy of Sciences* **88**(20): 9051–9055.

Li JZ, et al. 2008. Worldwide human relationships inferred from genome-wide patterns of variation. *Science* **319**(5866): 1100–1104.

Chapter 7

Emelyanov VV. 2001. Evolutionary relationship of rickettsiae and mitochondria. *FEBS Lett* **501**(1): 11–18.

Stewart WN, Stewart WM, Stewart WN, Rothwell GW. 1993. *Paleobotany and the evolution of plants*. Cambridge University Press.

Wallin IE. 1923. The mitochondria problem. *Am Nat* **57**(650): 255–261.

Zimorski V, Ku C, Martin WF, Gould SB. 2014. Endosymbiotic theory for organelle origins. *Curr Opin Microbiol* **22**: 38–48.

Chapter 8

Daeschler EB, Shubin NH, Jenkins FA. 2006. A Devonian tetrapod-like fish and the evolution of the tetrapod body plan. *Nature* **440**(7085): 757–763.

Gould SJ. 1990. *Wonderful life: The Burgess Shale and the nature of history*. WW Norton & Company.

Meyer A. 1995. Molecular evidence on the origin of tetrapods and the relationships of the coelacanth. *Trends in Ecology & Evolution* **10**(3): 111–116.

Polet D. 2011. The biggest bugs: An investigation into the factors controlling the maximum size of insects. *Eureka* **2**(1): 43–46.

Wang DY, Kumar S, Hedges SB. 1999. Divergence time estimates for the early history of animal phyla and the origin of plants, animals and fungi. *Proceedings of the Royal Society of London, Series B: Biological Sciences* **266**(1415): 163–171.

Chapter 9

Alvarez LW, Alvarez W, Asaro F, Michel HV. 1980. Extraterrestrial cause for the cretaceous-tertiary extinction. *Science* **208**(4448): 1095–1108.

Bond DP, Grasby SE. 2017. On the causes of mass extinctions. *Palaeogeogr, Palaeoclimatol, Palaeoecol* **478**: 3–29.

Field DJ, et al. 2018. Early evolution of modern birds structured by global forest collapse at the end-cretaceous mass extinction. *Current Biology* **28**(11): 1825–1831.

Hull P. 2015. Life in the aftermath of mass extinctions. *Current Biology* **25**(19): R941–R952.

Preston D. 2019. The day the dinosaurs died. *New Yorker*, April 8.

Wang SC, Dodson P. 2006. Estimating the diversity of dinosaurs. *Proceedings of the National Academy of Sciences* **103**(37): 13601–13605.

Chapter 10

Begun DR. 2016. *The real planet of the apes: A new story of human origins.* Princeton University Press.

Dart RA, Salmons A. 1925. *Australopithecus africanus:* The man-ape of South Africa. *Nature* **115**: 195–199.

Johanson DC, Taieb M. 1976. Plio-pleistocene hominid discoveries in Hadar, Ethiopia. *Nature* **260** (5549): 293–297.

Leakey LS. 1959. A new fossil skull from Olduvai. *Nature* **184**(4685): 491–493.

White TD, et al. 2009. *Ardipithecus ramidus* and the paleobiology of early hominids. *Science* **326** (5949): 64–86.

Wood B. 1991. *Koobi Fora research project: Hominid cranial remains.* Oxford University Press.

Chapter 11

Antón SC. 2003. Natural history of *Homo erectus. American Journal of Physical Anthropology* **122** (S37): 126–170.

Dubois E. 1896. On pithecanthropus erectus: A transitional form between man and the apes. *The Journal of the Anthropological Institute of Great Britain and Ireland* **25**: 240–255.

King W. 1864a. On the Neanderthal skull, or reasons for believing it to belong to the Clydian period and to a species different from that represented by man. *Notices Abstr Br Assoc Advance Sci* **1863**: 81–82.

Leakey LS. 1961. New finds at Oduvai Gorge. *Nature* **189**(4765): 649–650.

Tobias PV, Day MH, Howell FC, Von Koenigswald G, Napier JR, Robinson JT. 1965. New discoveries in Tanganyika: Their bearing on hominid evolution [and comments and reply]. *Curr Anthropol* **6**(4): 391–411.

Chapter 12

Ashley Montague MF. 1941. The concept of race in the human species in the light of genetics. *J Hered* **32**(8): 243–248.

Gould SJ, Gold SJ. 1996. *The mismeasure of man.* WW Norton & Company.

MacEachern S. 2012. *The concept of race in contemporary anthropology.* Race and Ethnicity: The United States and the World.

Mayblin L. 2013. Never look back: Political thought and the abolition of slavery. *Cambridge Review of International Affairs* **26**(1): 93–110.

Witherspoon DJ, Wooding S, Rogers AR, Marchani EE, Watkins WS, Batzer MA, Jorde LB. 2007. Genetic similarities within and between human populations. *Genetics* **176**(1): 351–359.

Yu N, et al. 2002. Larger genetic differences within Africans than between Africans and Eurasians. *Genetics* **161**(1): 269–274.

Chapter 13

Allen EA. 1975. Against sociobiology. *New York Review of Books*, November 13, **22**: 43–44.

Benshoof L, Thornhill R. 1979. The evolution of monogamy and concealed ovulation in humans. *J Soc Biol Struct* **2**(2): 95–106.

Borries C, Savini T, Koenig A. 2011. Social monogamy and the threat of infanticide in larger mammals. *Behav Ecol Sociobiol* **65**(4): 685–693.

Foster KR. 2009. A defense of sociobiology. In: *Anonymous Cold Spring Harbor symposia on quantitative biology.* Cold Spring Harbor Laboratory Press, 403–418.

Frankel C. 1979. Sociobiology and its critics. *Commentary* **68**(1): 39.

Larson P et al. 2021. Investigating the emergence of sex differences in jealousy responses in a large community sample from an evolutionary perspective. Scientific Reports 11, 6485.

Pagel M. 2016. Animal behaviour: Lethal violence deep in the human lineage. *Nature* **538**(7624): 180–181.

Pruetz J et al. 2017. Intragroup lethal aggression in West African chimpanzees (*Pan troglodytes verus*): Inferred killing of a former alpha male at Fongoli, Senegal. *Int J Primatol* **38**(1): 31–57.

Wilson EO. 1976. Dialogue, the response: Academic vigilantism and the political significance of sociobiology. *Bioscience* **26**(3): 183–190.

Zentner M, Mitura K. 2012. Stepping out of the caveman's shadow: Nations' gender gap predicts degree of sex differentiation in mate preferences. *Psychological Science* **23**(10): 1176–1185.

Chapter 14

Emery NJ. 2006. Cognitive ornithology: The evolution of avian intelligence. *Philosophical Transactions of the Royal Society B: Biological Sciences* **361**(1465): 23–43.

Fiorito G, Scotto P. 1992. Observational learning in octopus vulgaris. *Science* **256**(5056): 545–547.

Hawkes K, Finlay BL. 2018. Mammalian brain development and our grandmothering life history. *Physiol Behav* **193**: 55–68.

Heldstab S et al. 2016. Manipulation complexity in primates coevolved with brain size and terrestriality. *Scientific Reports* **6**(1): 1–9.

Herculano-Houzel S. 2009. The human brain in numbers: A linearly scaled-up primate brain. *Frontiers in Human Neuroscience* **3**: 31.

Roth G. 2015. Convergent evolution of complex brains and high intelligence. *Philosophical Transactions of the Royal Society B: Biological Sciences* **370**(1684): 20150049.

Sherwood CC, Gómez-Robles A. 2017. Brain plasticity and human evolution. *Annu Rev Anthropol* **46**: 399–419.

Vitti J. 2010. *The distribution and evolution of animal consciousness*. Harvard University Press.

Chapter 15

Adami C, Ofria C, Collier TC. 2000. Evolution of biological complexity. *Proceedings of the National Academy of Sciences* **97**(9): 4463–4468.

Ball P. 2015. Complex societies evolved without belief in all powerful deity. *Nature News*, March 4, https://doi.org/10.1038/nature.2015.17040.

Griffon D, Andara C, Jaffe K. 2015. Emergence, self-organization and network efficiency in gigantic termite-nest-networks build using simple rules. arXiv: 1506.01487.

Loye D, Eisler R. 1987. Chaos and transformation: Implications of nonequilibrium theory for social science and society. *Behav Sci* **32**(1): 53–65.

Mazzocchi F. 2008. Complexity in biology: Exceeding the limits of reductionism and determinism using complexity theory. *EMBO Rep* **9**(1): 10–14.

Chapter 16

Brown TA, Jones MK, Powell W, Allaby RG. 2009. The complex origins of domesticated crops in the fertile crescent. *Trends in Ecology & Evolution* **24**(2): 103–109.

Cook CJ. 2013. Long run health effects of the Neolithic revolution: The natural selection of infectious disease resistance. Working draft.

Denham WW. 1974. Population structure, infant transport, and infanticide among Pleistocene and modern hunter-gatherers. *Journal of Anthropological Research* **30**(3): 191–198.

Diamond J. 1987. The worst mistake in history of the human race. *Discover Magazine*, May: 64–66.

Lahr MM, et al. 2016. Inter-group violence among early Holocene hunter-gatherers of West Turkana, Kenya. *Nature* **529**(7586): 394–398.

Spielmann KA. 1989. A review: Dietary restrictions on hunter-gatherer women and the implications for fertility and infant mortality. *Hum Ecol* **17**(3): 321–345.

Chapter 17

Goodwin J. 1999. *Lords of the Horizon*. St. Martins' Press.

Hourani A. 2013. *A History of the Arab peoples: Updated edition*. Faber & Faber.

Richerson PJ, Boyd R. 2001. Institutional evolution in the Holocene: The rise of complex societies. In *The Origin of Human Social Institutions*, edited by W.G. Runciman. *Proceedings of the British Academy* **110.**

Service, E. 1975. *Origins of the state and civilization: The process of cultural evolution*. Norton.

Turchin P et al. 2013. War, space, and the evolution of old world complex societies. *Proceedings of the National Academy of Sciences* **110**(41): 16384–16389.

Chapter 18

Armstrong K. 2011. *A history of god: The 4,000-year quest of Judaism, Christianity and Islam*. Ballantine Books.

Banerjee K, Bloom P. 2013. Would Tarzan believe in god? Conditions for the emergence of religious belief. *Trends Cogn Sci.* **17**(1): 7–8.

Freud S 1927. *The future of an illusion*. Norton.

Harrod JB. 2014. The case for chimpanzee religion. *Journal for the Study of Religion, Nature & Culture* **8**(1): 8–45.

Hume D. 1779. *Dialogues concerning natural religion*. London.

Chapter 19

De Bellaigue C. 2017. *The Islamic enlightenment: The struggle between faith and reason, 1798 to modern times*. Liveright Publishing.

Hanna N. 2007. Literacy and the "great divide" in the Islamic world, 1300–1800. *Journal of Global History* **2**(2): 175.

Israeil J. 2001. *Radical Enlightenment: Philosophy and the making of modernity 1650–1750*. Oxford University Press.

Jacob, M. 2016. *The Enlightenment: A brief history with documents*. Bedford/St. Martin's.

Thomas, H. 1999. *The slave trade: The story of the Atlantic slave trade, 1440–1870*. Simon and Schuster.

Chapter 20

Allen RC. 2001. The great divergence in European wages and prices from the middle ages to the First World War. *Explorations in Economic History* **38**(4): 411–447.

Dublin T. 1995. *Transforming women's work: New England lives in the Industrial Revolution.* Cornell University Press.

Dubofsky M, McCartin JA. 2017. *Labor in America: A history.* John Wiley & Sons.

Mokyr, J. 2016. *A culture of growth: The origins of the modern economy.* Princeton University Press.

Overton M. 1996. *Agricultural revolution in England: The transformation of the agrarian economy 1500–1850.* Cambridge University Press.

Chapter 21

Ekelund Jr RB, Hébert RF. 2013. *A history of economic theory and method.* Waveland Press.

Ericksen RP. 1990. *The role of American churches in the McCarthy era.* Kirchliche Zeitgeschichte.

Grandin G. 2015. Capitalism and slavery. *The Nation*, May 1.

Kirman A. 1993. Ants, rationality, and recruitment. *The Quarterly Journal of Economics* **108**(1): 137–156.

Marx K, Lenin VI, Eastman M. 1990. *Capital, the communist manifesto and other writings.* New York: Modern Library.

Van Zanden JL et al. 2014. The changing shape of global inequality 1820–2000: exploring a new dataset. *Rev Income Wealth* **60**(2): 279–297.

Chapter 22

Chapman P. 2014. *Bananas: How the United Fruit Company shaped the world.* Open Road Grove/Atlantic.

Frankopan P. 2015. *The Silk Roads: A New History of the World.* Bloomsbury Publishing.

Hochschild A. 1999. *King Leopold's ghost: A story of greed, terror, and heroism in colonial Africa.* Houghton Mifflin Harcourt.

Lovell, J. 2011. *The Opium War: Drugs, dreams, and the making of modern China.* Overlook Press.

Yergin D. 2011. *The prize: The epic quest for oil, money and power.* Simon and Schuster.

Chapter 23

Armstrong K. 2001. *The battle for God.* Ballantine Reader's Circle.

Bach D. 2016. *State and society in francophone Africa since independence.* Springer.

Fletcher IC, Levine P, Mayhall LEN. 2012. *Women's suffrage in the British Empire: Citizenship, nation and race.* Routledge.

Paxton RO. 1998. The five stages of fascism. *The Journal of Modern History* **70**(1): 1–23.

Rader M. 1943. The conflict of fascist and democratic ideals. *The Antioch Review* **3**(2): 246–261.

Smith T. 1978. A comparative study of French and British decolonization. *Comparative Studies in Society and History* **20**(1): 70–102.

Chapter 24

Herz B, Herz BK, Sperling GB. 2004. *What works in girls' education: Evidence and policies from the developing world.* Council on Foreign Relations.

Mcnab RM, Mohamed AL. 2006. Human capital, natural resource scarcity and the Rwandan genocide. *Small Wars and Insurgencies* **17**(3): 311–332.

Somani T. 2017. Importance of educating girls for the overall development of society: A global perspective. *Journal of Educational Research and Practice* **7**(1): 10.

Szayna Tset al. 2017. Conflict trends and conflict drivers: An empirical assessment of historical conflict patterns and future conflict projections. Rand Corporation Research.

Index

Printed in the United States
by Baker & Taylor Publisher Services